T0314274

Resiliency of Power Distribution Systems

Resiliency of Power Distribution Systems

Edited by
Anurag K. Srivastava, Chen-Ching Liu, and Sayonsom Chanda

Registered Offices
John Wiley & Sons, Inc., 111 River Street, Hoboken, NJ 07030, USA
John Wiley & Sons Ltd, The Atrium, Southern Gate, Chichester, West Sussex, PO19 8SQ, UK

Editorial Office
The Atrium, Southern Gate, Chichester, West Sussex, PO19 8SQ, UK

For details of our global editorial offices, customer services, and more information about Wiley products visit us at www.wiley.com.

Wiley also publishes its books in a variety of electronic formats and by print-on-demand. Some content that appears in standard print versions of this book may not be available in other formats.

Library of Congress Cataloging-in-Publication Data

Names: Srivastava, Anurag K., editor. | Liu, Chen-Ching, editor. | Chanda, Sayonsom, editor.
Title: Resiliency of power distribution systems / edited by Anurag K. Srivastava, Chen-Ching Liu, and Sayonsom Chanda.
Description: Hoboken, NJ : Wiley, 2024. | Includes bibliographical references and index.
Identifiers: LCCN 2022044390 (print) | LCCN 2022044391 (ebook) | ISBN 9781119418672 (cloth) | ISBN 9781119418733 (adobe pdf) | ISBN 9781119418726 (epub)
Subjects: LCSH: Electric power distribution. | Electric power systems–Reliability.
Classification: LCC TK3001 .R385 2024 (print) | LCC TK3001 (ebook) | DDC 621.319–dc23/eng/20221017
LC record available at https://lccn.loc.gov/2022044390
LC ebook record available at https://lccn.loc.gov/2022044391

Cover Design: Wiley
Cover Image: © Zhao jian kang/Shutterstock

Set in 9.5/12.5pt STIXTwoText by Straive, Chennai, India
Printed and bound by CPI Group (UK) Ltd, Croydon, CR0 4YY

C9781119418672_031123

Contents

About the Editors

Anurag K. Srivastava is a Raymond J. Lane Professor and Chairperson of the Computer Science and Electrical Engineering Department at the West Virginia University. He is also an adjunct professor at Washington State University and a senior scientist at Pacific Northwest National Lab. He received his PhD degree in electrical engineering from the Illinois Institute of Technology in 2005. His research interest includes data-driven algorithms for power system operation and control including resiliency analysis. In past years, he has worked in a different capacity at the R'seau de transport d'électricit' in France; RWTH Aachen University in Germany; PEAK Reliability Coordinator, Idaho National Laboratory, PJM Interconnection, Schweitzer Engineering Lab (SEL), GE Grid Solutions, Massachusetts Institute of Technology and Mississippi State University in the USA; Indian Institute of Technology Kanpur in India; as well as at Asian Institute of Technology in Thailand. He is serving as co-chair of the IEEE Power & Energy Society's (PES) tools for power grid resilience TF and a member of CIGRE C4.47/C2.25 Resilience WG. Dr. Srivastava is serving or served as an editor of the *IEEE Transactions on Smart Grid*, *IEEE Transactions on Power Systems*, *IEEE Transactions on Industry Applications*, and *Elsevier Sustainable Computing*. He is an IEEE Fellow and the author of more than 300 technical publications including a book on power system security and 3 patents.

Chen-Ching Liu is an American Electric Power Professor and Director, Power and Energy Center, at Virginia Tech. During 1983–2017, he was on the faculty of the University of Washington, Iowa State University, University College Dublin (Ireland), and Washington State University. Dr. Liu is a leader in the areas of power system restoration, resiliency and microgrids in distribution systems, and cyber security of the power grid. Professor Liu received an IEEE Third Millennium Medal in 2000 and the Power and Energy Society Outstanding Power Engineering Educator Award in 2004. In 2013, Dr. Liu received a *Doctor Honoris Causa* from Polytechnic University of Bucharest, Romania. He chaired the IEEE Power and Energy Society Fellow Committee, Technical Committee on Power System

Analysis, Computing and Economics, and Outstanding Power Engineering Educator Award Committee. Professor Liu is the US Representative on the CIGRE Study Committee D2, Information Systems, and Telecommunication. He was elected a Fellow of the IEEE, a Member of the Virginia Academy of Science, Engineering, and Medicine, and a Member of the US National Academy of Engineering.

Sayonsom Chanda is a senior researcher of Energy Systems Integration at the National Renewable Energy Laboratory in Golden, Colorado, USA. He is a tech evangelist for grid modernization, digital transformation in electric utilities, and citizen science. He is the founder of clean-tech start-up companies like Plexflo and Sync Energy. He has received his master's and PhD degree in Electrical Engineering from Washington State University. Earlier in his career, Sayonsom worked as a senior analyst at National Grid and a research engineer at Idaho National Laboratory. He has served as the VP of IEEE Young Professional Society. Dr. Chanda has three patents in cloud computing for the power grid.

List of Contributors

Anuradha M. Annaswamy
Department of Mechanical Engineering
Massachusetts Institute of Technology
Cambridge
MA
USA

Prabodh Bajpai
Department of Sustainable Energy Engineering
I.I.T.
Kanpur
U.P.
India

Sandford Bessler
Digital Safety & Security Department
AIT Austrian Institute of Technology
GMBH
Vienna
Austria

Ted Brekken
School of Electrical Engineering and Computer Science
Oregon State University
Corvallis
OR
USA

KokKeong Chai
School of Electronic Engineering and Computer Sciences
Queen Mary University of London
London
UK

Sayonsom Chanda
National Renewable Energy Laboratory
Golden
CO
USA

Yue Chen
School of Electronic Engineering and Computer Sciences
Queen Mary University of London
London
UK

Adam Hahn
Washington State University
Pullman
WA
USA

John (JD) Hammerly
The Glarus Group
Spokane
WA
USA

Nikos Hatziargyriou
Electric Energy Systems Laboratory
School of Electrical and Computer Engineering
National Technical University of Athens
Athens
Greece

Junho Hong
University of Michigan-Dearborn
Department of Electrical and Computer Engineering
Dearborn
MI
USA

Oliver Jung
Digital Safety & Security Department
AIT Austrian Institute of Technology GMBH
Vienna
Austria

Hyung-Seung Kim
Myongji University
Department of Electrical Engineering
Yongin
South Korea

Alexis Kwasinski
Department of Electrical and Computer Engineering
University of Pittsburgh
Pittsburgh
PA
USA

EngTseng Lau
School of Electronic Engineering and Computer Sciences
Queen Mary University of London
London
UK

Seung-Jae Lee
Myongji University
Department of Electrical Engineering
Yongin
South Korea

Zhiyi Li
The College of Electrical Engineering
Zhejiang University
Hangzhou
Zhejiang
China

Siyang Liao
School of Electrical Engineering and Automation
Wuhan University
Wuhan
China

Chen-Ching Liu
The Bradley Department of Electrical and Computer Engineering
Virginia Polytechnic Institute and State University
Blacksburg
VA
USA

Ahmad R. Malekpour
Department of Mechanical Engineering
Massachusetts Institute of Technology
Cambridge
MA
USA

Pierluigi Mancarella
Department of Electrical and Electronic Engineering
The University of Melbourne
Parkville
Melbourne
Australia

Mathaios Panteli
Department of Electrical and Computer Engineering
University of Cyprus
Nicosia
Cyprus

Frédéric Petit
European Commission
Ispra
Lombardy
Italy

Julia Phillips
The Perduco Group
Beavercreek
OH
USA

Jalpa Shah
Sensata Technologies
Eaton Corporation Inc.
Eden Prairie
MN
USA

Mohammad Shahidehpour
The Robert W. Galvin Center for Electricity Innovation
Illinois Institute of Technology
Chicago
IL
USA

Anurag K. Srivastava
Lane Department of Computer Science and Electrical Engineering
West Virginia University
Morgantown
WV
USA

Gerald Stokes
Stony Brook University
Long Island
NY
USA

Yuanzhang Sun
School of Electrical Engineering and Automation
Wuhan University
Wuhan
China

Dimitris Trakas
Electric Energy Systems Laboratory
School of Electrical and Computer Engineering
National Technical University of Athens
Athens
Greece

Mani Vadari
Modern Grid Solutions
Redmond
WA
USA

Venkatesh Venkataramanan
National Renewable Energy Laboratory
Golden
CO
USA

Ying Wang
School of Electrical Engineering
Beijing Jiaotong University
Beijing
China

Jian Xu
School of Electrical Engineering and Automation
Wuhan University
Wuhan
China

Yin Xu
School of Electrical Engineering
Beijing Jiaotong University
Beijing
China

Foreword

Electrification has been recognized by the National Academy of Engineering as the engineering achievement having the greatest impact on the quality of life in the twentieth century. The centralized system designs and associated organizational structures that have provided reliable power delivery for the last century, however, face increasing challenges in meeting the key characteristics of power delivery that modern society demands: resilience, reliability, security, affordability, flexibility, and sustainability. The US Department of Energy Grid Modernization Initiative has identified emerging trends that drive the need for transformational changes in the grid. These trends posing challenges to the existing grid infrastructure include the following: a changing mix of types and characteristics of electric generation (in particular, distributed and clean energy); growing demand for a more resilient and reliable grid (especially due to weather impact, cyber, and physical attacks); growing supply- and demand-side opportunities for customers to participate in electricity markets; the emergence of interconnected electricity information and control systems; and aging electricity infrastructure.

These emerging trends have significantly affected electric power distribution systems, as an increasing level of penetration of distributed energy resources (DERs, such as demand response, energy storage, and microgrids) changes the traditional one-directional power flow paradigm into a bi-directional one, fundamentally altering how protection systems should be designed and operate. The increased variability of both electricity demand and supply introduced from DERs requires advanced sensing, communications, and control technologies – also known as smart grid technologies – to improve situational awareness and manage a balanced supply-and-demand operation in real time. Furthermore, exogenous events such as extreme weather and physical/cyber attacks have presented increasing threats to resilient and secure grid operations. According to the National Oceanic and Atmospheric Administration, the United States has sustained 219 weather and climate disasters since 1980 with the total cost of these events exceeding US$1.5 trillion. Further, a 2017 industry survey of utility professionals identified cyber and physical security as one of the five most important issues facing electric utilities, with the other four being DER policy, rate design reform, aging grid infrastructure, and the threat to reliability from integrating variable renewables and DERs.

The 2017 report entitled *Enhancing the Resilience of the Nation's Electricity System*, published by the National Academies of Sciences, Engineering, and Medicine, documents its national-level study findings on the future resilience and reliability of the nation's electric power transmission and distribution system. The report assesses various human and natural events that can cause outages with a range of consequences and concludes that the risks of physical and cyber attacks pose a serious and growing threat. The study recommendations include the following: conducting a coordinated assessment of the numerous resilience metrics being proposed for transmission and distribution systems and seeking to operationalize the metrics within the utility setting; supporting research, development, and demonstration of infrastructure and cyber monitoring and control systems; exploring the extent to which DERs can be used to prevent wide-area outages and improve restoration and overall resilience; and improving the cybersecurity and cyber resilience of the grid.

This timely book delves into critical topics of grid modernization and its associated challenges and technological solutions and examines some of the recommendations in the National Academies report, with a primary focus on the resiliency of distribution systems in the context of the new paradigm of smart grid technology and DERs. It also addresses improving the security of power system communications, control, and protection systems. Beginning with chapters on concept, framework, and metrics of resiliency, the book moves into measuring and visualizing resiliency; enabling, improving, and optimizing electricity network and cyber resiliency; and resilient operations employing control systems and distributed assets. It concludes by presenting some examples of practical implementation. Thus, this book addresses distribution system resiliency from concept to research and development, and through implementation and operation.

I am honored to write the foreword for this book, as it spans the interests of many US Department of Energy programs I have been involved in. These programs range from DER integration to smart grids, and to my current program focus areas of microgrid and resilient distribution system. This book encompasses what has been developed, what is being developed, and areas in need of further development. It also identifies exemplary development and implementation. I fully expect this volume to become a valuable resource for research, planning, implementation, and education on the key topics affecting the resiliency of distribution systems, a subject of national importance.

Dan T. Ton
Program Manager
Office of Electricity Delivery and Energy Reliability
US Department of Energy

Part I

Foundation

1

Concepts of Resiliency

Sayonsom Chanda[1], Anurag K. Srivastava[2], and Chen-Ching Liu[3]

[1] *National Renewable Energy Laboratory, Golden, CO, USA*
[2] *Lane Department of Computer Science and Electrical Engineering, West Virginia University, Morgantown, WV, USA*
[3] *The Bradley Department of Electrical and Computer Engineering, Virginia Polytechnic Institute and State University, Blacksburg, VA, USA*

1.1 Introduction

Resilience is an evolving concept for the power grid. There are many related system concepts – such as reliability, security, and system hardening – which further makes it complex to clearly and uniquely define resilience. This chapter attempts to lay a clear foundation about the various definitions and interpretations of power systems resiliency and a structured taxonomy. All modern critical infrastructure are greatly impacted by poor power quality issues and frequent discontinuity of service. *Higher-reliability and better-quality* electricity service is indispensable to sustain any strong and progressive modern economy and minimize financial losses. However, electric power infrastructure is encountering unique challenges:

- Climatological Challenges. Global climate change is attributed to number of weather-related significant power outages[1] at the US Atlantic coast. The number of significant power outages increased from average of 22.3 between 1990 and 2000 to 76.4 between 2001 and 2015 – which corresponds to an alarming 342% increase. In 2018, power delivery infrastructure in large regions in the states of Texas, Florida, Louisiana, and the territories of Puerto Rico and Virgin Islands were completely devastated by three successive Category 4 and above hurricanes in a span of two months.
  In other countries, devastating tsunamis (Japan, India, and Indonesia), and earthquakes (Nepal and Mexico) caused extensive infrastructure damage and power outages.
- Power Resources. Push for sustainability and the aforementioned ecological concerns led to transition to clean and renewable energy. These distributed energy resources (DERs) are interfaced to the power grid using power electronics causing harmonics. Also, distribution systems have to deal with intermittency. Reduced system inertia is also an associated result.

1 Power outage that impacted at least 50,000 customer businesses or homes.

Resiliency of Power Distribution Systems, First Edition.
Edited by Anurag K. Srivastava, Chen-Ching Liu, and Sayonsom Chanda.

- Aging Infrastructure. The power grid infrastructure was installed and has been operational (except for occasional repairs) for several decades. Several studies have shown that aging infrastructure, like transformers, are more prone to failures. Transmission and distribution system poles which may have been installed many decades ago, may not conform to modern ASCE standards for withstanding extreme wind speeds.
- Changing Demographics. By 2030, more than 60% of the 9 billion human population will live in urban regions. Urban regions have higher per capita power consumption and will likely to further stress system.
- Cybersecurity. Earlier adoption of digital devices and associated cyber infrastructure typically were not designed with security in mind, but security features was later added (as "patches") to them as an afterthought following many cybersecurity breaches. Smart grid substations with remote monitoring, controllability, and automation are vulnerable to cyber attacks.

These challenges must be addressed in order to meet the surge in demand for *higher-reliability, better-quality* electricity service.

1.2 Resilience of Complex Systems

It is important to approach this topic by breaking it down to its constituent components. First, a clarification of the definition of complex systems will be presented; then, a brief discourse on its origins and its importance will be discussed, before introducing the concept of resilience, as it applies to such systems.

1.2.1 What are Complex Networks and Complex Systems?

We are all exposed to multiple complex networks every day. Power grids, airplane networks, interstate and road networks, railway freight networks, supply chain networks, and the Internet are complex networks.

Definition 1.1 *A complex network is a graph comprising many nodes (which can independently act as source or sink or a modifier or a temporary buffer for storing or processing a logically consistent form of matter and/or information) and edges (through which matter and/or information is transported from one node to another) [1].*

Examples. The Internet complex network is made of millions of individual computers acting as nodes, with streams of data flowing via wireless or optical fiber communication channels, acting as edges (for the flow of information). In case of the power grid, load and generation buses are the nodes, while transmission or distribution lines are edges (for the flow of electrons). A detailed summary of industrial complex networks and a formal mathematical way of studying such networks has been presented in [2–4].

Definition 1.2 *A system is any machine with a large number of interdependent moving parts which might comprise even more granular moving parts, some of which can function autonomously, but constrained such that all of the moving parts must coordinate to solve a singular problem or perform a unique service.*

Examples. The human body is composed of multiple systems, such as the cardiac system or the digestive system. A simple example in the physical world is an elevator, which is a system. The original power grid with only generators, transmission lines, transformers, and loads can be considered as a complex network, working as a system.

Definition 1.3 *A complex system comprises a large number of interdependent complex networks and systems, each of which can function independently, but can be controlled and coordinated to increase the efficiency of each of its constituent networks and systems, and the outcome of such coordination is a distinctly higher societal or evolutionary value [5].*

Examples. The human body can be considered as the most complex system in the known universe. An airplane is a complex system, comprising of an interacting traffic communication system, with an internal communication complex network that connect with the mechanical engine, gyroscopes, radars, and auxiliary power generators within the aircraft. The evolving smart grid is also a complex network, which is the focus of this text.

1.2.2 Why Is Understanding Resilience of Complex Systems Important?

Complex systems often emerge and are not designed.

Unlike a carefully designed bridge, the complex power system topology grew without consideration for its cascading outages or cyber-attacks, and thus, the smart grid developing upon the legacy power system inherits the complex system flaws that emerge out of *ad hoc* growth. Since complex systems can contain structural flaws which aid in rapid propagation of disruptive factors. Once individual systems and networks are integrated into the complex system, the effectiveness of each component (or the reason for their existence) can be curtailed by the lack of defense mechanism of the complex system as a whole. That is why understanding the resilience of complex systems is extremely important.

A complex system, characterized by interdependent complex networks and systems, can be susceptible to the following, unlike a single wire connecting a battery to a light bulb:

- Butterfly Effect. Agents for damage, destruction, or disturbance can target a weakly guarded aspect of our lives to cause a widespread damage in another domain.
- Amplification Effect. Flaws or vulnerabilities in one complex network can go unnoticed for long periods of time because other interdependent complex networks often compensate shortcomings of another network. This leads to accumulation and amplification of factors that introduced the vulnerabilities, leading to stronger threats to the overall functioning of the complex networks.
- Domino Effect. Adversities can propagate from one domain of modern society to another across large geographical boundaries.

The abovementioned adverse implications of being reliant on interconnected complex networks can be conveniently remembered using the acronym "BAD" – from the first letters of each of the effects. It is well known that simplicity of networks lead to robustness, but complex systems, can lead to extremely efficient, high-functional societies.

However, the proliferating presence and negative influence of BAD actors in modern society is neither viable nor practical reason to retract from interconnected, interdependent, or inter-operating systems. Instead, it is upon us to deliver a carefully designed, engineering response – by leveraging latest tools and technologies to deploy materials and resources to empower existing infrastructure to minimize adversity and its consequences by responding to any known or unknown threat with *resilience.*

The origin of the word *resilience* dates back to the early seventeenth century in south-eastern Europe, where the present participle form of a Latin word *resilire* was being used

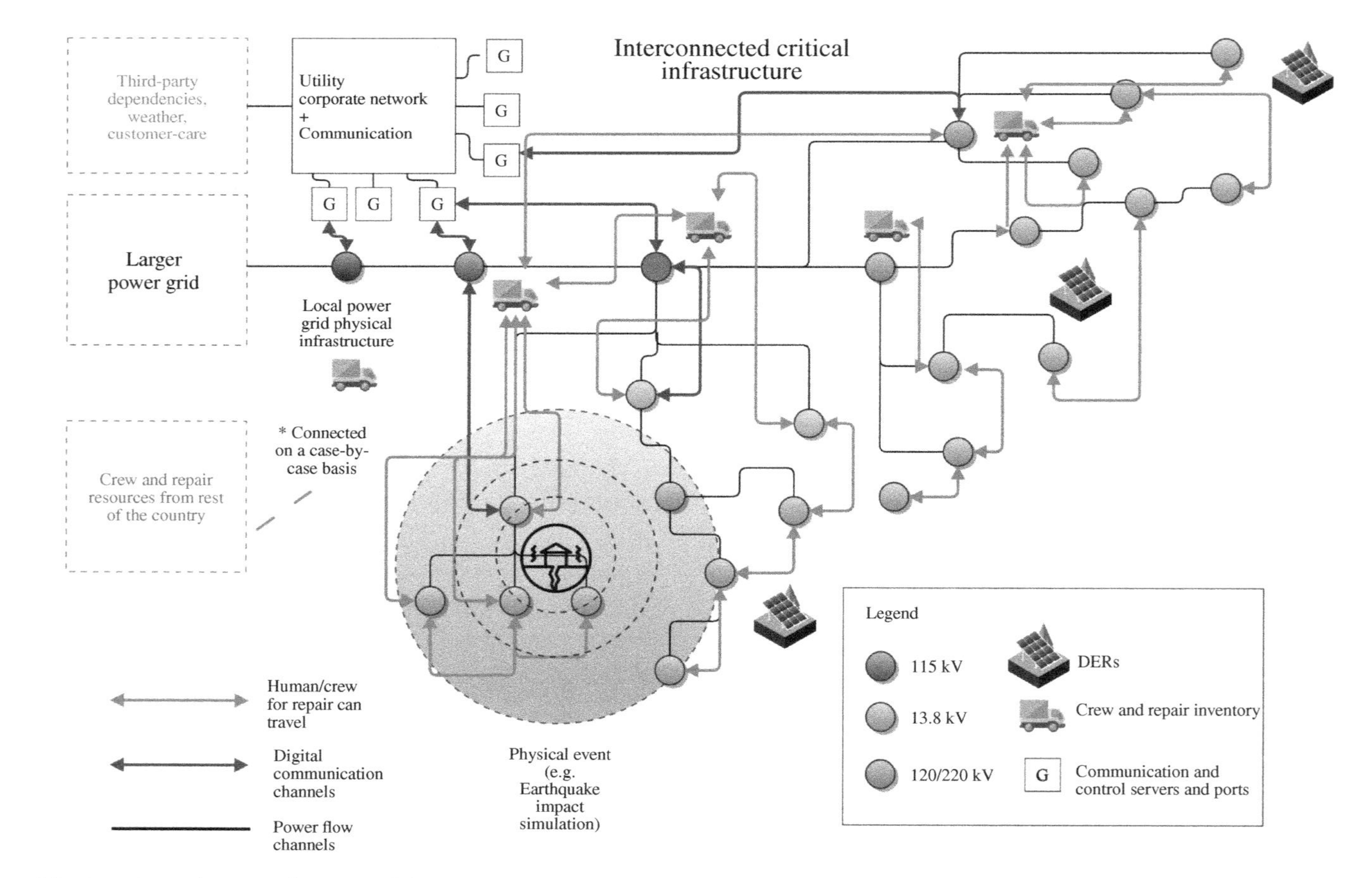

Figure 1.1 Interdependent complex network infrastructures in the smart grid.

to describe the ability of an object to rebound to its original state after having endured damaging forces. Five centuries later, the meaning of the word has not changed, though it has gained new definitions and contexts across many social and scientific domains, ranging from ecology to economics and engineering.

Definition 1.4 *Study of resilience is the scientific approach of optimizing the robustness with the control and structural complexity of the system, such that BAD effects are minimized.*

Figure 1.1 shows a normal interdependence of multiple complex networks – the power system, the communication, and the road transport system – that comprise the distribution system. In case a physical event (e.g. an earthquake) affects the region, it will impact all the three complex networks to various degrees. However, if each of the complex networks can be coordinated effectively, the overall functioning of the distribution system can be maximized despite the damages. That is the goal of resilient distribution systems.

In the next section, a review of closely related terms that partially overlap with the concept of resiliency (and help achieve resilience) is discussed.

1.3 Related Terms and Definitions for Power System

The most common terminologies associated with power system operations has been put forth by North American Reliability Corporation (NERC) [6]. In this section, we summarize the common terminologies, as a staging ground to highlight the uniqueness of the notion of resiliency.

1.3.1 Protection

Since a vast majority of the power system infrastructure, such as transmission and distribution lines, is exposed to the elements of nature and weather phenomena, faults such as line-line, line-ground, and over-current due to short circuits caused by insulation breakdowns are everyday occurrences [7].

Power system protective devices – such as electromechanical or digital relays – primarily protect the healthy parts of the power system from the currents induced by faults through the isolation of faulted parts from the rest of the electrical network.

Power system protection devices (over/under/rate-of-change-of current, frequency, voltage relays; differential relays, distance protection relays, and machine-specific relays) are the most fundamental component of ensuring high reliability and for preventing systemic damage from small, localized events.

1.3.2 Vulnerability

Vulnerabilities of any infrastructure complex network are the most probable points of failures at both component level and systems level. Vulnerabilities exist due to physical limitations of protection systems (e.g. strength of metal casing in case of component-level vulnerabilities or melting point of a fuse wire) or due to implementation (e.g. vulnerabilities due to weak tightening of screws connecting conductors to bushings or software bugs

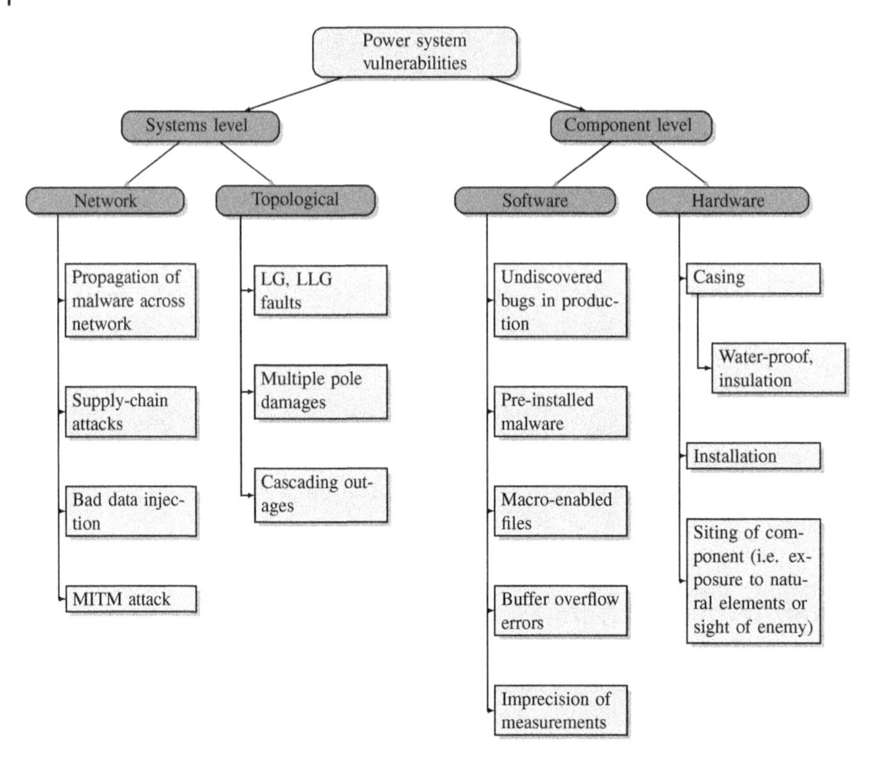

Figure 1.2 Summary of power systems vulnerabilities.

in the program used to control the systems operations). The summary of power system vulnerabilities is depicted in Figure 1.2.

In power systems, some researchers [8] have studied impacts of vulnerabilities that cause voltage instability and devised strategies that minimize costs of impact of other interdictions that disrupt normal power system operations [9, 10]. Implementation of strategic power infrastructure defense (SPID) design methodology for future power systems to respond faster to power system vulnerabilities was discussed by [11]. There also exists different metrics – such as Anticipate–Withstand–Recover (AWR) metrics – for quantifying the vulnerabilities of power system components [12, 13].

1.3.3 Threat

Events which increases the probability of failure of the power system as a whole or an individual component is considered as a *threat*. Threats can either expose and exploit existing known or unknown power system *vulnerabilities*, or create new ones. Like reliability, threat in power systems has been treated as a probabilistic measure by many researchers [14, 15]. Threat has also been studied based on how soon it is detected (temporal) [16, 17] or by the size of the power system it impacts (spatial) [18] or by the means of the attack – such as terror-based [19, 20] or cyber [21].

In spatial studies, a threat is considered *high-impact* if a large area or large number of customers are affected by the event and *low-impact* if it is local in scope. Power system events such as loss of large generators, or natural phenomena such as hurricanes and earthquakes can be used categorized as power system threats in such analyses.

In temporal studies of threat – if both the threat and the vulnerability are known long before the event takes place, and a power system *protection* is typically equipped to mitigate the event by an automated sequence of actions (such as breaker trip or recloser operations) and it is considered a *low-impact threat*. Else if, either the threat or the vulnerability can be known sufficiently long before the event takes place, and a power system *protection* can be configured to handle the threatening event (by installing special protection schemes [SPSs], it is considered a *medium-impact threat*. However, if both the threat and vulnerability become only apparent immediately before, during, or after the events through its consequences or an investigation, such threats are considered *high-impact threat*. Distribution-system faults or cyber-attack, quantified by measures such as Common Vulnerabilities and Exposures (CVEs) and Common Vulnerability Scoring System (CVSS) [22], can be grouped into one of the three categories in temporal studies of threat.

Other power system threats include irrational malicious human behavior, such as acts of terrorism, inadequate installation of power protection devices in distribution and transmission systems, and aging equipment, which have higher probability of failure and higher mean-time to repair (MTTR).

1.3.4 Reliability

According to NERC [6], power system reliability is the degree of performance of the elements of the bulk electric system that results in electricity being delivered to customers adequately, and within accepted power quality standards. Power system reliability is typically measured by using probabilistic indices such as LOLP (loss of load probability), and ENS (Energy Not Served). Utilities annually report to regulatory authorities metrics such as system average interruption frequency index (SAIFI), SAIDI (system average interruption duration index), and momentary average interruption frequency index (MAIFI) (among many others [23, 24]), in order to showcase their conformance of the NERC definition of reliability.

1.3.5 Security

Power system security refers to the degree of risk in the grid's ability to survive *imminent* contingencies without interruption to customer service [25]. Power system security assessment has been a standard industry practice for nearly three decades, ever since the mainstream use of state estimation. Many researchers [26] have proposed on-line screening filters for both static and dynamic security analysis, which is now been being used by several transmission and distribution system companies. Power system security assessment also uses optimal power flow schemes to recommend optimal preventive and restorative strategies – such that small signal and transient stability of the power system is not impacted. The advent of synchrophasors, high-resolution power system data, and security calculations for the power system are also being implemented in near real-time [27].

There is a strong conceptual interdependence of security in power systems and its reliability and stability [28, 29]. This is because it is not possible for a power system to be secure without being small-signal stable and nor can it be reliable without being secure. However, it is important to note that security is a time-varying attribute of the power system, while reliability is measured over longer periods of time.

1.3.6 Restorability

Power system restoration of distribution systems has been an active area of research, since it was realized that high reliability of bulk power distribution systems did not directly translate into improved power delivery to end-consumers at lower-voltage levels. The first restoration algorithms estimated the "restorability" of radial networks using [30, 31]. Efficiency of power system operations can be studied on the ability of the network to restore the disrupted loads following the disruption based on the impact on power system stability [32], cost–benefit analysis [33, 34], or via speed of restoration [35, 36]. The objective of restorability studies is to minimize the ENS metric that is used for computation for SAIDI and SAIFI. In case of common power system outages (e.g. tree falling on a distribution line, car crash in a utility pole), restorability is not impacted by the ability of crew or personnel to mobilize and repair the damage. In completely islanded, mission-critical systems such as hospital microgrids or shipboard systems, expert agents have been designed by several researchers to improve the restorability to outages [37–39].

Restorability, unlike reliability, takes into factors like time taken and path (i.e. redundancy) and other network topological parameters. Hence, several researchers have studied restorability of distribution systems as graph models, and developing spanning tree-based algorithms to minimize the loss of load expectation (LOLE). Automation of restoration by means of installation of smart reclosers, automatic transfer switches, and remote-controlled breakers have also increased restorability of power distribution systems in the recent years [40].

1.3.7 Storm Hardening

Storm hardening is physically changing the parts of the power grid infrastructure which are most vulnerable to exposure to the impact of a physical event (weather-based or man-based attack). The objective of storm hardening is to make power poles, lines, and other equipment less susceptible to extreme wind, flooding, or flying debris [41]. It involves upgrading the system to use cutting-edge technology, upgrade insulation, strengthen or waterproof or wind proof cables, transformers, and repair ducts, install newer components that are rated to higher tolerance to sudden stress. Storm hardened parts of the power grid are then often linked to each other through advanced communication or automated systems to create pockets of extremely resilient distribution systems that can be useful in supplying critical loads or provide blackstart capabilities following a major physical event.

1.4 Need for Grid Resiliency

1.4.1 Managing Extreme Weather Events

Number of weather events including hurricanes, Nor'easters, and wildfire have been increasing over years. According to the NFPA 110 standard, critical loads like hospitals are required to maintain around 72–96 hours of emergency fuel supply for backup diesel generators, depending on the facility's capacity and scale. Any outage lasting more than four days would put hundreds of lives in danger, not to mention the cryogenic assets stored in the hospital premises. To cite one of the many blows to such contingency plans was Hurricane Zeta, the 26th hurricane of the year 2020. On 29 October 2020, Zeta caused power outages lasting more than a fortnight in parts of Alabama, Mississippi, and Georgia. At its peak, the superstorm left two million in the dark. Several more events like Texas Polar vortex in 2021 led to the largest forced power outage in US history. This further shows the important of resilience-focused grid planning and design (Figure 1.3).

1.4.2 Cyber and Other Events

In 2015, cyber-attack on the power grid of Ukraine resulted in power outages for roughly 230,000 consumers for multiple hours. This was one of the first cyber event resulting into large impact on the power grid, even though most of the electric utilities reported cyber events happening very often.

COVID-19 is another important event with impact on the power grid operation. The disease affected hundreds of millions of people in a short amount of time, with high mortality rate. The impact of the sudden onset of the pandemic meant novel resilience challenges for electric utilities worldwide.

Impact on Load Profiles. COVID-19 upended traditional feeder-level load profiles used by utilities to manage their operations. In the United States, residential electricity sales increased 6% after the lockdowns orders were issued, while commercial and industrial demand went down by 10% [42].

1.4.3 Role of Resiliency-Focused Grid Planning

By running power system analytics with resiliency as one of its crucial optimization factors – utilities could have quickly identified parts of the network that can be designated

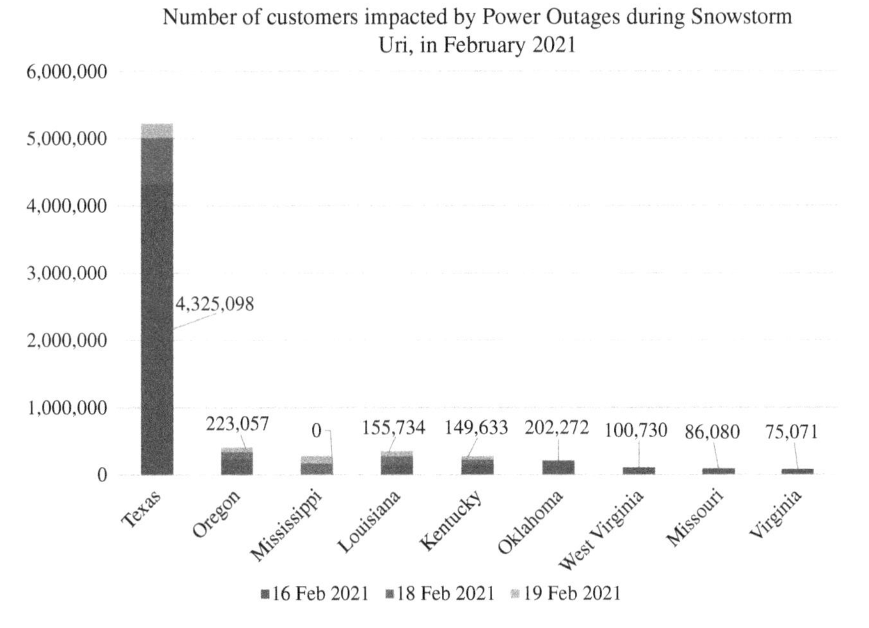

Figure 1.3 Number of customers impacted by the Texas Freeze of 2021. The numbers alongside the bar chart show the customers affected on the first day of the event.

to operate as microgrids. The resiliency-driven power flow solutions can help the network operators identify strategic locations for the point-of-common-coupling for the utility and the new microgrid (Figure 1.9).

The network studies conducted by resiliency-enabling tools and software can concurrently identify costs and create workflows that expedite the necessary purchase orders for distributed generation with correct sizing, accessories, and integration equipment. It can also generate site acceptance criteria and testing protocols for engineers to evaluate before fully commissioning the microgrid. Researchers have claimed that there can be up to 74% time reduction in the feasibility planning stages of microgrid design [43]. This way utilities could have ensured extreme resilience for critical facilities in vulnerable regions by deploying functional microgrids faster. This can only be possible by automating several design steps of a typical project using resiliency-enabling tools. Since these resiliency-focused microgrids are tested via simulations from conception, they are derisked and compliant, using only equipment that meets IEEE 1547-2018, IEEE P2030.7, Clean Air Act, and IEC 61727 standards. Thus, they require fewer field validations before meeting the acceptability criteria – and can be quickly deployed during the emergencies, such as the one posed by the COVID-19 pandemic.

1.5 Resiliency of Power Distribution Systems

1.5.1 Existing Definitions and Interpretations

In US Department of Energy's 2015 Quadrennial Report [44], resilience of power distribution systems has been considered as one of the highest priorities during the elaborate and expensive process of modernizing the power grid. The report attributes the ability of the power system or its components to adapt to changing conditions and withstand and rapidly recover from disruptions as *resilience*.

Some commonly discussed definitions of resilience of power distribution systems are as follows:

Definition 1.5 *According to North American Energy Resilience Model (NAERM) (2019), resilience is defined as the availability of potential solutions that enable systematic identification of threats to the power grid infrastructure, development of hardening options that reduce exposure to weather-based threats, and situational awareness and sophisticated analytics to minimize the impact of threats as they evolve in real time [45].*

Definition 1.6 *Resiliency is a power system attribute that will enable it to operate despite high-impact, low-frequency events [46].*

Definition 1.7 *Resiliency of a power distribution system is its ability to withstand impact of unfavorable events, recover rapidly from any damage incurred through the event, adapt the system to minimize the damages in successive events of similar strength, and prevent the system from further damages in future unfavorable events [47].*

Definition 1.8 *Power system resilience is a quantitative way of understanding the system's boundary conditions and their changes during disturbances [48].*

Definition 1.9 *The resilience of a system presented with an unexpected set of disturbances is the system's ability to reduce the magnitude and duration of the disruption. A resilient system downgrades its functionality and alters its structure in an agile way [49].*

Definition 1.10 *The resilience of a system can be defined as a total amount of energy served to critical load during the preset time period and can be measured as a cumulative service time to critical load weighted by priority [50, 51].*

There has been a body of literature developing steadily over the last five years, around the concept of resilience – in general, as well as specifically for power distribution systems [52, 53]. Several researchers have proposed generic fragility models [54], while others have focused on resiliency under specific contexts – such as hurricanes [55, 56], or wildfires [57]. Despite isolated differences and uniqueness of perspectives of individual researchers, the following points summarize the importance of enabling higher resilience in power distribution systems:

- The Why?. Resilience will help minimize the costs or loss of lives incurred due to large-scale power outages across vast geographical areas due to low-frequency, high-impact events.
- The What?. Higher resiliency will enable the networks to serve critical facilities and customers despite simultaneous multiple outages across the larger power grid and protect it from catastrophic damages (in terms of stability, service continuity, infrastructure, or financial losses).
- The How?. Resilience can be enabled by strengthening all interdependent networks, integrating diverse geographical resources with flexibility to island, and operating off-grid for long periods of time by having reliable and renewable DERs with flexible storage capacities. This is possible by leveraging latest sensing and forecasting technologies, data-driven algorithms for monitoring malicious traffic, weather, and geo-magnetic disturbances. Improved power storage and flexibility introduced via demand-side management and energy markets will be crucial in stable postcontingency operation as well as deploying resilience with a viable investment strategy (i.e. utility's will have a business-case for deploying higher resilience). At the personnel level, cybersecurity awareness will be crucial to prevent spread of malware. Also optimized location of first-responder crew and inventory (of replacement transformers, poles, conductors, breakers, switches, hand pumps to de-flood submerged transformers and such equipment) ahead of forecasted strong weather events.

Table 1.1 summarizes the key differences between resilience and the other notions associated with stable power system operations in Section 1.5.

A vast majority of contemporary research on resiliency also agree that a resilient power distribution system is characterized by its *flexibility*, which comes from a synergy of

- the system's ability to maintain voltage stability in deeply fragmented condition
- redundant feeders, breakers, reclosers, and distribution poles and lines – with high insulation strength and rated to high wind speeds if they are installed overhead
- adequate storage capacity
- DERs with fast ramping rates
- automation enabled through remote controlled switches and smart algorithms

Table 1.1 Key differences between resiliency and reliability, security, and storm hardening

Point	Resiliency	Reliability	Security	Storm Hardening
1. *Objective*	Higher tolerance to extreme events and rapid recovery to damaged components			Strengthen and proof individual components of power grid to physical events
2. *Quantification Approach*	No standard	SAIDI, SAIFI	Voltage stability	Wind rated strength (ASCE) [58], MTTR
3. *Context specificity*	Very high (Refer to Section 1.6)	None	Depends on power system loading conditions	Depends on geographical conditions
4. *Enabling process*	Continuous process, with an active feedback loop			Periodic upgrade to increase the infrastructure's strength to endure events that induce stress beyond rated strengths of existing infrastructure
5. *Scope*	Cyber, physical, transport complex networks	Physical network	Cyber, physical network	Physical network
6. *Depends on*	Low-frequency, high-impact weather, geographical conditions, control algorithms, implemented cyber-security, repair crew availability, strength of the event reliability metrics of the system	Utility maintenance schedules, local animals, foliage	System inertia and small signal stability	Strength of low-frequency, high-impact weather events

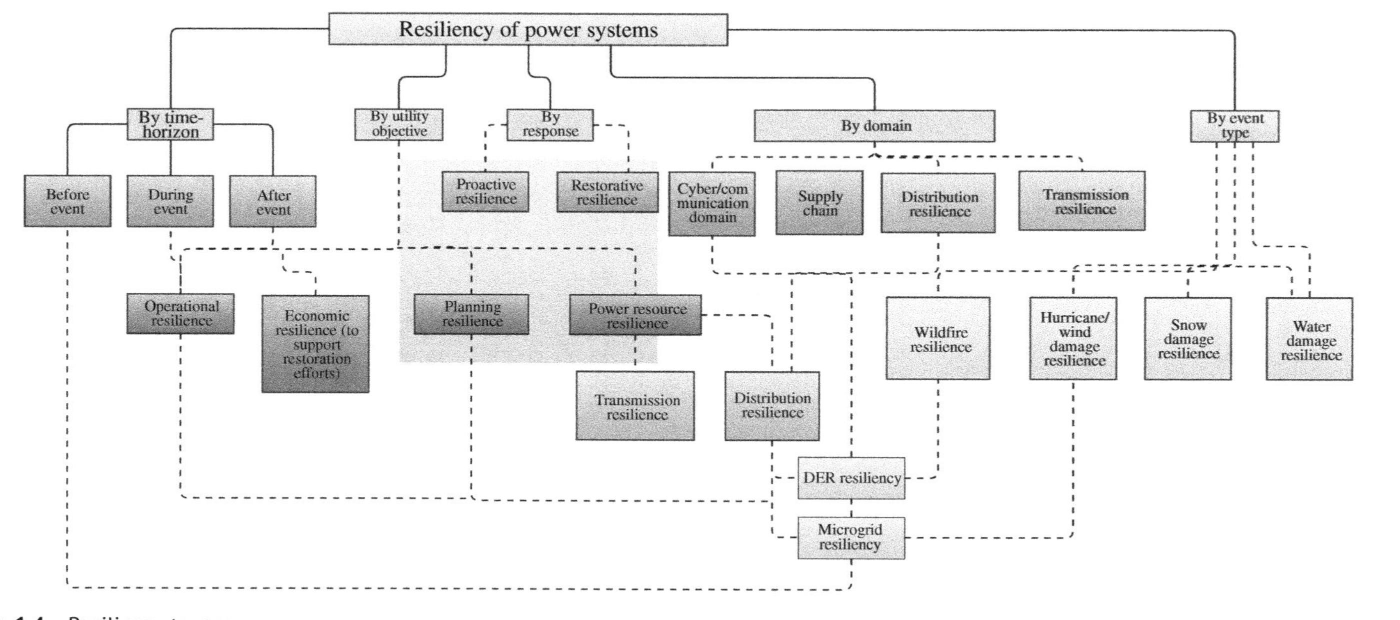

Figure 1.4 Resiliency taxonomy.

It is worth clarifying that the objective of resilience is not to make the power system disaster-proof as forces of nature will occasionally overwhelm man-made defenses. Instead, the objective of resilience is that a cyber-physical system should remain operable and rapidly recoverable, even though all its components are susceptible to severe damages. Thus, resilience is not a brute-force, completely physical hardware-based solution, like storm-hardening. Instead, resilience depends on actionable intelligence that can be gathered through smart use of modern metering devices, advanced communication tools, and digital computers. It may be argued that the ultimate resiliency will be achieved when a complex cyber-physical system will be able to forecast threats without active supervision, self-identify, and autonomously undergo upgrades to improve upon its existing resiliency to de-risk itself from the impending threat and repeat this process indefinitely (Figure 1.4).

1.6 Taxonomy of Resiliency

Resiliency is a challenging topic in power systems due to the wide variety of definitions and interpretations, as discussed in Section 1.5. The variability in interpreting resilience stems from the fact that a large number of entities participate in financing, planning, supplying, operating, maintaining, and repairing the power grid. Also, the complete concept of resilience varies greatly based on the context of its analysis, and the role of the person in the utility or organization whose resiliency is being evaluated.

Countries across the world differ in their policies toward interpreting resiliency. In poor communities (in certain sub-Saharan African regions), resiliency of the community can depend on occasional access to electric power to charge their communication devices, or meet limited, basic human needs. However, in a sharp contrast, for developed economies like the United States which has the most reliable electric power in the world [44], resiliency implies striving toward lesser discontinuity to power supply, even in extreme operating conditions. Thus, a taxonomy of resiliency is required to help facilitate the comprehension of the context.

Taxonomy of resiliency is aimed at creating a larger framework in which resiliency of electric power distribution systems can be analyzed. The proposed categorization is summarized with their respective enabling tools, advantages, and disadvantages in Table 1.2. Each category of resiliency represents a perspective of looking at a system's resilience and contributes partially to the complete concept of resilience. Except for event-specific resilience, all other classifications of resilience deal with any generic threat that can come from the system's physical environment, or the cyber network dependencies.

The two interdependent "dimensions" along which resilience can be studied are the following:

- Economic. In this dimension, financial losses incurred by the utility (or community) due to an unfavorable event that caused a power outage, or damage to power grid infrastructure, or both are computed. Some utilities in areas frequently affected by adverse weather, such as Florida Power and Lights serving Miami-Dade county in the United States, proactively charge a surplus fee to keep their replacement inventory up to date and engage consumers in sharing the costs associated with high resiliency. For areas that have

Table 1.2 Summary of taxonomy of resilient power systems.

Applications/enabling tools	Advantages	Disadvantages
By time-horizon		
	Short-term resilience	
Critical facilities without known load priority	Continuity of supply to entire facilities of critical loads	Resources may be quickly depleted
Traffic and egress lights	Consumers will notice only a flicker when power goes out	Voltage drop and suboptimal performance of certain machinery
Data centers to quickly save and backup data	No planning computation required	Voltage instability [61]
In events of cyber-attacks that last only a brief period of time		
	Contingency resiliency	
– Critical facilities with known load priorities	– Sustainability of resources over projected period of contingency	Many nonpriority loads are lost
– Interruptions due to outages are not experienced by mission-critical facilities	– Advanced planning and computation required in DMS	
– Restoration of path finding in very large distribution systems is computationally intensive and is guaranteed to yield suboptimal results in definite time (NP-complete)		
	Long-term resilience	
– Critical facilities with known or unknown load priorities	– Sustainability of nonrenewable energy resources for maximum time possible	Many nonpriority and some priority loads are lost
– Customer-level disconnect switches, fuses, and breakers	– Power quality may be reduced	
– Customer-owned renewable energy resources		
– Diesel gen-sets with regular refills of fuel resource		
By utility objective		
	Power resource resilience	
	Distribution systems	
Diesel generators	Ability to withstand long periods of outages caused due to catastrophic infrastructure damage in transmission systems	Cost-intensive to implement

(Continued)

Table 1.2 (Continued)

Applications/enabling tools	Advantages	Disadvantages
Inverters and battery packs	Business-case is hard to arrive at due to high-frequency/low-impact nature of the events this resiliency is designed to protect	
Fuel cells	It can be rendered ineffective if the end-user (power sink) is also damaged (buildings razed by earthquake) and not capable of using power	
Roof-top solar installation		
	Transmission systems	
Interconnection of large-scale wind farms, solar farms, nuclear generators, and conventional power systems	Redundancy of bulk power resource protects leads to increased diversity in the economy's energy portfolio. More diversity implies reduced risk for large regions of the power grid from prolonged interruptions in the supply chain due to political instability or war	Cost-intensive to implement
FACTS devices and HVDC link	Power system stability concerns	
Dependencies on multiple organizations, projects, and resources reduce reliability, especially during early stages of the technology		
	Preventative resilience	
Storm hardening of infrastructure	Prevention of cascading outages	Some loads will be disconnected in anticipation of outage
Advanced forecasting technologies and simulation to plan and prepare for damages	Fastest possible restoration time is ensured	Conservative approach
Load shedding in anticipation of storm to minimize chance of causing grid instability		
Strategic position of crews in anticipation of damage to minimize downtimes		

(Continued)

Table 1.2 (Continued)

Applications/enabling tools	Advantages	Disadvantages
	Economic resilience	
Charging customers a nominal fee in high-risk areas	Utilities will not have to take major hits to their operating budget in storm recovery	Customers may not be convinced to pay premium rates for an event which has not happened
Insurance (if available)	Finances better maintenance of grid	Chances of overstocking of inventory and cash reserves
Participation in energy markets	Inventory is ready and stocked in preparation of event	
	Service resilience	
Surplus inventory	Fastest possible restoration time is ensured	Chances of overstocking of inventory and cash reserves
Strategic position of crews in anticipation of damage to minimize downtimes	Less financial impact from customer revenue as power is restored very quickly	Dependent on transportation complex network, hence not guaranteed
Use of innovative and modern tools (such as aerial drones and unmanned automated vehicles) for quickly detecting damaged poles in difficult to reach terrain		
Robust, fuel-efficient crew vehicles capable of navigation through debris, lifting weight, and cranes		
	By event	
Specific strategies in regions with highest probability for • Wildfires • Hurricanes, tropical storms • Storm surge, flooding • Snow storms • Foliage or animal related damages	Targeted investments yield highest return on investment for utilities	Focus on protection against one/few types of event leaves the system vulnerable to unprecedented events
	By response	
	Proactive response	
Shedding load to minimize cascading power outage (Grid Islanding and Resynchronizing relays)	Higher reliability metrics	Conservative approach
Strategic position of crews in anticipation of damage to minimize downtimes	Better maintenance and upkeep of utility owned infrastructure	Cost-intensive
Microgrid formation and stable operation (Microgrid controller and management system installations)		

(Continued)

Table 1.2 (Continued)

Applications/enabling tools	Advantages	Disadvantages
	Restorative response	
Restoration algorithms	Reduced outage times leading to more consumer revenue	Risky approach, as many damages may have been avoidable through proactive response
Improved and optimizing manual restoration process		
Automation Infrastructure to enable		
	By domain	
	Cyber-system resilience	
Regular security upgrades	Defense against cyber-attacks and cyber-bullying	No cyber-defense is guaranteed to be attack proof as adversaries can find new mechanisms
Monitoring of network traffic	Improved communication over system components and greater visibility	Improved cyber-system resilience may not be able to detect dormant malware installed in the past
Prevent unauthorized USB/network access by enforcing least privilege to all employees and devices	Ensures systems are up to date with latest software upgrades, so peak performance is achieved at the component level	Improved cyber-system resilience cannot fix potential/future damages due to previous data breaches
Stronger passwords/firewalls	Dependence on third-party vendors	
Ability to operate critical loads without cyber-control		
Encryption		
	Supply chain resilience	
More dependence on renewable energy resources	Ensures high system reliability even during geopolitically unstable conditions	Cost-intensive
Energy market formation	Competition among energy vendors may bring down cost of electricity	Difficult to increase energy resource diversity due to geopolitical constraints
Stricter quality control of suppliers and vendors	Diversity in energy resources introduces grid instability issues	
Diversification in domestic and international supply of raw materials		

(Continued)

Table 1.2 (Continued)

Applications/enabling tools	Advantages	Disadvantages
	By voltage level	
	Distribution system resilience	
DERs including domestic renewable energy resource installations	Improved resilience also leads to higher SAIDI, SAIFI metrics for utilities to report to regulators	Business-case is hard to arrive at due to high-frequency/ low-impact nature of the events this resiliency is designed to protect
Flexible energy storage	Reduced maintenance costs	
Advanced restoration algorithms	Customer participation in peak-energy demand management reduces financial burden on the power grid companies to install bulk power systems	
Improved outage management systems	Utility revenue increases due to more customer hours as a consequence of quick restoration or total avoidance of power loss	
Microgrid formation capability		
Advanced metering systems		
Storm hardened distribution infrastructure		
	Transmission system resilience	
Vulnerability analysis	Improved voltage stability	Cost-intensive for implementation
Higher than N-2 contingency tools	Greater leverage of restructured grid policies and meshed interconnectedness of networks	Computation costs are very high due to large number of nodes to be analyzed
PMU, cascading outage prevention	Ensuring pathways to supply power to substations in very high risk regions	

not experienced tangible adversities (yet) in recent times, the business case for utilities to offer higher resilience is not very clear due to budgetary constraints, or other immediate priorities. Economic resilience changes according to the variety of business models for investing, owning, and operating grid infrastructure.

- Engineering. The engineering aspect of resiliency is frequently discussed from multiple time-frames for analysis (postevent, ongoing-after, or preevent), or based on number of critical loads that were not lost or quickly recovered following an unfavorable event or cyber-attack, or based on the nature of the attack itself ("resilient with respect to what?"). This dimension of engineering-based resilience coincides with the temporal nature of the problem. As shown in Figure 1.5, response of a power distribution systems to adversity

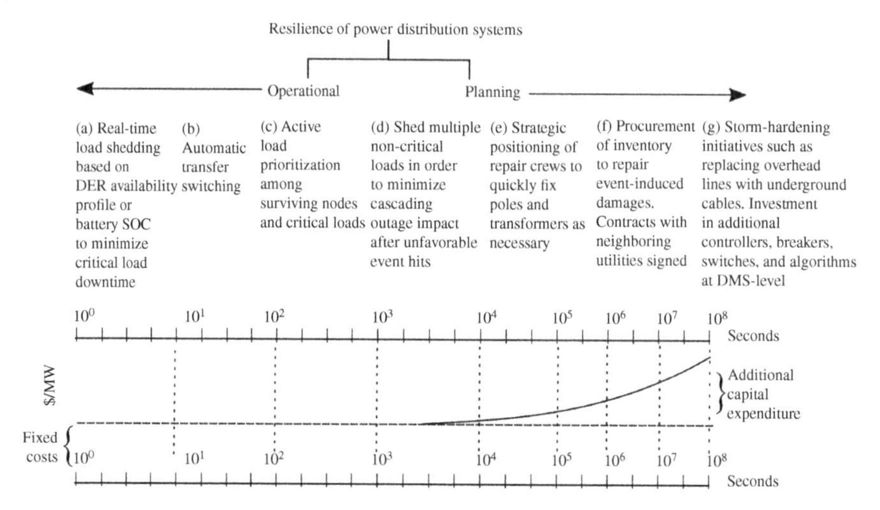

Figure 1.5 Operational and planning resiliency.

can be plotted on a time-scale. A well-engineered distribution network (that has most of the flexibility characteristics discussed in Section 1.5) will be able to ensure least LOLP during the event, in near real-time, and contributes to the *operational resilience* of the network [59, 60]. On the other hand, if the resiliency in place cannot help the system recover in near real time or save critical loads during or immediately after an event, the loads are recovered and strengthened for future events by strategies included as part of *planning resilience.*

Planning resilience or "resilience design" to improve existing networks requires significant capital expenditure. Operational resilience reaps the benefits of resilience design and does not increase the capital costs for the utilities; instead, it minimizes financial loss for the utility. The objective of planning resilience is to drive the network response to a state of operational resilience.

A utility can benefit from identifying their point of intersection of the engineering and economic dimensions of enabling resilience for developing a business case for resiliency. It is worth noting that high resiliency in one category does not imply improved resiliency in another category. The taxonomy can also be used to list the strengths and weaknesses of a system, helping utilities assess which areas to leverage for strategic advantage in case of an impending or ongoing unfavorable event, and which areas to build on during the recovery after the unfavorable event has subsided in strength.

Most resiliency-related engineering efforts are focused toward serving the critical loads in the aftermath of a large-scale disruptive event, in the near-term. Some other forms of resilience can be the following:

- A distribution system is said to be *long-term resilient* if it is able to provide a reduced service to a critical facility(ies) during the entire duration of a power outage lasting more than one week.
- A power network is said to be *contingency resilient* if it is able to restore normal electric supply to a functional critical facility(ies) by means of reconfiguration of the network

topology, by accessing and altering states of breakers and switches, manually or remotely, using a primary or secondary or reserve energy resource.
- A power distribution system is said to be characterized by *economic resilience* if it has adequate financial reserves to maintain a steady inventory of backup equipment, fund recovery operations of damaged infrastructure in minimum time, or have mechanisms in place to recover/share repair costs with customers or federal authorities following an unfavorable event impacting the network's normal operations.
- A power distribution system is said to be characterized by *preventative resilience* if it is able to reliably and proactively identify vulnerable distribution system infrastructure, make continuous improvements to existing installations such that worst-case response of the infrastructure is minimized or averted at both systems level and component level.
- A power distribution system is said to be characterized by *service resilience* if it can maintain a safe, reliable, and accessible inventory of replacement equipment and maintain a diverse fleet of transportation mechanism, and personnel who can quickly reconstruct damaged parts of the distribution systems, reconfigure settings in the control rooms, and assist customers through the period of recovery from an outage.

The taxonomy of resiliency can help design new conceptual or computing tools for understanding resiliency of the power grid, guide on-going or planned resiliency-oriented projects, and shape business-case for utilities who need to upgrade their resilience to be better prepared against uncertainties of the power grid and future losses.

1.7 Tools for Enabling Resiliency

The problems faced by the power grid are not because of nonconformance of any power system equipment connected to the grid, but because today's requirements and standards are no longer adequate to address the contemporary and emerging challenges. Due to its diversity in definitions, resilient distribution systems have no standards metrics, tools for implementation, nor any accountability toward regulators.

In order to achieve resilient power distribution systems at a scale that encompasses countries and continents, multiple organizations, and millions of customers, the scope of work spans governments, industrial partners, as well as participation from individuals. Since the problem impacts a significant portion of the population, identification of key issues by the state and federal governments is the desirable first step.

In the United States, several countries of the European Union, Russia, Brazil, Japan, India, Australia, and other nations – governments have introduced initiatives to reinforce the strength of infrastructure within the last decade [62–67].

These initiatives are funded through policies and programs, which are called upon to explore the challenges at greater depth, and then design new tools to deal with the identified problems using existing technology, or come up with breakthrough technology to solve the problem. Generic technologies are developed through specific projects, from where data are generated and provided as a feedback for success of the solutions being proposed in addressing the problem. When specific projects succeed, the results are generalized and published as standards, against which future operators and utilities will be

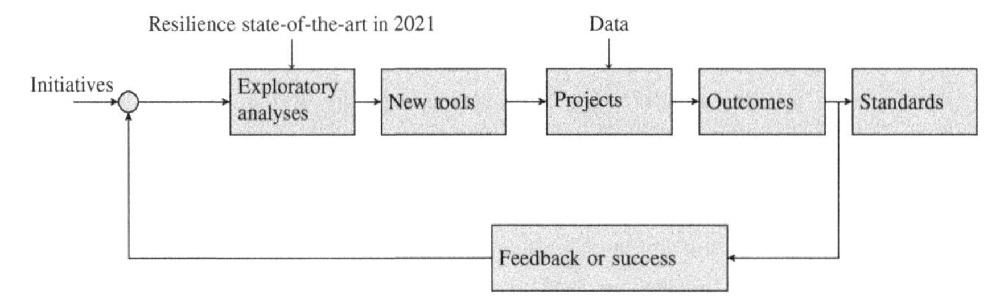

Figure 1.6 Development and adoption cycle of new technologies in power systems.

held accountable for. Figure 1.6, summarizes how new operating standards emerge from contemporary challenges and the position of resiliency in that context.

1.7.1 Pyramid Approach for Performing Resiliency Cost-Benefit Exploratory Analyses

Upon review of hundreds of research articles on power system resiliency, it can be summarized that resiliency research and development projects emphasize on being able to answer the following triads of questions-and-answers during a low-frequency/high-impact event:

1.7.1.1 Three Hows

- How many ways a critical load can be connected to a source of power during an ongoing event?
  Solutions. Depending upon the fragmentation of the network and available technologies, a critical loads can be connected to (i) main power grid; (ii) alternate substation; (iii) DERs; (iv) community-scale energy storage facilities; (v) individually owned diesel generator.
- How can the existing infrastructure be upgraded to enable and deploy resilience?
  Solutions. Resiliency-oriented infrastructure upgrades can happen through (i) expensive storm-hardening practices such as converting overhead lines to underground lines to prevent wind and storm related damages (ii) installing redundant feeders, lines, and transformers (iii) software tools that leverage artificial intelligence and machine learning to proactively detect threats and prepare the power system in advance.
- How to enable and coordinate resilience control-actions effectively?
  Solutions. Modern data-sensing tools such as Phasor measurement units (PMU) deployed at both transmission and distribution systems and innovative algorithms for automating restoration and optimizing storage controllers to maximize the number of critical loads picked up after a forced outage. Advanced Metering Infrastructure (AMI) and its network of smart meters is actively (automatically and instantly) identifying which customers have been restored and which still need to be recovered – thus further reducing outage times. These advances are further strengthened by their integration with advanced building management systems (BMSs), and emerging home automation technologies – that not only aid in planning resiliency by improving home energy consumption efficiency but also assist in enabling automation resiliency by curtailing noncritical loads at the domestic-level during extreme events.

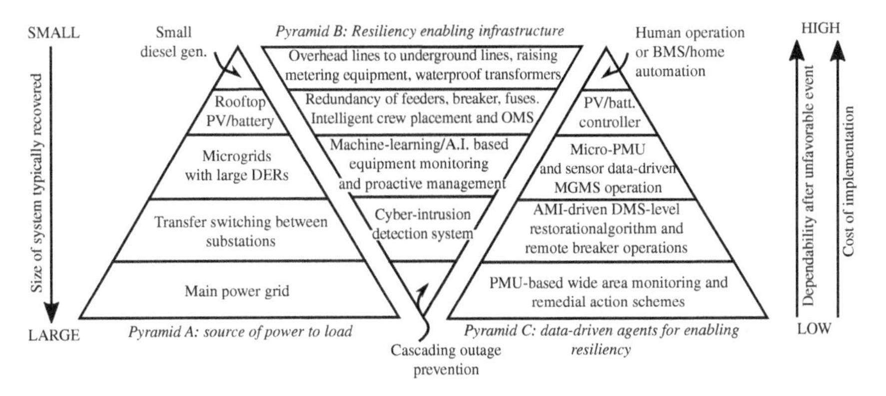

Figure 1.7 Resiliency pyramids.

1.7.1.2 Three Whats

- What fraction of the power system can be rapidly recovered or prevented from catastrophic damage?
- What is the reliability of the system after the damaging event has passed?
- What is the cost of enabling and implementing resilience of power distribution systems?

Each of the "three hows" and their resolutions can be arranged on top of each other in three separate columns. Since well-known standard quantifiable quantities are chosen as "the whats," which can be used as a vertical axis to arrange the stack "the hows" in ascending or descending order along "the what" axes. Every stack in each pyramid correspond to a stack in the same level but in another pyramid, such that the two resilience-enabling strategies require complementary enabling tools, same costs for implementation, and have similar impacts. It may be emphasized that each of the pyramids are interdependent, and resilience is not achievable without one or the other Figure 1.7.

The pyramid model can be also be assembled along three axes, in the shape of a "Y," as shown in Figure 1.8. As one travels inwards toward the center along each axis, a more dependable, but more expensive form of resiliency is enabled. It is possible that a distribution system can have penetrated different depths along each of the axis. The Y chart is a useful tool in estimating and visualizing the existing state of resiliency of a distribution system concisely, and the future investments necessary for enabling higher resiliency in the distribution network.

Considering the United States as an example of a developed economy in pursuit of higher resiliency of its critical infrastructure, a fiscal deficit in hundreds of billions of dollars exist between the need and availability of funds required for enabling nationwide resilience. According to American Society of Civil Engineers (ASCE), the investment required for upgrading the US power grid would cost approximately $673 billions [68].

The US federal government has invested $16 billions for specific smart grid projects and new transmission lines under the American Recovery and Reinvestment Act of 2009 [69] and $ 220 millions through rounds of Grid Modernization Initiatives between 2016 and 2020 [70, 71], to meet the visions outlined in the high-level Multi-Year Program Plan (MYPP) of the Department of Energy (DOE) [72]. In order to optimally use the monetary resources for

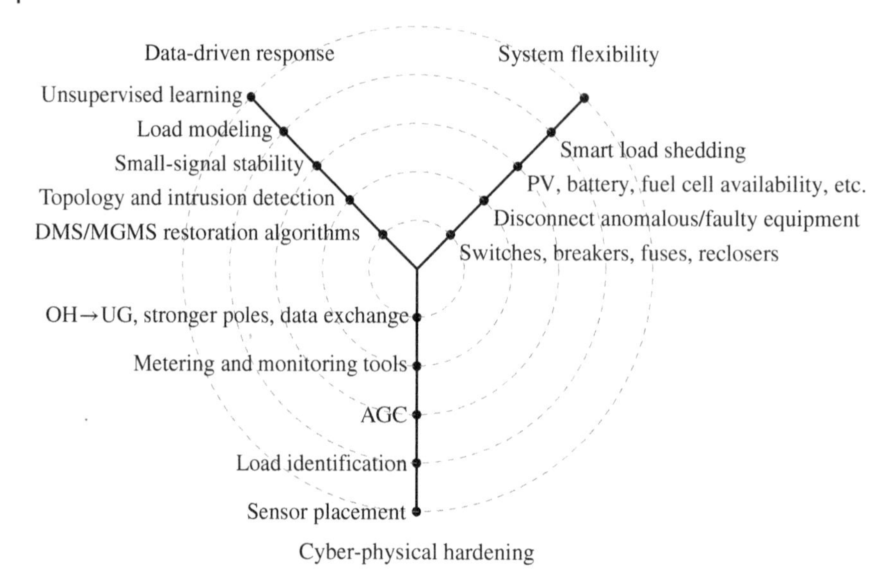

Figure 1.8 Resiliency Y-chart.

maximum results, proper evaluation of the status of resiliency of networks would make it easier to channelize efforts toward modernizing the power grid.

In this section, a deeper discourse is provided on how advancements in measurement devices, computation power, and accessible and actionable artificial intelligence available through machine learning and smart data visualization can empower utilities in creation of new tools and technologies for resilient, and consequently more stable, secure, and reliable, power distribution systems.

1.7.2 Data-Driven Approaches for Higher Operational Resiliency

One of the greatest advantages of modern power systems is the availability of high fidelity, high resolution data that help operators have unprecedented insights and forward-thinking analytics.

The key benefit of data-driven approaches for higher operational resiliency can be conceptualized by considering the following use case of a faulty distribution transformer and a potential outage scenarios.

1.7.2.1 Contemporary Approach

An important distribution system transformer faults due to an internal problem which has been developing for weeks. Customers and smart metering reports an outage in a region, and field crews are dispatched to switch and replace. Despite DERs and microgrids that are available to respond, the lack of prediction and prescription does not allow their full capabilities to be utilized. The transformer will be replaced, but the customer loses reliable power and generation revenue.

1.7.2.2 Data-Driven Approach

With descriptive and diagnostics enabled, the fault can be quickly located and the exact reason for the problem diagnosed. Field crews are more accurately dispatched, and therefore a more efficient use of resources. However, simple diagnostic data base cannot prevent the power outage but lay the foundation for anticipatory analysis using machine-learning.

Based on available measured data, there was a developing fault within the transformer (incipient fault detection). The application, in near real time, can be used to analyze a set of scenarios and determine the best reconfiguration and resource dispatch and the appropriate customer rates applied for participation. The action to reduce transformer load is taken quickly, without major operator or field crew intervention and the transformer can then be easily repaired or replaced.

One of the examples of enabling resiliency is the case of *smart vegetation tracking*. Vegetation management is a mandatory exercise all utilities engage in to maintain the reliability of the systems. However, in context of resiliency, traditional vegetation management approaches are falling short in performance. As evidenced by all hurricanes in past and recent history, power outages in distribution networks are a consequence of noncompliant trees (i.e. trees that have violated the 15-feet right of way from the power lines) tripping on the conductors during extreme weather events. Today's satellite imagery technology – such as the ones used by Google Maps for navigation – can help utilities detect trees most likely to cause a power outage and use that data to proactively trim high-risk vegetation. This is a data-driven way of enabling resiliency, and the workflow of this has been shown in Figure 1.9.

The benefits of the data-driven approach to the consumer are reduced outage time during extreme events (which is the main objective of resilient distribution systems), and new markets for the DER (which help create the business-case for resilience). The utility benefits from predictive analysis by enabling repair rather than run to failure. Field crew dispatch time is minimized with appropriate analysis.

Resiliency metrics are quantified representations of the system's ability to withstand, protect, respond, and recover to large-scale outage events.

Resiliency, irrespective of how it may be classified or interpreted, is not a fixed quantity for a given system and varies with availability of resources, customer demands, operating conditions, and several other variables. This further underlines the importance of having a data-driven approach toward enabling and maintaining resiliency in the network. Higher data resolutions and data fidelity, coupled with the ability to analyze these data streams faster, will be key toward further improvements in leveraging resiliency.

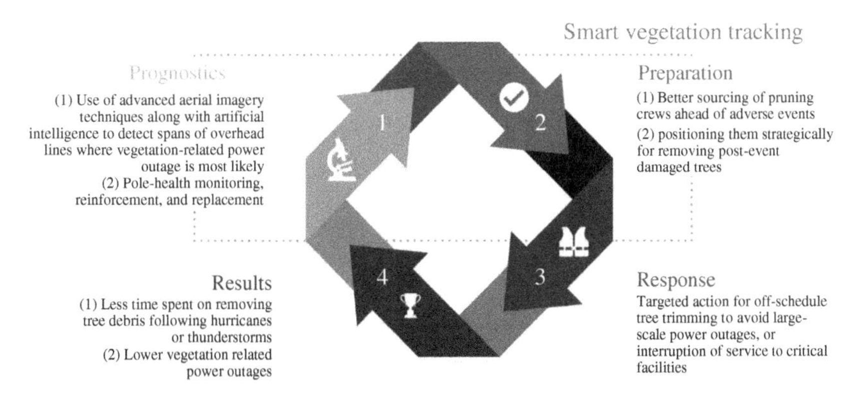

Figure 1.9 Example of data-driven resiliency, using smart vegetation tracking.

Several authors have proposed resiliency metrics of power distribution systems. However, most of these measurements are computed for power system state at one one-instant of time. For example, a multicriteria decision-making-based resiliency metric, proposed in [47], provides a mapping between resource availability within a distribution network and access of critical loads to the available resources [73].

If a successive measurement is to be made using the existing process for deployment of the resiliency metric to the distribution management system (DMS), the algorithm would need to assume that changes in the operating conditions would not change to an extent where a small-signal instability is caused in the distribution system and that power flows will converge for all feeders.

Using the precise distribution-level PMU voltage and angle data at multiple distribution nodes data resolution with integrated renewable resources, the modified metric can be integrated into the on-line DMS for enabling resiliency-driven control actions. This modified resiliency metric can be used as an input and optimization reference to help maintain stability in, especially, low-inertia, islanded distribution system, operating as a microgrid [74].

1.8 Summary

In this chapter, resiliency of power systems has been explored according to multiple prevalent concepts, definitions, and interpretations of the topic. The differences between resiliency and other closely associated terms such as reliability, security, and vulnerability have been clearly identified. It has been established that investments in resilience have multiple benefits. Investments in energy efficiency, smart grid technologies, storage, and distributed generation can contribute to enhanced resiliency and low-carbon electric grid, as well as provide operational flexibility for grid operators. Innovative technologies have significant value for the electricity system. Resilient power systems provide the ecosystem where new technologies and data applications can enable new services and customer choices, which further aid to enhance the resiliency of the power systems [75].

A future crisis in this century, whether it is a climate-change catastrophe, another pandemic, war, or an economic meltdown, will probably take another form without any precedence.

Since electric utilities will need to continue delivering reliable power, resiliency-driven adaptive load profiles in action will go a long way in helping utilities deal with the jolts with more efficiency and flexibility [76–81].

Resilience is built on simplicity – simple to set up, simple to use, and simple to fix when broken. Simplicity does not mean a lack of sophistication; on the contrary, it means a higher level of abstraction of complex processes, such that the creativity of human minds and engineering excellence can have more space to manifest.

As utilities continue to iterate on deploying greater resiliency in the modern power grid, resilience algorithms should complement human knowledge and past experiences in dealing with crises and enable action by delivering insights and information in correct proportions with precision, without any complex paraphernalia.

Utilities must continue to innovate to empower customers to refuel with new kinds of fuels for their cars (electric, hydrogen). Regulators require utilities to conform to new principles of climate-friendliness and decarbonization, and on top increased more affordable and more reliable power systems that do not fail under strong weather conditions or cyber-attacks. It is counterintuitive, but our experts know that the aging critical infrastructure can become more fragile with more digital devices if the complexity is not strengthened by the superior simplicity of managing the digitization of the power grid.

Thus, a widespread conceptual understanding of resilience will enable utilities in leveraging the benefits of resilient systems, and leverage the innovations to potentially increase their revenues beyond the sale of electric kilowatt-hours.

References

1 Biswas, S., Singh, M.K., and Centeno, V.A. (2021). Chance-constrained optimal distribution network partitioning to enhance power grid resilience. *IEEE Access* 9: 42169–42181.

2 Strogatz, S.H. (2001). Exploring complex networks. *Nature* 410 (6825): 268.

3 Liu, Y.-Y., Slotine, J.-J., and Barabási, A.-L. (2011). Controllability of complex networks. *Nature* 473 (7346): 167.

4 Aziz, T., Lin, Z., Waseem, M., and Liu, S. (2021). Review on optimization methodologies in transmission network reconfiguration of power systems for grid resilience. *International Transactions on Electrical Energy Systems* 31 (3): e12704.

5 Noebels, M., Quirós-Tortós, J., and Panteli, M. (2021). Decision-making under uncertainty on preventive actions boosting power grid resilience. *IEEE Systems Journal* 16 (2): 2614–2625.

6 NERC (2020). NERC Reliability Standards Development Plan 2021–2023. *Tech. Rep.*

7 Kadir, S.U., Majumder, S., Chhokra, A.D. et al. (2021). Reinforcement learning based proactive control for transmission grid resilience to wildfire. *arXiv preprint arXiv:2107.05756.*

8 Kim, T., Wright, S.J., Bienstock, D., and Harnett, S. (2015). Vulnerability analysis of power systems. *arXiv preprint:1503.02360.*

9 Salmeron, J., Wood, K., and Baldick, R. (2009). Worst-case interdiction analysis of large-scale electric power grids. *IEEE Transactions on Power Systems* 24 (1): 96–104.

10 Arroyo, J.M. (2010). Bilevel programming applied to power system vulnerability analysis under multiple contingencies. *IET Generation, Transmission and Distribution* 4 (2): 178–190.

11 Liu, C.-C., Jung, J., Heydt, G.T. et al. (2000). The strategic power infrastructure defense (SPID) system. A conceptual design. *IEEE Control Systems* 20 (4): 40–52.

12 Vellaithurai, C., Srivastava, A., Zonouz, S., and Berthier, R. (2015). CPIndex: Cyber-physical vulnerability assessment for power-grid infrastructures. *IEEE Transactions on Smart Grid* 6 (2): 566–575.

13 Kandaperumal, G., Pandey, S., and Srivastava, A. (2021). AWR: Anticipate, withstand, and recover resilience metric for operational and planning decision support in electric distribution system. *IEEE Transactions on Smart Grid* 13 (1): 179–190.

14 Allan, R. and Billinton, R. (2000). Probabilistic assessment of power systems. *Proceedings of the IEEE* 88 (2): 140–162.

15 Jia, L., Pannala, S., Kandaperumal, G., and Srivastava, A. (2022). Coordinating energy resources in an islanded microgrid for economic and resilient operation. *IEEE Transactions on Industry Applications* 58 (3): 3054–3063.

16 Kropp, T. (2006). System threats and vulnerabilities [power system protection]. *IEEE Power and Energy Magazine* 4 (2): 46–50.

17 Bompard, E., Huang, T., Wu, Y., and Cremenescu, M. (2013). Classification and trend analysis of threats origins to the security of power systems. *International Journal of Electrical Power & Energy Systems* 50: 50–64.

18 Flick, T. and Morehouse, J. (2010). *Securing the Smart Grid: Next Generation Power Grid Security*. Elsevier.

19 Rose, A., Oladosu, G., and Liao, S.-Y. (2007). Business interruption impacts of a terrorist attack on the electric power system of los angeles: customer resilience to a total blackout. *Risk Analysis* 27 (3): 513–531.

20 Noorazar, H., Srivastava, A., Pannala, S., and Sadanandan, S.K. (2021). Data-driven operation of the resilient electric grid: a case of COVID-19. *The Journal of Engineering* 2021 (11): 665–684.

21 Kosut, O., Jia, L., Thomas, R.J., and Tong, L. (2010). Malicious data attacks on smart grid state estimation: attack strategies and countermeasures. *IEEE International Conference on Smart Grid Communications (SmartGridComm)*, pp. 220–225.

22 Hahn, A. (2010). Smart grid architecture risk optimization through vulnerability scoring. *IEEE Conference on Innovative Technologies for an Efficient and Reliable Electricity Supply (CITRES)*, pp. 36–41.

23 Billinton, R. and Allan, R.N. (1992). *Reliability Evaluation of Engineering Systems*. Springer.

24 Brown, R.E. (2017). *Electric Power Distribution Reliability*. CRC Press.

25 Morison, K., Wang, L., and Kundur, P. (2004). Power system security assessment. *IEEE Power and Energy Magazine* 2 (5): 30–39.

26 Balu, N., Bertram, T., Bose, A. et al. (1992). On-line power system security analysis. *Proceedings of the IEEE* 80 (2): 262–282.

27 Biswas, S.S., Vellaithurai, C.B., and Srivastava, A.K. (2013). Development and real time implementation of a synchrophasor based fast voltage stability monitoring algorithm with consideration of load models. *Proceedings of IEEE Industry Applications Society Annual Meeting*.

28 Kundur, P., Paserba, J., Ajjarapu, V. et al. (2004). Definition and classification of power system stability IEEE/CIGRE joint task force on stability terms and definitions. *IEEE Transactions on Power Systems* 19 (3): 1387–1401.

29 Marceau, R., Endrenyi, J., Allan, R. et al. (1997). Power system security assessment: a position paper. *Electra* 175: 49–77.

30 Fink, L.H., Liou, K.-L., and Liu, C.-C. (1995). From generic restoration actions to specific restoration strategies. *IEEE Transactions on Power Systems* 10 (2): 745–752.

31 Ma, T.-K., Liu, C.-C., Damborg, M., and Chang, M. (1994). An integrated model-and rule-based approach to design of automatic switching for subtransmission lines. *IEEE Transactions on Power Systems* 9 (2): 750–756.

32 Liu, C.-C. and Tomsovic, K. (1986). An expert system assisting decision-making of reactive power/voltage control. *IEEE Transactions on Power Systems* 1 (3): 195–201.

33 Deng, Y., Cai, L., and Ni, Y. (2003). Algorithm for improving the restorability of power supply in distribution systems. *IEEE Transactions on Power delivery* 18 (4): 1497–1502.

34 Yutian, L., Rui, F., and Terzija, V. (2016). Power system restoration: a literature review from 2006 to 2016. *Journal of Modern Power Systems and Clean Energy* 4 (3): 332–341.

35 Miu, K.N., Chiang, H.-D., Yuan, B., and Darling, G. (1998). Fast service restoration for large-scale distribution systems with priority customers and constraints. *IEEE Transactions on Power Systems* 13 (3): 789–795.

36 Hou, Y., Liu, C.-C., Sun, K. et al. (2011). Computation of milestones for decision support during system restoration. *2011 IEEE Power and Energy Society General Meeting*, pp. 1–10.

37 Sakaguchi, T. and Matsumoto, K. (1983). Development of a knowledge based system for power system restoration. *IEEE Transactions on Power Apparatus and Systems* (2): 320–329.

38 Srivastava, S.K., Butler-Purry, K.L., and Sarma, N. (2002). Shipboard power restored for active duty. *IEEE Computer Applications in Power* 15 (3): 16–23.

39 Srivastava, S.K., Xiao, H., and Butler-Purry, K.L. (2002). Multi-agent system for automated service restoration of shipboard power systems. *CAINE*. CiteSeer, pp. 119–123.

40 Xu, Y., Liu, C.-C., Schneider, K.P., and Ton, D.T. (2016). Placement of remote-controlled switches to enhance distribution system restoration capability. *IEEE Transactions on Power Systems* 31 (2): 1139–1150.

41 Richard, J. and Rossini, A. (2017). Grid Hardening and Resiliency. *Tech. Report, Leidos, VA, 2019, Tech. Rep.*

42 Burleyson, C.D., Rahman, A., Rice, J.S. et al. (2021). Multiscale effects masked the impact of the COVID-19 pandemic on electricity demand in the United States. *Applied Energy* 304: 117711.

43 Hussain, A., Bui, V.-H., and Kim, H.-M. (2019). Microgrids as a resilience resource and strategies used by microgrids for enhancing resilience. *Applied Energy* 240: 56–72.

44 DOE. Quadrennial Technology Review: An Assessment of Energy Technologies and Research Opportunities, Washington D.C., September 2015. Quadrennial Report. *Tech. Rep.*

45 Staid, A. (2021). North American Energy Resilience Model (NAERM). *Tech. Rep. No. SAND2021-3075PE* Albuquerque, NM (United States): Sandia National Lab. (SNL-NM).

46 EPRI (2013). Enhancing Distribution Resiliency: Opportunities for Applying Innovative Technologies. *Tech. Rep. 1026889* Electric Power Research Institute. 20

47 Chanda, S. and Srivastava, A.K. (2016). Defining and enabling resiliency of electric distribution systems with multiple microgrids. *IEEE Transactions on Smart Grid* 7 (6): 2859–2868.

48 Hollnagel, E., Woods, D.D., and Leveson, N. (2007). *Resilience Engineering: Concepts and Precepts*. Ashgate Publishing, Ltd.

49 Arghandeh, R., von Meier, A., Mehrmanesh, L., and Mili, L. (2016). On the definition of cyber-physical resilience in power systems. *Renewable and Sustainable Energy Reviews* 58: 1060–1069.

50 Gao, H., Chen, Y., Xu, Y., and Liu, C.C. (2016). Resiliency-oriented critical load restoration using microgrids in distribution systems. *IEEE Transactions on Smart Grid* 7 (6): 2837–2848.

51 Xu, Y., Liu, C.C., Schneider, K. et al. (2018). Resiliency-oriented critical load restoration using microgrids in distribution systems. *IEEE Transactions on Smart Grid* 7 (6): 2837–2848.

52 Wang, Y., Chen, C., Wang, J., and Baldick, R. (2016). Research on resilience of power systems under natural disasters–a review. *IEEE Transactions on Power Systems* 31 (2): 1604–1613.

53 Hosseini, S., Barker, K., and Ramirez-Marquez, J.E. (2016). A review of definitions and measures of system resilience. *Reliability Engineering & System Safety* 145: 47–61.

54 Shafieezadeh, A., Onyewuchi, U.P., Begovic, M.M., and DesRoches, R. (2014). Age-dependent fragility models of utility wood poles in power distribution networks against extreme wind hazards. *IEEE Transactions on Power Delivery* 29 (1): 131–139.

55 Bjarnadottir, S., Li, Y., and Stewart, M.G. (2012). Hurricane risk assessment of power distribution poles considering impacts of a changing climate. *Journal of Infrastructure Systems* 19 (1): 12–24.

56 Bjarnadottir, S., Li, Y., and Stewart, M.G. (2014). Risk-based economic assessment of mitigation strategies for power distribution poles subjected to hurricanes. *Structure and Infrastructure Engineering* 10 (6): 740–752.

57 Mitchell, J.W. (2013). Power line failures and catastrophic wildfires under extreme weather conditions. *Engineering Failure Analysis* 35: 726–735.

58 Dagher, H. (2006). *Reliability-Based Design of Utility Pole-Structures.* American Society of Civil Engineers.

59 Chanda, S., Srivastava, A.K., Mohanpurkar, M.U., and Hovsapian, R. (2016). Quantifying power distribution system resiliency using code based metric. *2016 IEEE International Conference on Power Electronics, Drives and Energy Systems (PEDES).* IEEE, pp. 1–6.

60 Chanda, S., Srivastava, A.K., Mohanpurkar, M.U., and Hovsapian, R. (2018). Quantifying power distribution system resiliency using code based metric. *IEEE Transactions on Industrial Applications* 54 (4): 3676–3686.

61 Zhang, X., Tu, H., Guo, J. et al. (2021). Braess paradox and double-loop optimization method to enhance power grid resilience. *Reliability Engineering & System Safety* 215: 107913.

62 Molyneaux, L., Wagner, L., Froome, C., and Foster, J. (2012). Resilience and electricity systems: a comparative analysis. *Energy Policy* 47: 188–201.

63 Grove, A. and Burgelman, R. (2008). An electric plan for energy resilience. *The McKinsey Quarterly (December).*

64 Hussey, K. and Pittock, J. (2012). The energy–water nexus: managing the links between energy and water for a sustainable future. *Ecology and Society* 17 (1): 9 p.

65 Manfren, M., Caputo, P., and Costa, G. (2011). Paradigm shift in urban energy systems through distributed generation: methods and models. *Applied Energy* 88 (4): 1032–1048.

66 Leung, G.C., Cherp, A., Jewell, J., and Wei, Y.-M. (2014). Securitization of energy supply chains in china. *Applied Energy* 123: 316–326.

67 Panteli, M. and Mancarella, P. (2015). The grid: stronger, bigger, smarter?: Presenting a conceptual framework of power system resilience. *IEEE Power and Energy Magazine* 13 (3): 58–66.

68 Li, Z. and Guo, J. (2006). Wisdom about age [aging electricity infrastructure]. *IEEE Power and Energy Magazine* 4 (3): 44–51.

69 ARRA (2009). American recovery and reinvestment act. *U.S. Congress*, 7.

70 GMLC (2015). *Grid Modernization Laboratory Consortium*. U.S. Department of Energy.

71 Ton, D.T. and Wang, W.P. (2015). A more resilient grid: the us department of energy joins with stakeholders in an R&D plan. *IEEE Power and Energy Magazine* 13 (3): 26–34.

72 USDOE. (2016) Multi-year program plan.

73 Kandaperumal, G., Linli, J., Pannala, S. et al. (2021). RT-RMS: A real-time resiliency management system for operational decision support. *2020 52nd North American Power Symposium (NAPS)*. IEEE, pp. 1–6.

74 Sarker, P.S., Sadanandan, S.K., and Srivastava, A.K. (2022). Resiliency metrics for monitoring and analysis of cyber-power distribution system with IoTs. *IEEE Internet of Things Journal*.

75 Acharya, A., Pannala, S., Srivastava, A.K., and Bhavirisetty, S.R. (2021). Resiliency planning and analysis tool for the power grid with hydro generation and DERs. 2021 North American Power Symposium (NAPS). IEEE, pp. 1–6.

76 Kandaperumal, G., Pandey, S., and Srivastava, A.K. (2021). Enabling electric distribution system resiliency through metrics-driven black start restoration. 2021 IEEE Industry Applications Society Annual Meeting (IAS). IEEE, pp. 1–8.

77 Kandaperumal, G., Majumder, S., and Srivastava, A.K. (2022). Microgrids as a resilience resource in the electric distribution grid. In: *Electric Power Systems Resiliency: Modeling, Opportunity, and Challenges*, Elsevier Academic Press.

78 Qin, C., Jia, L., Bajagain, S. et al. (2023). An integrated situational awareness tool for resilience-driven restoration with sustainable energy resources. *IEEE Transactions on Sustainable Energy* 14 (2): 1099–1111.

79 Menazzi, M., Qin, C., and Srivastava, A.K. (2023). Enabling resiliency through outage management and data-driven real-time aggregated DERs. *IEEE Transactions on Industry Applications* 59 (5): 5728–5738.

80 Kadir, S.U., Majumder, S., Srivastava, A.K. et al. (2023). Reinforcement learning based proactive control for enabling power grid resilience to wildfire. *IEEE Transactions on Industrial Informatics*, pp. 1–10.

81 Stankovic, A.M., Tomsovic, K.L., De Caro, F. et al. (2023). Methods for analysis and quantification of power system resilience. *IEEE Transactions on Power Systems* 38 (5): 4474–4787.

2

Measuring Resiliency Using Integrated Decision-Making Approach

Sayonsom Chanda[1], Prabodh Bajpai[2], and Anurag K. Srivastava[3]

[1] *National Renewable Energy Laboratory, Golden, CO, USA*
[2] *Department of Sustainable Energy Engineering, I.I.T., Kanpur, U.P., India*
[3] *Lane Department of Computer Science and Electrical Engineering, West Virginia University, Morgantown, WV, USA*

2.1 Introduction

Resiliency has been defined through numerous working definitions in [1–3], which can be summarized as "Resilience includes the ability to withstand and recover from deliberate attacks, accidents, or naturally occurring threats or incidents." The key features of resiliency are resourcefulness, robustness, and rapid recovery. Resourcefulness relates to contingency planning for resources prepared to mitigate the crisis, robustness relates to the ability to maintain critical operations during extreme contingencies, and rapid recovery relates to the shortest possible time required to bring the system back to normal operations [4].

The critical infrastructure of a community when affected by a power outage possesses a great national security risk and leads to huge financial losses [5]. There are multiple factors those results in power outages, such as weather [6], insufficient generation, and transmission failure [7]. The increase in the number of extreme weather events coincident with the increasing demand for reliable power mainly resulted in greater emphasis on the resiliency of the power distribution system (PDS) [8]. During extreme weather events, resiliency efforts should focus on rectifying power interruptions to critical loads (CLs), such as airports, hospitals, city halls, and the other buildings deemed important to the community as providing power to all the connected loads during severe events will be economically infeasible [9, 10]. While most of such power outages are caused by faults as a result of trees (or their branches) falling on distribution lines and poles, major power outages tend to be caused by extreme weather events, like US Hurricane Sandy in 2012 [11], a flood of 2010–2011 in Queensland (Australia) [12], and severe ice storm in China in 2008 [13]. With a higher expectancy of power interruptions due to weather and progressive climate change, there has been an increased community-wide emphasis on "developing highly resilient critical infrastructure," which is critical to the safety and affluence of the community [14]. As more acts continue to be mandated based on the US Presidential Policy Directive-21 [1], traditional methods of measuring PDS efficiency and reliability might be inadequate to address these.

Resiliency of Power Distribution Systems, First Edition.
Edited by Anurag K. Srivastava, Chen-Ching Liu, and Sayonsom Chanda.

Some researchers have proposed formulating resiliency metrics considering the entire infrastructure of a community or a city [3, 15]. Their method comprises a survey of all the resources available to the entire community for their index. However, their method is not focused on PDS, and thus cannot be effectively used for electrical engineering purposes. Sandia National Laboratories (SNL) has proposed a "Resilience Framework" where a "resilience metric framework is defined as the probability of consequence X given threat Y." There are further reviews on several methods of determining the resilience of energy infrastructure as proposed by Watson and O'Rourke [2, 16]. However, such models are probabilistic, and the computation of the probability of an event occurring depends on several other factors. Proper evaluation of the status of the resiliency of PDS would not only make it easier to channelize funds and efforts toward making the power grid better in the long term but also improve the decisions taken by a distribution network operator (DNO) immediately following an event that triggers multiple contingencies. Recent advances in distribution automation [17], campus-based or community-based microgrids [18], relaying, automatic transfer switching, and "micro"-phasor measurement units [19] can be used to improve the resiliency of the system. However, these investments are expensive [20] and require a thorough cost–benefit analysis to justify their effectiveness. Moreover, the number of factors impacting resiliency considered in the published literature is limited and not very inclusive. Also, the operational constraints (power flow voltage limits, generation limits, thermal limits) of a PDS provide additional challenges.

A microgrid is able to contribute to grid resiliency by restoring CLs in active distribution systems due to the impact of natural forces [21]. Being an energy resource, a microgrid is able to supply its local loads during fault or emergency conditions in an islanded operation. Being a community resource, a microgrid is able to support the load beyond its geographical boundaries (e.g. by transferring the extra power to loads outside the microgrid). Networked microgrids will be more common in the future and can be defined as "a high-level structure, formed at the medium voltage (MV) level, consisting of several low voltage (LV) microgrids generation (MG) and distributed generation (DG) units connected to adjacent MV feeders" [22, 23]. In case of networked microgrids, loads are served in neighboring microgrids in the same cluster. Enabling resiliency using microgrids is in infancy and there are limited publications [3, 6, 7]. A preventive reinforcement approach is offered in [25] to the microgrid operators to develop the resiliency of generation and demand scheduling against disruptions in multiple energy carrier microgrids. To enable resilience in PDS, after major faults due to natural disasters, optimum autonomous communication requirements in microgrids have been examined in [26]. In a review of smart grid technologies being presently employed to improve the resilience of PDS, it has been noted that utility depends greatly on "experience" to take restorative actions for hardening and enabling resiliency in their systems [27].

An algorithm to quantify the resiliency of different configurations possible within a distribution network is presented here, thus enabling better cost–benefit analysis, planning and operation of modern and future, more resilient distribution systems. Mostly, resiliency studies of critical infrastructure have used complex network theory and are directed toward their structure and organization [28–31]. However, the physical constraints (power flow voltage limits, generation limits, thermal limits) of a PDS provide additional challenges. The resilience of a PDS is also affected by the control decisions taken by the operator or the reconfiguration method. Resiliency and robustness of the power systems complex network have been studied by several researchers [24, 32–35]; however, they have not included the comparative effectiveness of different network configurations supplying all CLs during planning and contingent operations of a PDS.

This chapter presents the resiliency metrics based on several indicators to measure the resiliency of the PDS in Section 2.2 and quantify it using the integrated decision-making approach in Section 2.3. Further, an algorithm for enabling the resiliency of a given distribution network through feasible network (FN) solutions and their resiliency quantification is presented in Section 2.4. The maximum positive impact on resiliency while planning as well as operating stages is demonstrated through several simulation studies in Section 2.5 followed by a conclusion in Section 2.6.

2.2 Feature to Measure Resiliency of Power Distribution System

Definitions of resilience are subjective, but it is imperative to find means to quantify it because it is hard to improve upon the existing resiliency of the system if there is no way to measure it. The resiliency metric proposed in this chapter is based on the principle that all causative factors to enhance resiliency will be strongly interrelated to the indicators described in this section. The resiliency of a power system depends on the ability of a system to maintain continuous supply to CLs, in events of "low frequency, high impact" contingencies such as a hurricane, storms, floods, cyberattacks, etc. To enable resiliency in PDS, there can be numerous ways such as storm-hardening of infrastructure, improved cyber-security, and implementation of more intelligent control algorithms. However, above these factors, improved management of system redundancy will be critical to improving the resiliency of a PDS. So, it requires that there should be multiple paths to connect the CL to several sources, and the number of feasible paths plays a critical role in determining the resiliency of the distribution system.

To define the following indicators a generic PDS is considered as a graph with b branches and n nodes. By selecting one path for each CL, all possible path combinations to provide power supply to all CLs are found and after eliminating path combinations with loop (PCWL) they become possible networks (PNs). After eliminating similar networks, PNs become unique networks (UNs) and each unique network satisfying the power flow convergence becomes the FN.

(1) Switching operations indicator (SOI): This indicator is defined as the total number of changes in the state of the switches, that is, from normally open (NO) to closed and from normally closed (NC) to open. These switching operations create different FNs without any loop connecting all CLs to a source.
(2) Branch count indicator (BCI): This indicator is defined for nth PN by the ratio of the total number of connected branches for each PCWL in nth PN to the number of all CLs in nth PN. Finally, the BCI for mth FN is defined in Eq. (2.1) as the average value of all corresponding similar PNs:

$$\mathrm{BCI}_m = \frac{\sum_{k=1}^{N_m} \frac{\text{Branches in PCWL for } n\text{th PN}}{\text{Number of CLs in } n\text{th PN}}}{N_m} \tag{2.1}$$

where m is for mth FN being considered and N_m is the total number of similar PNs for the mth FN.

(3) Branch overlapping indicator (BOI): It is defined as the total number of common branches in each PCWL of a PN. Again the average value of all corresponding similar PN is considered to define the BOI for the mth FN in Eq. (2.2):

$$\mathrm{BOI}_m = \frac{\sum_{n=1}^{N_m} \text{Common Branches in } n\text{th PN}}{N_m} \tag{2.2}$$

(4) Path availability indicator (PAI): This indicator is the ratio of the total number of paths available for all CLs connecting to all sources to the total number of CLs in an FN. PAI for mth FN is expressed in Eq. (2.3):

$$\mathrm{PAI}_m = \frac{\text{Path connecting all CLs to all sources in } m\text{th FN}}{\text{Number of CLs in } m\text{th FN}} \tag{2.3}$$

(5) Source availability indicator (SAI): It refers to the ratio of the number of available sources used to supply all CLs to the number of all CLs in a PN. The average value of all similar PNs is considered for the corresponding FN in Eq. (2.4). This indicator should have a high value for high resiliency:

$$\mathrm{SAI}_m = \frac{\sum_{n=1}^{N_m} \frac{\text{Sources supplying all CLs in } n\text{th PN}}{\text{Number of CLs in } n\text{th PN}}}{N_m} \tag{2.4}$$

(6) Reliability and loss factor indicator (RLFI): This indicator includes the impact of the type of source and its location with respect to CL together. This is based on the reliability or probability of availability of the source supplying to the CL and the losses in a distribution corresponding to this supply. If the CL is drawing power from the main grid, both the probability of availability and loss factor should be highest because the main grid will be more reliable compared to DGs but physically far away from the CL. However, the reliability of all DGs to supply a CL may be assumed the same but a high loss factor should be considered if power is drawn from a DG located in another MG than from a DG located in the same MG, where CL is located. RLFI for mth FN is expressed by Eq. (2.5):

$$\mathrm{RLFI}_m = \frac{\sum_{n=1}^{N_m} \text{Reliability factor} \times \text{Loss factor for } n\text{th PN}}{N_m} \tag{2.5}$$

(7) Aggregated centrality indicator (ACI): This indicator confines the information about the importance of the node for the given topology of the network. To determine this value, the central point dominance (or betweenness centrality) of each node in a path is determined using Eq. (2.6) to identify the importance of each node for the connectivity of the network:

$$C_B(q) = \sum_{x \neq q \neq y} \frac{\delta_{xy}(q)}{\delta_{xy}} \tag{2.6}$$

where δ_{xy} is the total number of shortest paths from node x to y, and $\delta_{xy}(q)$ is the number of those paths that pass through node q. The central point dominance of the full network is illustrated by the ACI of mth FN as expressed by Eq. (2.7):

$$\mathrm{ACI}_m = \frac{\sum_{q=1}^{Q_m} \zeta_q^m \times C_B(q)}{Q_m} \tag{2.7}$$

where ζ_q^m is the order of node q and Q_m is the total number of nodes, respectively, in the mth FN.

A metric based on the above-mentioned indicators is used to measure and quantify the resiliency of PDS through an integrated decision-making approach considering the interdependency of these indicators. Basically resiliency computation of a network is the mathematical modeling of interaction and simultaneous effect of numerous indicators of resiliency metrics, which can be computationally expensive for medium to large distribution networks.

The problem of finding the number of simple paths between any two nodes is an non-polynomial (NP)-hard problem [36]. However, for a PDS, the problem of finding the number of paths becomes even more complex due to an increase in the number of possible paths with additional switches possible in each section. It requires $\Gamma(n + b)$ space and runs in $\Gamma(nb + s)$ and $\Gamma(2s{:}nb + \log n)$ time to compute on unweighted and weighted networks, respectively, where Γ is a Landau notation [37], used to denote the increase in the growth rate of the function when the number of nodes increases, s is the number of switching states possible in the sections of the PDS feeder(s). Since for distribution system weighted graphs will be more applicable to use, the computation time is even larger for a real PDS [38]. Hence, it is a complex problem and one of the limitations in the rapid calculation of the resiliency of a PDS.

There are many ways of path search in a graph that depends upon the technique in which branch to search is selected. Numerous path discovery algorithms (depth-first search, breadth-first search, Tarjan's Algorithm [39]) have been proposed to effectively determine the number of simple paths (i.e. a sub-graph without repetition of vertices) between two nodes in a graph. A depth-first search algorithm is a recursive algorithm that uses the idea of backtracking. It involves exhaustive searches of all the nodes by going ahead, if possible, else by backtracking. Backtrack means that when you are moving forward and there are no more nodes along the current path, you move backward on the same path to find nodes to traverse. All the nodes will be visited on the current path till all the unvisited nodes have been traversed after which the next path will be selected. After finding all possible paths, their different combinations lead to PNs after eliminating loops formation. The operationally feasible PNs become FNs satisfying the limitations of power flow constraints, IEEE standards, safety codes, and local jurisdiction specifications.

The indicators discussed in this section cannot be evaluated by additive measures because they have collaborating characteristics and are interdependent. Thus, non-additive tools for summative measures of different criteria are supposed to be better tools for quantification of the resiliency matrices. Therefore, resiliency quantification problem of a PDS has been framed as a multi-criteria decision-making (MCDM) problem. The use of MCDM helps to account for all these different types of attributes without any homogeneity or a normalization requirement which speeds up the computation process [40].

2.3 Integrated Decision-Making Approach

The resiliency of a distribution system indicates its ability to provide power to all CLs during contingency. Therefore, resiliency values for different configurations of a given network help the planners as well as operators to choose the most resilient network configuration. The methodology discussed in this chapter first illustrates the resiliency quantification process with an integrated decision-making approach in this section and then discusses the algorithm to find the FN solutions to meet all CLs and the most resilient network in the following Section 2.4. The integrated decision-making approach uses Choquet Integral (CI) as an aggregation operator to quantify resiliency metrics in a unique numerical solution taking into account multiple interdependent indicators presented in Section 2.2.

CI is a practical approach to quantify resiliency as a single numerical solution for easier interpretation based on several criteria [41]. The algorithm of resiliency quantification using CI is illustrated in Figure 2.1, where seven network metrics values are used as input, and the parameters of the model are input weights and interaction index. When a single criterion is considered at a time, an adequate answer to the problem cannot be found due to interdependence among them. Therefore, CI is used as an aggregation operator that can be applied to multiple criteria together to find a satisfactory and single numerical result. It uses the concept of measure and is able to take into account multiple interdependent criteria collectively. There are various aggregation methods like weighted arithmetic mean, maximum and minimum, etc. but the CI provides greater flexibility. The use of CI in MCDM has been proposed by many authors for several complex MCDM problems [41] and thus has been chosen as the method for evaluating the resiliency of all FNs to supply CLs in a PDS [42].

The concept of CI is based on the measure which acts as indices on parameters and is used for aggregation of partial values to interpret the importance of each criterion when

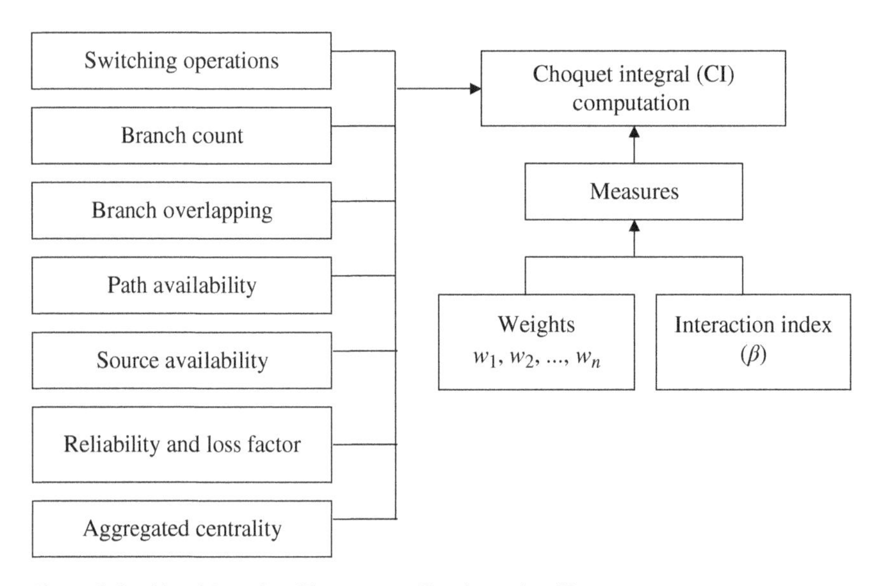

Figure 2.1 Algorithm of resiliency quantification using CI.

taken separately as well as when taken together with other criteria. Measure Δ_i indicates the importance of individual criteria c_i, and combinations of criteria taken together in a Set. A criteria c_i is important if the value of Δ_i is high. However for the importance of c_i, it may not be enough to look at only the value of Δ_i but also have to consider the value of Δ_{ij}, $\Delta_{ijk,}$ etc. for considering the importance of multiple criteria taken together. Now if Δ_i and Δ_j are high but Δ_{ij} does not have much difference from Δ_i and Δ_j then the importance of criteria c_i and c_j taken separately is the same as c_i and c_j taken together. So the interest in considering them both together should be less. On the other hand, if c_i and c_j have low values but Δ_{ij} is very large then c_i and c_j are not as much important as when both are taken together.

CI is defined by several different definitions [41]: A measure Δ on a set of criteria C is a function $\Delta(C) \rightarrow [0, 1]$, satisfying the assumptions (i) $\Delta(\Theta) = 0$, where Θ represents an empty set, and (ii) $I \subset J \subset C$ implies $\Delta(I) \leq \Delta(J)$, I and J are nonempty sets representing different alternatives. In another definition, if measure Δ on a set of criteria $C = [c_{1,}\ c_{2,\ldots,} c_n]$, then the CI of a function f: $C \rightarrow R+$ with respect to Δ is defined by Eq. (2.8):

$$\mathrm{CI}_\Delta(f) = \int f \mathrm{d}\Delta = \sum_{i=1}^{n} (f(x_i) - f(x_{i-1}))\Delta(I_{(i)}) \tag{2.8}$$

$$\text{such that } 0 \leq f(x_1) \leq f(x_2) \cdots \leq f(x_n), I_{(i)} \in C \text{ and } f(x_0) = 0$$

This equation is used to aggregate the impact of different criteria to compute the resiliency metric. The proposed resiliency quantification process considers C as a set of the seven criteria mentioned in Section 2.2. In power networks, the criteria to determine resiliency are not independent of each other, and f determined from one criterion may impact the values from other criteria. For example, let us assume that there are two criteria "I" and "J" and two FN alternatives, $N1$ and $N2$. Also, let us assume that both the criteria, "I" and "J," are equally important to evaluate the resiliency with other possible criteria: $\Delta(I) = \Delta(J) = 0.25$. If for network $N1$, criteria "I" and "J" complement each other to increase the resiliency, there would be a "positive synergy" between the interacting factors, and $\Delta_{N1}(I, J) = 0.661 > \Delta_{N1}(\mathrm{I}) + \Delta_{N1}(J) = 0.5$. On the contrary, if for network $N2$, criteria "I" and "J" contradict each other to decrease the resiliency, there would be a "negative synergy" between the interacting factors, and $\Delta_{N2}(I, J) = 0.425 < \Delta_{N2}(I) + \Delta_{N2}(J) = 0.5$.

The proposed methodology using CI takes into account a positive effect or a negative effect of each criterion on resiliency separately using weights. Thus, Δ had to be determined for each network alternative. In order to identify the impact of each criterion on the overall resiliency, the interaction between each criterion needs to be determined for the network. Weights assigned to these criteria are user-defined and may be directly assigned based on the computation of the resiliency factors described in Section 2.2, and assigned by pairwise comparison for the final computation of Eq. (2.8). For pairwise comparison, each of the criteria is taken in the order of one-at-a-time, two-at-a-time, …, seven-at-a-time to determine the measure. The weights are assigned in a typical Analytical Hierarchical Process (AHP) matrix to determine the overall interaction of resiliency-determining factors. If a positive synergy exists between two interacting parameters, positive weights $\in(0\ 9]$ are assigned, and negative weights are assigned for negative synergy interactions.

The criteria being considered are inter-dependent and to model it, a special measure β (called the "interaction index" in certain literature [43]) is defined on 2^C of the finite set of criteria C and satisfies the following expressions

$$\Delta_\beta(I \cup J) = \Delta_\beta(I) + \Delta_\beta(J) + \beta\Delta_\beta(I)\Delta_\beta(J) \tag{2.9}$$

Whenever $I \cap J = \emptyset$ and $\beta \in (-1, \infty)$ since C is a finite set, the β measure of $\Delta_\beta(C)$ can be written as [44]

$$\Delta_\beta(C) = \sum \Delta_i + K \tag{2.10}$$

where

$$K = \beta \sum_{i_1=1}^{n-1} \sum_{i_2=1+i_1}^{n} \Delta_{i_1} \cdot \Delta_{i_2} + \ldots \beta^{n-1} \Delta_{i_1} \cdot \Delta_{i_2} \ldots \Delta_{i_n}$$

$$\Delta_\beta(C) = \frac{1}{\beta}\left[\prod_{i=1}^{n}(1 + \beta\Delta_i) - 1\right] \text{ by definition, } \Delta_\beta(C) = 1$$

$$\beta + 1 = \prod_{i=1}^{n}(1 = \beta\Delta_i) \tag{2.11}$$

The interaction index is determined using Eq. (2.11). After the measure for all the criteria combination is computed using Eq. (2.8), all the FNs can be ranked in the order of their resiliency values.

2.4 Algorithm to Enable Resilient Power Distribution System

The following three logical statements serve as a premise for justified reasoning for choosing the seven indicators that affect the operational resiliency of a PDS.

- The increasing number of switches increases resiliency, but the increasing number of switching operations to connect the source to the sink decreases the resiliency of a network.
- The resiliency of a network is directly proportional to the number of paths that connect a source node to a sink node.
- Increasing the ratio of the number of sources to the number of CL increases the resiliency of a network.

These three axioms lead to the following three objectives, which need to be implemented in order to enable a resilient PDS through reconfiguration of the network

- Minimization of the number of switching;
- Maximization of the utilization of power under different contingent conditions;
- Maximization of number of the energized loads.

The major aim of a resilient PDS is to minimize any downtime and the approach [24] used here proactively determines alternative paths and their corresponding resiliency values when a load needs to be connected to a source using a path that is not its regular path. It is well-understood and agreed upon (by virtue of a large number of restoration algorithm works in literature) that a load can be restored using multiple paths. However, the approach suggested here would enable an operator to take the most resilient decision. It is not easy to take decisions during contingency, and not all power system state-estimating sensors may be working correctly in a post-emergency environment. Thus, the method proposed here prepares a list of paths an operator can choose if an emergency occurs. Switching operations to restore CLs by the operator is mathematically guaranteed to have the least probability of further failures in the system. To determine the operationally FN solutions to meet all CLs during the planning and operational contingency, it is important to start by determining the CLs and sources available in the network. Then find all possible path combinations taking one path for each CL. To maintain the radial nature of the PDS, path combinations with the loop are eliminated using the loop elimination technique [45].

The algorithm proposed in Figure 2.2 emphasizes the restoration of CL (loads assigned high priority) prior to restoring other loads in the network for enabling resiliency through PN solutions and their resiliency quantification. It is also worth noting that the resiliency indices determined from Figure 2.2 are functions of operating conditions at a given time, and are not strictly time dependent. If a system continues to operate without any change in demand or supply or network configuration, the values of resiliency indices determined will not change over time; however, as the operating conditions vary according to days and seasons, the values of the indices also change correspondingly.

It is important to start the resiliency determination process by determining the number of CLs and sources available in the network, as well as to find the operationally FN solutions to supply power to all CLs during the planning and operational contingency. Paths containing more than one CL and/or more than one source are eliminated. By selecting one path for each CL, all possible path combinations to provide power supply to all CLs are listed. To maintain the radial nature of the PDS, path combinations with the loop are eliminated using the loop elimination technique [41]. Each PCWL gives different PN configurations resulting in UNs after eliminating similar networks based on switch configurations.

UNs satisfying the power flow convergence (using backward-forward sweep technique for unbalanced distribution systems) become FNs satisfying all operating constraints. The forward-backward sweep method is designed to solve the differential-algebraic system generated by the maximum principle that characterizes the solution. It is easy to program and runs quickly. Using the resiliency quantification process discussed in Section 2.4, the network resiliency of each FN is calculated. All FN configurations supplying all CLs are listed in the hierarchical order of their resiliency matrices as an output of this algorithm.

The network resiliency of each FN is calculated and finally compared to get the most resilient network configuration. The DGs are modeled as PV buses and assumed to be capable of meeting the active and reactive power demands of all normal and CLs of the given network. For each FN capable of feeding all CLs, seven factors affecting the resiliency of the network are considered and the network resiliency is quantified using CI. A distribution network planner may use this information to find the location and number of sectionalizing and tie-line switches taking into account cost and other related factors. A DNO may take appropriate control actions to enable higher resiliency in the PDS during contingency based on the hierarchical order of resiliency matrices.

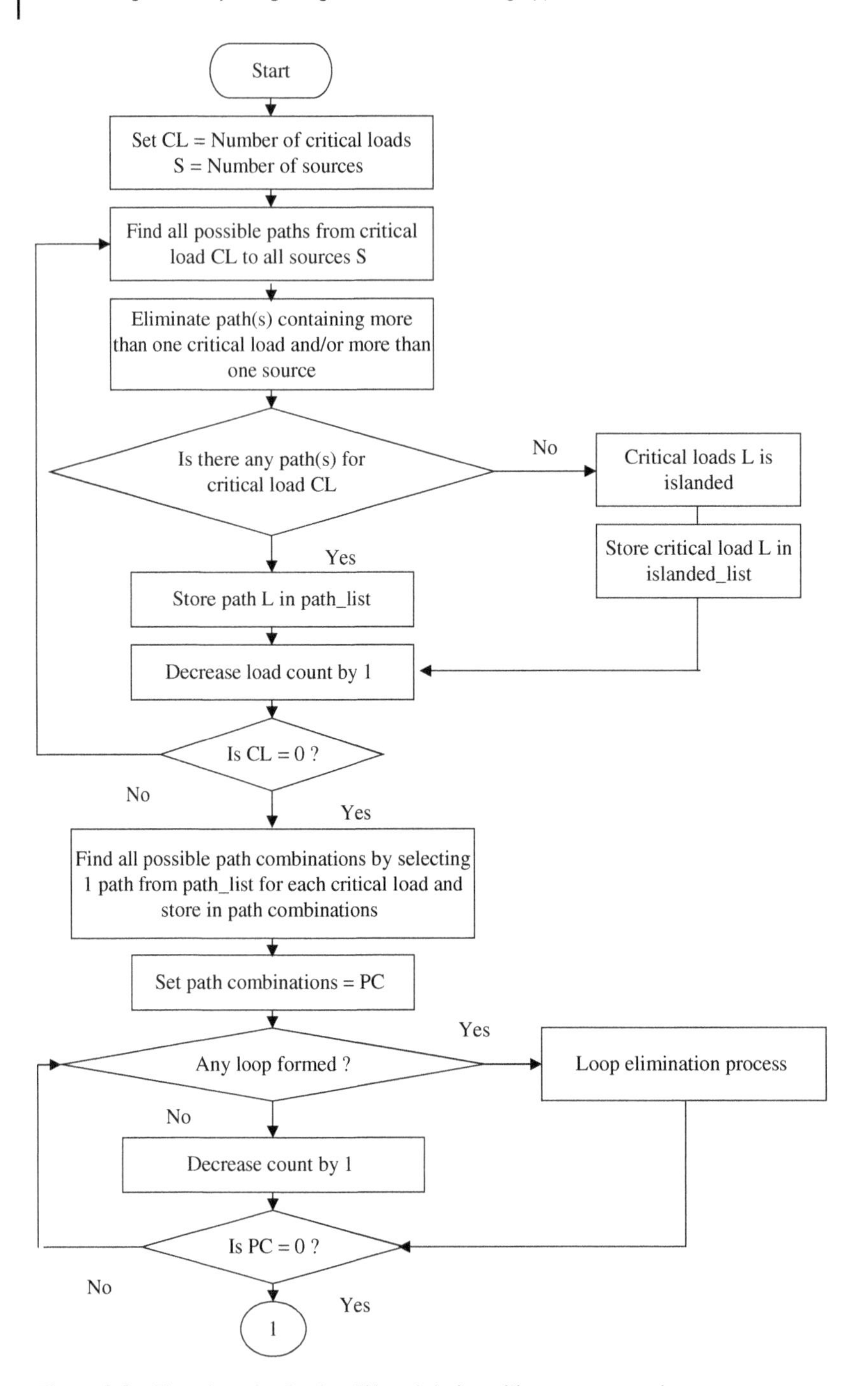

Figure 2.2 Flowchart for finding FN and their resiliency computation.

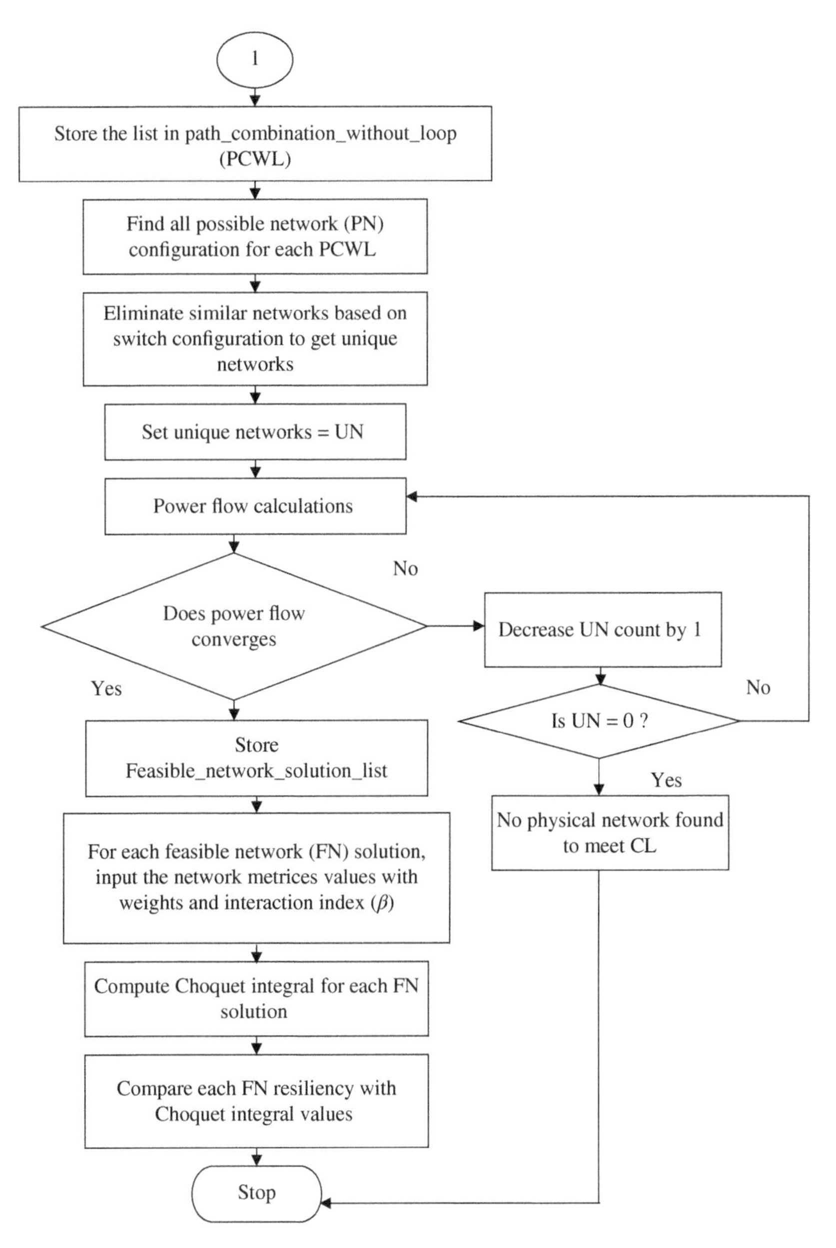

Figure 2.2 (*Continued*)

2.5 Case Study

The proposed algorithm to compute resiliency has been validated using two proximally located Consortium for Electric Reliability Technology Solutions (CERTS) microgrids [4]. The line data and bus data have been obtained from [5]. The DGs and load data are listed in Table 2.1.

Table 2.1 DG and load data.

Source	Node	Capacity (kVA)	Microgrid	
DG1	*N*8	262	1	
DG2	*N*16	262	2	
Priority	Load node	*P* (kW)	*Q* (kVAR)	Microgrid
Normal	*N*5	48.8	36.6	1
Critical	*N*7	84.5	64.5	
Normal	*N*9	77.3	58.9	
Normal	*N*11	79.9	59.9	
Normal	*N*15	46.6	34.5	2
Normal	*N*17	52.5	39.4	
Critical	*N*19	81.1	60.8	
Normal	*N*21	69.8	52.3	

The DGs have been located at nodes 8 and 16, capable of serving the 165.6 kW CL demand of the network. The remaining capacity of the generators is used to feed the remainder of the normal loads in the same feeder as CLs. The critical loads CL1 and CL2 are identified at nodes 7 and 19, as shown in Figure 2.3. There are six NC sectionalizing switches and it is possible to install three tie-lines with NO switches (T_1, T_2, and T_3) in the network between nodes 7-11, 17-21, and 11-21.

The proposed algorithm may be used to calculate the network resiliency for all FN configurations supplying all CLs in a network that is useful in the planning as well as operational stage of a PDS. The application is demonstrated through several case studies. Common objectives for the planning task, in general, include minimization of losses and cost and maximization of reliability and resiliency. This leads to finding the optimal location and size of DG, optimal location and number of re-configurable switches, etc. Several cases based on possible combinations of tie-line switches are illustrated for the planning of a resilient network by placing additional switches in a multiple microgrid network. Multiple case studies are also included based on different contingencies during operation. DNO will have the list of FN solutions to be operational and their hierarchy order of resiliency metrics that help the DNO to quickly choose the most resilient network configuration.

In the first step, all the paths are determined using a depth-first search algorithm [6]. It is observable that some of the PNs are similarly based on switch configurations, so they are reduced to nine *unique* networks listed in Table 2.2 with all sectionalizing and tie-line switches status. In case the CL is being supplied from the main grid, probability of availability (PoA) and Penalty Factor (PF) are considered 0.98 and 0.9, respectively. The PoA of each DG is assumed same as 0.95. PF is assumed to be 1, if both DG and CL are in same MG, and PF is assumed to be 0.8, if DG and CL are in different MGs. All nine unique networks are operationally feasible with ±0.05 p.u. bus voltage violation limit, therefore, the network metrics are quantified for all the nine FNs and listed in Table 2.3. Detailed computation of network metrics for FN1 is shown in the following example.

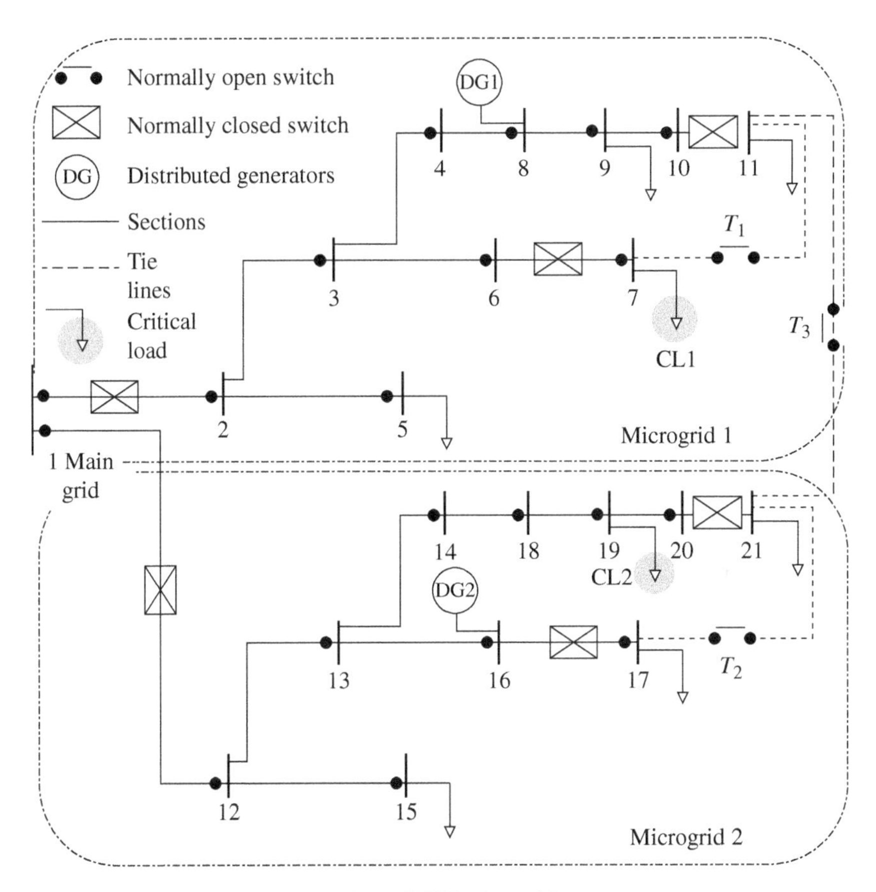

Figure 2.3 Test system: two proximal CERTS microgrids.

Table 2.2 Unique network configurations with corresponding similar PNs and switch configurations.

Unique network configurations	Similar PNs	Sectionalizing switches (NC)					Tie-line switches (NO)			
		r2-1	12-1	7-6	11-10	16-17	20-21	11-7	21-11	21-17
$N1$	$N1, N4, N9, N12$	C	C	C	C	C	C	O	O	O
$N2$	$N2, N10$	C	O	C	C	C	C	O	C	O
$N3$	$N3, N11$	O	C	C	C	C	C	O	C	O
$N4$	$N5, N8$	C	C	O	C	C	C	C	O	O
$N5$	$N6$	C	O	O	C	C	C	C	C	O
$N6$	$N7$	O	C	O	C	C	C	C	C	O
$N7$	$N13, N16, N18$	C	O	O	C	C	O	C	C	C
$N8$	$N14, N17, N19$	O	C	O	C	C	O	C	C	C
$N9$	$N15$	O	O	O	C	C	O	C	C	C

Table 2.3 Network metrics for feasible networks (FNs).

	FN1	FN2	FN3	FN4	FN5	FN6	FN7	FN8	FN9
Branch count effect (BCE)	4.25	5	5	4.25	5	5	4.5	4.5	5
Overlapping branches (OB)	0	0	0	0	3	3	0	0	0
Switching operations (SO)	0	2	2	2	4	4	6	6	7
Repetition of sources (RoS)	0.88	0.75	0.75	1	0.50	0.50	0.83	0.83	1
Path redundancy (PR)	3	3	1.5	3	3	3	3	3	2
PoA and PF	0.84	0.70	0.70	0.81	0.72	0.72	0.66	0.66	0.58
Aggregated central point dominance (ACPD)	0.55	0.62	0.62	0.58	0.62	0.62	0.61	0.60	0.58

Network metrics for all the nine FNs are listed in Table 2.3. Detailed calculation of all network metrics values only for FN1 is shown here. There are four similar PNs ($N1$, $N2$, $N4$, $N9$, and $N12$) for FN1 as their switch configurations (shown in Table 2.2) are the same. So q is equal to 1, and N_q is equal to 4 for FN1. The sequence of nodes for the PCWL, corresponding to these four PNs is listed in Table 2.6.

Nodes in PCWL for $N1$, $N4$, $N9$, and $N12$ are 9, 8, 9, and 8, respectively, and the number of CLs is the same (i.e. 2) for all. So $\text{BCE}_1 = (9/2 + 8/2 + 9/2 + 8/2)/4 = 4.25$.

The number of common branches in each PN ($N1$, $N4$, $N9$, $N12$) is zero. So $\text{OB}_1 = 0$.

As shown in Table 2.2, all Sectionalizing Switches are closed and all tie-line switches are open for all PNs ($N1$, $N4$, $N9$, and $N12$). So there is no change in switch configuration from the normal position, therefore, $\text{SO}_1 = 0$.

The number of sources supplying all CLs in PNs ($N1$, $N4$, $N9$, and $N12$) are 1, 2, 2, and 2, respectively, and the number of CLs for all PNs is 2. So $\text{RoS}_1 = (1/2 + 2/2 + 2/2 + 2/2)/4 = 0.875$.

Each CL has three paths available to connect to all three sources in FN1. So $\text{PR}_1 = (3 + 3)/2 = 3$.

- Calculation for PoA and PF for each PN is listed in Table 2.7 based on values given in Section 2.5. PoA and PF $= (0.882 \times 0.882) + (0.882 \times 0.95) + (0.882 \times 0.95) + (0.95 \times 0.95)/4 = 0.8391 - 0.84$.

ACPD_1 depends on the order and C_B of each node in FN1. For FN1, $D = 21$. The order of each node, and corresponding betweenness centrality of each node d (i.e. $C_B(d)$) are shown in Table 2.8. The betweenness centrality of each node is computed from Eq. (2.6), using MATLABs graph analysis libraries. Exactly similar results can be obtained by creating an undirected graph of the multiple microgrid distribution systems using any network analyzing software, such as Cytoscape [7].

For the sake of demonstration of not FN paths, the network selection criteria have been made more stringent. Only the first seven unique networks show power flow convergence with ±0.04 p.u. bus voltage violation limit. The set of FNs is a subset of unique networks depending upon operating constraints. The pair-wise comparative weights assigned to these network metrics are shown in Table 2.4. Table 2.5 shows the application of the

Table 2.4 Pairwise comparative weights.

	BCE	OB	SO	RoS	PR	PoA and PF	ACPD	Weights
BCE	1	1	3	1	0.33	0.2	0.16	5.64×10^{-2}
OB	1	1	3	1	0.33	0.2	0.16	5.64×10^{-2}
SO	0.33	0.33	1	0.33	0.25	0.16	0.14	2.95×10^{-2}
RoS	1	1	3	1	0.33	0.2	0.16	5.64×10^{-2}
PR	3	3	4	3	1	0.33	0.16	1.21×10^{-1}
PoA and PF	5	5	6	5	3	1	0.25	2.33×10^{-1}
ACPD	6	6	7	6	6	4	1	4.47×10^{-1}

developed algorithm during an ongoing contingency. The CI values, after the aggregating impact of all the different criteria for all the nine FNs using network metrics input values and pairwise comparative weights, are shown in Table 2.9 for three different interaction index values (−0.5, 0, and 1). The interaction degrees are identified by the importance ratio between the maximum input value and the minimum input value of network metrics. First, one has to select which input value is important. If one selects a minimum input value, it means a super-additive measure is selected. If one selects a maximum input value, it means a sub-additive measure is selected. Second, one has to select how many times one put importance. The value is bigger; the output is closer to the maximum or minimum input. In this study, even with varying degree of interaction index (−0.5, 0, and 1), the order of PNs in ascending order of their resiliency values remained the same, that is, FN3, FN1, FN4, FN9, FN2, FN8, FN7, FN6, and FN5. It implies that the relative importance of maximum and minimum values of network metrics are adequately represented in pairwise comparative weights. Broadly PDS planning and operation scenarios are considered for simulation case studies in this paper. Two geographically proximate CERTS microgrids are considered with given DGs, CLs, tie-lines, and sectionalizing switches. Four different operational contingency cases are studied to find the most resilient network configuration (Table 2.10).

2.5.1 Planning Strategy Case

The distribution planning scenario is to be used to develop insights into the resiliency of a system prior to infrastructural investment and development. Drafts of different network designs, load assignments, and DER installation can be evaluated to proactively determine the PDS configuration is most resilient to physical attacks. In the system being studied in this paper, three tie-line switches between node 7 and node 11 (T_1), between node 17 and node 21 (T_2), and between node 21 and node 11 (T_3) are planned in the network with all possible combinations. A total of seven cases corresponding to all possible combinations of three tie-line switches are listed in Table 2.5. For each combination, all FN configurations supplying all CLs are listed along with their hierarchy order of resiliency metrics. FN with all closed (C) switches (T_1 and/or T_2 and/or T_3) and the highest resiliency is considered

Table 2.5 Operational contingency scenario results.

Cases	Faulted nodes	Feasible networks (FN)	FN in ascending order of resiliency	Decision
1	2	FN1, FN2, FN3, FN4, FN5, FN6, FN7, FN8, FN9	FN3, FN1, FN4, FN2, FN9, FN8, FN7, FN6, FN5	FN5
2	12	FN1, FN2, FN3, FN4, FN5, FN6, FN7, FN8, FN9	FN3, FN1, FN4, FN2, FN9, FN8, FN7, FN6, FN5	FN5
3.A	2, 12, 11	FN1	FN1	FN1
3.B	2, 12, 21	FN1, FN4	FN1, FN4	FN4
4	Sections 3-8, 13-19, 7-11	FN2, FN3	FN3, FN2	FN2

for the planning decision. Three cases (II, IV, and VI) are not providing any PN because the T_2 switch (alone or with T_1 or T_3 switch) creates loop formation, therefore, made open (O) in FN configuration. N_1 is the only FN without any tie-line switch and has less resiliency compared to all other networks possible in the planning decision. This information will help the planners to choose the location and number of additional switches after appropriate trade-off with cost and other factors.

2.5.2 Operational Contingency Case

Operational scenarios are important to demonstrate the usability of the proposed metrics to improve distribution network automation and reduce the chances of power interruption to CLs. The metric can be used as a future control signal to influence control actions that enable higher resiliency in PDS. Four different contingency cases have been formulated in the multiple CERTS microgrid studied in this paper. They are reported in this section to list all PN configurations supplying all CLs along with their hierarchy order of resiliency metrics (listed for each case in Table 2.6).

2.5.2.1 Case 1

In this case, it is assumed that a fault occurred at one of the main grid's feeder to MG-1. Fault at node 2 will island the MG-1 from the main grid and will result in the exclusion

Table 2.6 PCWL corresponding to four PNs.

PNs	Path 1 for CL1	Path 2 for CL2
N_1	7, 6, 3, 2, 1	19, 18, 14, 13, 12, 1
N_2	7, 6, 3, 2, 1	19, 18, 14, 13, 16
N_3	7, 6, 3, 4, 8	19, 18, 14, 13, 12, 1
N_4	7, 6, 3, 4, 8	19, 18, 14, 13, 16

Table 2.7 PoA and PF for each PN.

	Sources for CL1			Sources for CL2		
PNs	**PoA**	**PF**	**PoA and PF**	**PoA**	**PF**	**PoA and PF**
$N1$	0.98	0.9	0.882	0.98	0.9	0.882
$N2$	0.98	0.9	0.882	0.95	1	0.95
$N3$	0.95	1	0.95	0.98	0.9	0.882
$N4$	0.95	1	0.95	0.95	1	0.95

Table 2.8 Computation of ACPD for FN1.

Node (d)	Order Ω	$C_B(d)$	$\Omega \times C_B(d)$
1	2.00	0.179	0.357894737
2	3.00	0.274	0.821052632
3	3.00	0.316	0.947368421
4	2.00	0.289	0.578947368
5	2.00	0.205	0.410526316
6	1.00	0.000	0
7	2.00	0.100	0.2
8	2.00	0.189	0.378947368
9	2.00	0.333	0.666315789
10	2.00	0.263	0.526315789
11	1.00	0.000	0
12	2.00	0.268	0.536842105
13	3.00	0.584	1.752631579
14	3.00	0.568	1.705263158
15	3.00	0.582	1.744736842
16	2.00	0.153	0.305263158
17	2.00	0.137	0.273684211
18	2.00	0.147	0.294736842
19	1.00	0.000	0
20	1.00	0.000	0
21	1.00	0.000	0
ACPD($FN1$)			**11.5005/21 = 0.55**

Table 2.9 Choquet integral values with varying interaction degree.

Feasible networks (FN)	Choquet integral with		
	$\lambda = -0.5$	$\lambda = 0$	$\lambda = 1$
FN1	123.3	109.3	95.2
FN2	133.2	118.9	104.7
FN3	111.2	100.7	90.2
FN4	130.1	116.5	102.8
FN5	156.3	140.9	125.5
FN6	156.1	140.6	125.2
FN7	142.5	126.5	110.6
FN8	141.9	125.8	109.9
FN9	133.2	118.0	102.9

Table 2.10 Resiliency metric ranking table for critical load restoration.

Network	Switching operation (O: Open, C: Close)	CI resiliency metrics		
		Load 1	Load 2	Load 3
FN1	O:S3, S′1 C: S4, S′2, S5, S8	134.2	127.9	164.2
FN2	O:S5, C:S′1, S′2, S8	133.3	167.9	147.3
FN3	O:S5, S8, S′2 C: S7, S′1	131.2	148.9	154.2

of all PCWL passing through node 2. Out of 19 PNs, four networks (FN1, FN2, FN3, FN4) will not be operational. Due to the similar overlapping of PNs, all nine FN solutions will still be operational for supplying all CLs. DNO should choose the network with the highest resiliency, that is FN5, where CL1 and CL2 are both supplied by DG1.

If the most resilient topology is infeasible, the proposed algorithm selects the next feasible and most resilient topology. In case 1 under distribution planning scenario, FN1 is the only FN without any tie-line switch and has less resiliency compared to all other networks possible in the planning decision. In this case, only FN1 can be adopted and will be the most resilient feasible topology adopted for the system.

2.5.2.2 Case 2

In this case, it is assumed that a fault occurred at one of the main grid's feeder to MG-2. Fault at node 12 will island the MG-2 from the main grid and will result in the elimination of all PCWL passing through node 12. Out of 19 PNs, five networks ($N1, N5, N9, N13, N14$) will not be operational. Due to the similar overlapping of PNs, all nine FN solutions will still be operational for supplying all CLs. DNO should choose the network with the highest resiliency, that is, FN5.

2.5.2.3 Case 3

In this case, it is assumed that microgrids MG-1 and MG-2 both are in islanded mode due to fault in their feeder to the main grid and also fault in the tie-line between MG-1 and MG-2. This case is realized with two different scenarios. Scenario 1 is considered with fault at nodes 2, 12, and 11. This will reject all PCWL passing through these three nodes and only one PN ($N12$) will be operational. Therefore, the DNO will not have any choice and has to operate the corresponding FN, that is, FN1. In scenario 2, fault at nodes 2, 12, and 21 may be considered to realize the same case. This will reject all PCWL passing through nodes 2, 12, and 21 and only two PNs ($N8$, $N12$) will be operational. Thus, the DNO will have a choice between corresponding FNs (FN1 and FN4) and will choose the network with higher resiliency, that is, FN4.

2.5.2.4 Case 4

In this case, it is assumed that both the MGs are grid-connected and their interconnected tie-line is also healthy but the fault happens within MGs due to a storm hitting several areas, such that the lines between Bus-3 and Bus-8, lines between Bus 13 and Bus 19 and line between Bus 7 and Bus 11 are out of service. This will result in the elimination of all PCWL passing through paths joining these nodes. Out of 19 PNs, only two networks (N_2 and N_3) will be operational. This leads to corresponding two FN solutions (FN2 and FN3) to be operational for supplying all CLs. Therefore, the DNO will choose the network with higher resiliency, that is, FN2.

2.5.3 Application to IEEE 123 Node Distribution System

In order to demonstrate the scalability of the proposed approach for implementation in an operational setting of a larger system, IEEE 123 node PDS has been chosen for demonstration. IEEE 123 node PDS is an unbalanced and multi-phase radial distribution network with 11 three-phase switches [8]. There are enough switches in the feeder so that multiple paths for restoring a load can be tested, and the resiliency metric for each corresponding path can be easily determined. Some assumptions have been made in this paper for the IEEE 123-bus test system:

- Assumed all loads are modeled as constant PQ loads.
- Three CLs have been assumed to be at Node 66 (single phase load *Load* 1 – 75 kW and 35 kVAr), Node 37 (single phase load *Load 2* – 40 kW and 20 kVAr), and Node 82 (single phase load *Load 3* – 40 kW and 20 kVAr), as shown in Figure 2.4.
- Three sectionalizing switches S′1, S′2, and S′3 were installed between nodes 37 and 59, 39 and 66, and 82 and 86, respectively, as shown in Figure 2.4.
- DNO has control limited to the three-phase switches and the added sectionalizng switches.
- 500 kW renewable generation (with storage) has been modeled at node 149 in order to supply power in the system in event of failures upstream of the substation (node 150).

The total load of the system is 1420 kW, and scaled to fit a load profile from PJMs 2010 metered historical data [9]. Therefore, the load profile of the 123 node distribution system

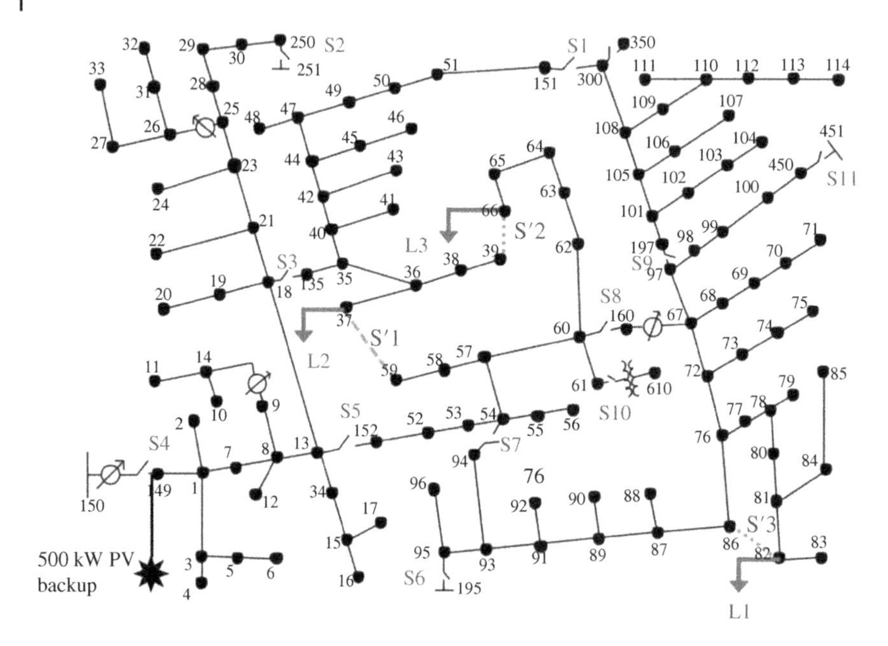

Figure 2.4 Modifications made to standard IEEE 123 node distribution system.

is normalized so that its peak value is 1 and multiplied by the value of the static load at each bus.

Due to the integration of a 500 kW solar backup power system in the PDS, the voltage profiles at several nodes go beyond the 1.0 ± 0.05 p.u. operational limits, as shown in Figure 2.5. The number of paths available to restore Load 1, Load 2, and Load 3 have a varying number of paths available for restoration, depending on the time of day (as shown in Figure 2.6). Since the resiliency of network is dependent on the number of paths available for restoration, it may be concluded from this simulation that resiliency of the network is a function

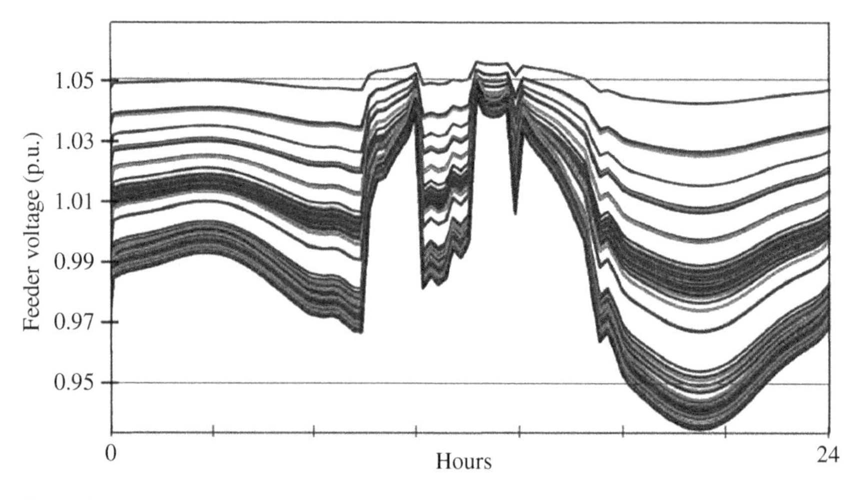

Figure 2.5 Voltage profiles of all nodes over a 24 hour load-profile.

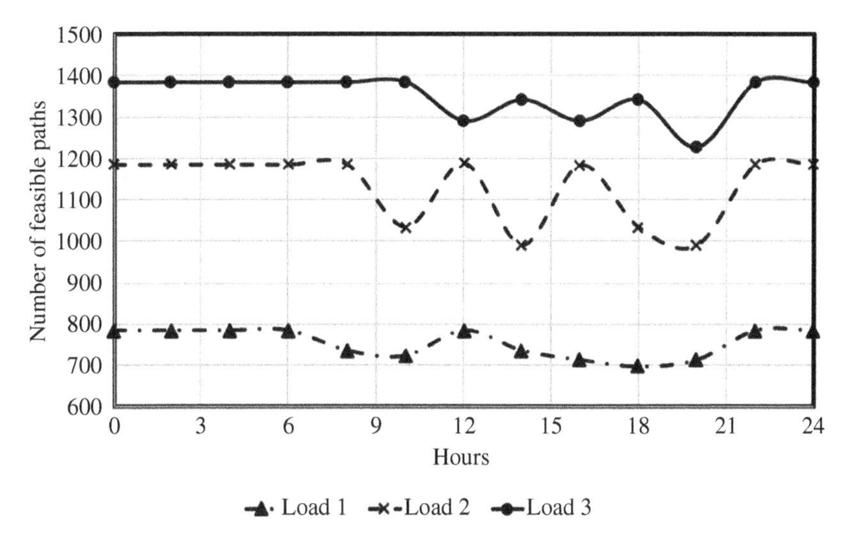

Figure 2.6 Diurnal variation of number of available paths for load restoration.

of the operational state, and the operator needs to be informed in real-time about the most resilient path for restoring a CL, with respect to the ongoing system conditions.

In Figure 2.6, all possible paths for restoring the loads, for all PNs have been computed.

Let us consider a scenario where the PDS is islanded from the grid and being sustained on the 500 kW solar generation backup at 10 a.m. on the day of the chosen load profile. Also, uncleared faults are being repaired by crews in Sections 35 and 36, 63 and 64, and 72 and 76 – such that the DNO is tasked with restoring all CLs (Load 1, Load 2, and Load 3) in the most resilient configuration before normal operation can be resumed. In this scenario, the proposed approach of the paper will present the DNO with the switching actions that can be executed in the system and the corresponding resiliency value for the path to restoring the load. The ranking of switching actions is shown in Table 2.10. For brevity (and clarity of understanding of the DNO), only the top three resilient configuration options are presented in the table, though the preference can be customized. The resiliency values are computed using Eqs. (2.8)–(2.11), with an interaction value of $\lambda = 0$, implying the factors affecting resiliency impact the overall resiliency value independently, without any positive or negative synergy. The ranking table is sorted according to the most resilient configuration for Load 1 but can be re-arranged according to the preference of the DNO. The impact of the switching actions on the paths restoring Loads 2 and 3 are also presented. In this way, the operator's decision-making for the switching operation can be aided during contingencies.

From Table 2.10, it can be interpreted that $FN1$ is the most resilient path to provide power to Loads 1 and 3, while $FN2$ is the most resilient path to provide power to Load 2.

2.5.4 Meshed or Ring Main Distribution Topology

Though PDS in North America and worldwide are predominantly radial in nature, the proposed algorithm is valid for meshed distribution systems as well. Meshed (or Ring Main) distribution topology is characterized by greater feeder redundancy and loop formation in the network. By taking out the radial constraints imposed in the proposed approach, such as loop elimination, the algorithm shown in Figure 2.2 will work for meshed or ring networks. A meshed network is characterized by the presence of loops formed by strategically placed switches in the PDS. Loops in a meshed network are achieved by making tie-line switches status as NC instead of being NO as in radial distribution feeders. In the multiple CERTS microgrid system case study shown in Figure 2.3, switches T_1, T_2, and T_3 would be NC for meshed topology. As a result, the only unique network configurations feasible would be restricted to $N7$, $N8$, $N9$, $N13$, $N14$, $N15$, $N16$, $N17$, $N18$, and $N19$, as shown in Table 2.2. The rest of the computation of network metrics (as in Table 2.3) and subsequent derivation of the resilient metrics of the FNs is identical for both radial and meshed distribution topologies. Results have not to be shown here to avoid redundancy in results. Thus, the approach is suitable for both radial and meshed distribution topologies.

2.5.5 Spatial Factors Impacting Resilience

In the example with the PV-integrated IEEE-123 bus system described before, we have emphasized on reconnecting components of the grid by intelligent switching or higher redundancy as means of enabling resiliency in the grid. However, in practice, all PDS are sited in regions with variable degrees of risk and ability to restore. Thus, though the approach described in Section 2.4 is universal, it needs further exploration in terms of its ability to incorporate spatial aspects of PDS resilience.

PDS in sub-urban and rural regions are characterized by radial networks, seasonal variations in demand, and lower population densities, while regions within city limits or downtown regions have a larger number of consumers, with higher requirements for reliable power. Such regions are characterized by a stronger meshed network, and further strengthened by more underground cabling in many of the developed metropolitan cities of the world. The location of the city also plays an important role in determining the weather factors to which the city is more vulnerable, such as cities in Florida and Texas are more prone to strong hurricane winds and flooding, however, cities in California are designed to be more risk averse about earthquakes. Thus, it is important to study resilience by factoring in the local geography and ability to restore.

For illustration, let the previously discussed IEEE-123 node distribution system be sited in four identified zones of New York with varying levels of threat and capability to restore infrastructure. The different regions of New York are shown in Figure 2.7.

We introduce a new weight sf representing the relative risk due to spatial factors, to improve the CI metrics and obtain the resiliency results for the modified IEEE 123 bus system.

2.5.5.1 Computation Costs

In order to check the feasibility of the proposed solution to real-world distribution systems, the computation cost for generating the resiliency value ranking table was also considered. The memory requirement for the computation of all feasible configurations based on a network positively depends upon the number of nodes in the network. Each node accounted for 19 bytes of memory while running the 123-node PDS simulation, and required 17.83 KB of random-access memory (RAM) for processing. For larger networks, the memory requirement is a function of $O(2^s \cdot ne + \log n)$, where n is the number of nodes and e is the number

Figure 2.7 Different risk regions in New York metropolitan area.

of branches in the PDS. Thus, it can be estimated for a typical distribution with 10,000 nodes, the memory requirement is 26.12 MB of RAM – which is well within the capacity of single-threaded modern digital computers installed in distribution automation control centers. The processing power is further enhanced if multiple threads of the computer are used, but that was not explored in this work.

2.6 Conclusion

A novel problem formulation and solution approach for measuring and enabling the resiliency of a PDS has been presented in this paper. The algorithm proposed in this paper can be used by DNOs to determine the most resilient paths to restore CLs, under all possible contingencies feasible in the network. An array of all feasible paths corresponding to any loss-of-source situation is ranked, and available to the distribution system operator. The paths are determined optimally to maximize the resiliency of the PDS. The simulations validate the conjecture that the resiliency of a network depends on the number of paths to connect a source node to a sink node, and the ratio of the number of sources to the number of CL. The simulation results also show that increasing the number of switches increases more switches means more points of failure.

The CI approach is not computationally expensive and can be applied to large batches of feasible paths to restore a load. Moreover, distribution planners may use this information to find the number and locations of switches to enable higher resiliency in a power distribution network.

The proposed algorithm has been validated using a model of two geographically proximal industry-standard CERTS microgrids for all PNs configurations. The approach has been further validated using IEEE 123 node distribution feeder. This paper also illustrates the application of the proposed algorithm in distribution network planning as well as in operational contingency scenarios. The solution approach will be extended in the future by including the duration of time taken to restore service to CLs following a power outage. Similarly in the planning scenario, the placement of DGs and additional lines needed to improve resiliency will be considered in future work.

References

1 Presidential Policy Directive 21 (PPD-21), *Critical Infrastructure Security and Resilience*, White House, February 12, 2013, Press Secretary Release.

2 Watson, J.P., Guttromson, R., Silva-Monroy, C., et al. (2014). *Conceptual framework for developing resilience metrics for the electricity oil and gas sectors in the United States.* Sandia National Laboratories, Albuquerque, NM (United States), Tech. Rep.

3 Petit, F.D., Fisher, R.E., and Veselka, S.N. (2013). *Resiliency Measurement Index.* USA: Argonne National Laboratory.

4 National Infrastructure Advisory Council (2010). *Framework for Establishing Critical Infrastructure Resilience Goals.* US Department of Homeland Security.

5 Brown, G., Carlyle, M., Salmerón, J., and Wood, K. (2006). Defending critical infrastructure. *Interfaces* 36 (6): 530–544.

6 Marnay, C., Aki, H., Hirose, K. et al. (2015). Japan's pivot to resilience: how two microgrids fared after the 2011 earthquake. *IEEE Power and Energy Magazine* 13 (3): 44–57.

7 Wang, Z., Chen, B., Wang, J., and Chen, C. (2016). Networked microgrids for self-healing power systems. *IEEE Transactions on Smart Grid* 7 (1): 310–319.

8 Bahramirad, S., Khodaei, A., Svachula, J., and Aguero, J.R. (2015). Building resilient integrated grids: one neighborhood at a time. *IEEE Electrification Magazine* 3 (1): 48–55.

9 Che, L., Zhang, X., and Shahidehpour, M. (2015). Resilience enhancement with DC microgrids. In: *IEEE Power & Energy Society General Meeting*. IEEE, Denver, CO, USA, pp. 1–5.

10 Wang, Z. and Wang, J. (2015). Self-healing resilient distribution systems based on sectionalization into microgrids. *IEEE Transactions on Power Systems* 30 (6): 3139–3149.

11 Force, H.S.R.T. (2013). *Hurricane Sandy Rebuilding Strategy*. Washington, DC: US Department of Housing and Urban Development.

12 van den Honert, R.C. and McAneney, J. (2011). The 2011 Brisbane floods: causes, impacts and implications. *Water* 3 (4): 1149–1173.

13 Zhou, B., Gu, L., Ding, Y. et al. (2011). The great 2008 Chinese ice storm: its socio economic ecological impact and sustainability lessons learned. *Bulletin of the American Meteorological Society* 92 (1): 47–60.

14 Vugrin, E.D., Warren, D.E., and Ehlen, M.A. (2011). A resilience assessment framework for infrastructure and economic systems: quantitative and qualitative resilience analysis of petrochemical supply chains to a hurricane. *Process Safety Progress* 30 (3): 280–290.

15 Fisher, R.E. and Norman, M. (2010). Developing measurement indices to enhance protection and resilience of critical infrastructure and key resources. *Journal of Business Continuity & Emergency Planning* 4 (3): 191–206.

16 O, Rourke, T.D. (2007). Critical infrastructure, interdependencies, and resilience. *Bridge Washington* 37 (1): 22.

17 Hsi, P.-H. and Chen, S.-L. (1998). Distribution automation communication infrastructure. *IEEE Transactions on Power Delivery* 13 (3): 728–734.

18 Abbey, C., Cornforth, D., Hatziargyriou, N. et al. (2014). Powering through the storm: microgrids operation for more efficient disaster recovery. *IEEE Power and Energy Magazine* 12 (3): 67–76.

19 Pinte, B., Quinlan, M., and Reinhard, K. (2015). Low voltage micro-phasor measurement unit (PMU). In: *IEEE Power and Energy Conference at Illinois (PECI)*. IEEE, Chicago, IL, USA, pp. 1–4.

20 Farhangi, H. (2010). The path of the smart grid. *IEEE Power and Energy Magazine* 8 (1): 18–28.

21 Schneider, K., Tuffner, F., Elizondo, M. et al. (2016). Evaluating the feasibility to use microgrids as a resiliency resource. *IEEE Transactions on Smart Grid* PP (99): 1.

22 Manshadi, S. and Khodayar, M. (2015). Resilient operation of multiple energy carrier microgrids. *IEEE Transactions on Smart Grid* 6 (5): 2283–2292.

23 Hatziargyriou, N. (2014). Operation of multi-microgrids, ch. 5, sec. 5.1. In: *Microgrids: Architecture and Control*, 165. UK: Wiley.

24 Khodaei, A. (2014). Resiliency-oriented microgrid optimal scheduling. *IEEE Transactions on Smart Grid* 5 (4): 1584–1591.

25 Manshadi, S.D. and Khodayar, M.E. (2015). Resilient operation of multiple energy carrier microgrids. *IEEE Transactions on Smart Grid* 6 (5): 2283–2292.

26 Chen, C., Wang, J., Qiu, F., and Zhao, D. (2016). Resilient distribution system by microgrids formation after natural disasters. *IEEE Transactions on Smart Grid* 7 (2): 958–966.

27 Wang, Y., Chen, C., Wang, J., and Baldick, R. (2016). Research on resilience of power systems under natural disasters – a review. *IEEE Transactions on Power Systems* 31 (2): 1604–1613.

28 Pandit, A., Jeong, H., Crittenden, J.C., and Xu, M. 2011). An infrastructure ecology approach for urban infrastructure sustainability and resiliency. In: *Power Systems Conference and Exposition (PSCE), 2011 IEEE/PES*. IEEE, Phoenix, AZ, USA, pp. 1–2.

29 Wang, X.F. and Chen, G. (2003). Complex networks: small-world, scale-free and beyond. *Circuits and Systems Magazine* 3 (1): 6–20.

30 Albert, R. and Barabási, A.-L. (2002). Statistical mechanics of complex networks. *Reviews of Modern Physics* 74 (1): 47.

31 Nguyen, D.T., Shen, Y., and Thai, M.T. (2013). Detecting critical nodes in interdependent power networks for vulnerability assessment. *IEEE Transactions on Smart Grid* 4 (1): 151–159.

32 Chanda, S. and Srivastava, A. (2015). Quantifying resiliency of smart power distribution systems with distributed energy resources. In: *Proceedings of the 24th IEEE International Symposium on Industrial Electronics*. Búzios, Rio de Janeiro, Brazil: IEEE, pp. 766–771.

33 Chanda, S. (2015). Measuring and enabling of resiliency using multiple microgrids. Master's thesis. Washington State University.

34 Arab, A., Khodaei, A., Khator, S.K. et al. (2015). Stochastic pre-hurricane restoration planning for electric power systems infrastructure. *IEEE Transactions on Smart Grid* 6 (2): 1046–1054.

35 Zonouz, S., Davis, C.M., Davis, K.R. et al. (2014). Socca: a security-oriented cyber-physical contingency analysis in power infrastructures. *IEEE Transactions on Smart Grid* 5 (1): 3–13.

36 Rubin, F. (1978). Enumerating all simple paths in a graph. *IEEE Transactions on Circuits and Systems* 25 (8): 641–642.

37 Knuth, D.E. (1976). Big omicron and big omega and big theta. *ACM Sigact News* 8 (2): 18–24.

38 Mateti, P. and Deo, N. (1976). On algorithms for enumerating all circuits of a graph. *SIAM Journal on Computing* 5 (1): 90–99.

39 Tarjan, R. (1972). Depth-first search and linear graph algorithms. *SIAM Journal on Computing* 1 (2): 146–160.

40 Zeleny, M. and Cochrane, J.L. (1973). *Multiple Criteria Decision Making*. University of South Carolina Press.

41 Labreuche, C. and Grabisch, M. (2003). The Choquet integral for the aggregation of interval scales in multicriteria decision making. *Fuzzy Sets and Systems* 137 (1): 11–26.

42 Bajpai, P., Chanda, S., and Srivastava, A.K. (2016). A novel metric to quantify and enable resilient distribution system using graph theory and Choquet integral. *IEEE Transactions on Smart Grid* 9 (4): 2918–2929. https://doi.org/10.1109/TSG.2016.2623818.

43 Takahagi, E. (2005). *λ Fuzzy Measure Identification Methods Using λ and Weights*. Japan: Senshu University.

44 Leszczýnski, K., Penczek, P., and Grochulski, W. (1985). Sugeno's fuzzy measure and fuzzy clustering. *Fuzzy Sets and Systems* 15 (2): 147–158.

45 Jabr, R. et al. (2013). Polyhedral formulations and loop elimination constraints for distribution network expansion planning. *IEEE Transactions on Power Systems* 28 (2): 1888–1897.

3

Resilience Indices Using Markov Modeling and Monte Carlo Simulation

Mohammad Shahidehpour[1] *and Zhiyi Li*[2]

[1] *The Robert W. Galvin Center for Electricity Innovation, Illinois Institute of Technology, Chicago, IL, USA*
[2] *The College of Electrical Engineering, Zhejiang University, Hangzhou, Zhejiang, China*

3.1 Introduction

Climate change has increased both the frequency and the severity of adverse weather throughout the world. In the year of 2019, the United States has suffered 14 weather-related disasters, including three flooding events, one wildfire event, and eight severe storm events, each of which incurred economic losses exceeding $1 billion [1]. Given the aging and already-stressed infrastructure, power distribution systems, which are fundamental for powering up the functioning and development of our society, have been continuously threatened by such extreme weather events. In fact, disruptions in power distribution systems are estimated to account for roughly 90% of outages in the United States, when adverse weather is proved as the primary cause of these outages [2]. Therefore, there exists an urgent need to prepare power distribution systems more adequately for meeting significant challenges posed by extreme weather events.

Extreme weather events are rare but disastrous, potentially triggering extensive disruptions within a very short time in power distribution systems. In order to alleviate the whims of Mother Nature, power distribution systems are expected to remain functional with minimal effect on electricity services supplied to end customers during and after the occurrence of extreme weather events. Accordingly, power distribution systems should be inherently capable of reacting effectively to and recovering promptly from severe disruptions resulting from extreme weather events. Such an inherent capability which exceeds the context of reliability is defined as resilience. In principle, both reliability and resilience are indispensable for power distribution systems to sustain electricity services under any environmental and operating conditions. Reliability is concerned with high-frequency but low-impact perturbations like $N-1$ contingencies, whereas resilience focuses on low-frequency but high-impact disruptions like $N-k$ contingencies [3].

Power distribution systems are conventionally designed in accordance with reliability requirements, but a highly reliable system is not necessarily resilient [4]. The reliability of a power distribution system is conventionally measured using historical outage data in terms of expected or long-run average values such as the system (or customer) average interruption frequency and the system (or customer) average interruption duration. However, these

Resiliency of Power Distribution Systems, First Edition.
Edited by Anurag K. Srivastava, Chen-Ching Liu, and Sayonsom Chanda.

statistics-based indices cast little light on the resilience of a power distribution system due primarily to insufficient information for representing extreme weather events that were rare in history. It is also rather difficult to predict the occurrence and duration of extreme weather events, as compared with typical disruptive events considered in reliability analyses. Considering that common approaches for reliability analysis fail to assess the resilience of a power distribution system, instituting a novel resilience analysis approach is essential for guiding electric grid modernization efforts to defeat extreme weather events.

Moreover, with the increasing integration of information and communication technologies, power distribution systems have been undergoing a transition to cyber-physical systems where physical processes in power grids and information flows in communication networks are tightly coupled for the overall distribution system functionality. Accordingly, the power grid (i.e. physical subsystem) and the communication network (i.e. cyber subsystem), which were isolated from one another, have been incorporated as two highly interdependent subsystems in the operation of a power distribution system. Such interdependencies are advantageous in enhancing the economics, security, and quality of electricity services, but they also represent driving forces for cascading failures across cyber and physical subsystems [5]. Hence, the modeling and analysis of cyber-physical interdependencies are integral to the resilience of a power distribution system.

In order to proactively enhance the resilience of power distribution systems against extreme weather events, this chapter will propose a comprehensive framework for analyzing the resilience of power distribution systems, and provide a detailed modeling and formulation with the consideration of cyber-physical interdependencies for decision making in practical energy management systems. The remainder of this chapter is organized as follows: Section 3.2 introduces extensive cyber-physical interdependencies in the operation of power distribution systems, especially the functionality of dependent subsystems when any subsystem suffers disruptions; Section 3.3 presents a generic resilience analysis framework for power distribution systems and proposes several resilience indices for the system performance evaluation following an extreme weather event; Section 3.4 states the detailed process for the implementation of the proposed resilience analysis framework; Section 3.5 shows simulation results for applying the proposed resilience analysis framework to a test power distribution system that would be potentially stricken by an extreme weather event; Section 3.6 concludes this chapter and points out to the role of microgrids in enhancing the resilience of power distribution systems.

3.2 Cyber-Physical Interdependencies in Power Distribution Systems

A power distribution system possesses a multitude of physically and functionally heterogeneous components which work in harmony for providing end customers with satisfactory electricity services. Figure 3.1 illustrates a typical power distribution system consisting of a hierarchy of cyber and physical subsystems that interact with each other. The physical subsystem is composed of equipment and infrastructure for delivering electricity services including buses and power lines in the power grid, while the cyber subsystem is composed of systems and devices for processing and exchanging information including remote

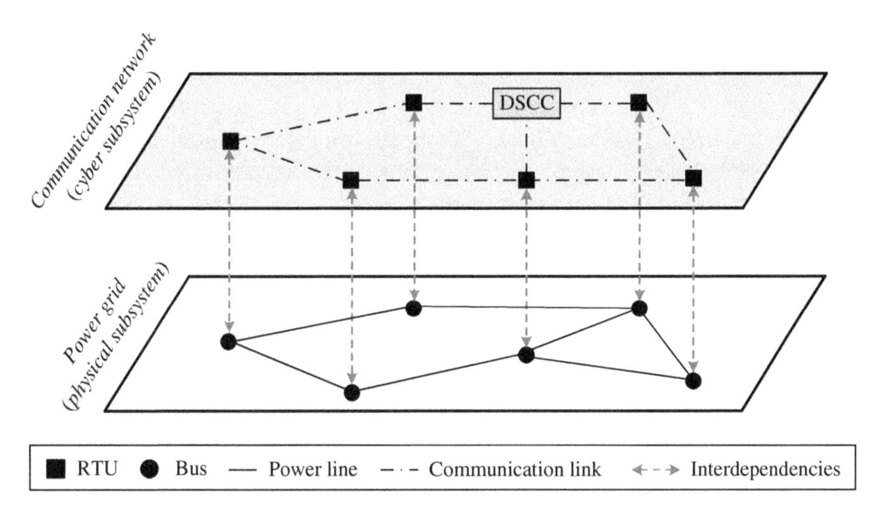

Figure 3.1 Cyber-physical power distribution system.

terminal units (RTUs), distribution system control center (DSCC), and the associated communication links in the communication network.

In a power distribution system, electricity flows unidirectionally from generation sources to load sites via buses and power lines, but information flows bidirectionally between RTUs and DSCC via communication links. There have also been obvious interdependencies between cyber and physical subsystems (as represented by vertical cross-layer links in Figure 3.1) when the full functionality of each subsystem hinges on the existence and sufficiency of services provided by the other subsystem. In particular, the cyber subsystem plays a key role in maintaining the controllability and observability of the physical subsystem which in turn empowers the cyber subsystem to fulfill its responsibility. For instance, each electrical bus is associated with an RTU for the monitoring and control of physical processes at a substation, but the RTU would fail when the supply of power is insufficient. Meanwhile, in order to avoid a single point of failure in the cyber subsystem, DSCC commonly hinges on multiple buses for a sustained supply of power and even equips uninterruptible power supply apparatus as a backup.

The operation of a power distribution system could be abstracted as a general control loop depicted in Figure 3.2. Under normal circumstances, DSCC continuously interacts with RTUs over the communication network for enabling operators' situational awareness and

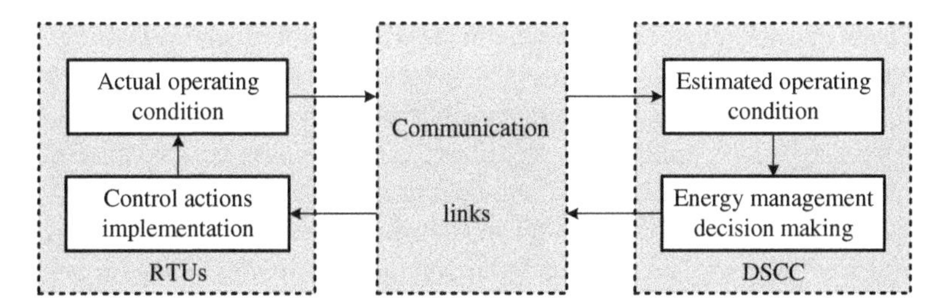

Figure 3.2 Close-loop control of power distribution systems.

further facilitating their decision-making in energy management systems [6]. RTUs acquire real-time measurements that reflect the actual operating condition of the physical subsystem, including analog data like voltage magnitudes of buses, and digital data like breaker statutes of incident power lines. Meanwhile, DSCC operators have a continuous observation of the physical subsystem's operating condition (e.g. by means of state estimation) after collecting and synchronizing the RTU measurements. The analyses of these measurements also allow DSCC operators to take corrective actions (e.g. adjust voltages or redispatch power generation units) in response to any changes in estimated operating conditions of the physical subsystem. DSCC instructs and supervises RTUs over the communication network to implement control actions, when any energy management decisions need to be realized in the physical subsystem. The closed-loop control procedure is performed on an ongoing basis representing the tie between operators' decision-making process and system operating conditions.

When cyber-physical interdependencies contribute to the overall functionality of a power distribution system, they unintentionally expose both cyber and physical subsystems to a broader range of disruptions. Figure 3.3 illustrates three types of disruptions caused by cyber-physical interdependencies rather than extreme weather events, where each bus is equipped with an RTU that is assigned with the same sequence number as the bus. Originally, all components in both cyber and physical subsystems are functional. In Figure 3.3a, initial failures of power lines 1-2 and 2-4 caused by an extreme weather event place buses 1 and 4 on outage since the two buses are isolated from the remaining physical subsystem (such a bus is labeled as an isolated bus). The outages further force RTUs 1 and 4 to be inoperable (such an RTU is labeled as an inoperable RTU) since the supply of power to the two RTUs cannot be sustained by the associated buses; the two RTUs therefore lose the capability of communication, which is equivalent to removing them from the cyber subsystem, so that the other two RTUs are also disconnected from DSCC in terms of communication accessibility.

In Figure 3.3b, initial failures of communication links 2-3 and 3-4 caused by an extreme weather event render RTU 3 inoperable since the RTU is isolated from the remaining cyber subsystem; DSCC therefore fails to communicate with the RTU for collecting real-time measurements from and implementing supervisory control commands at bus 3 (such a bus is labeled as an idle bus).

In Figure 3.3c, initial failures of communication links 1-2, 2-3, 2-4, and 3-4 caused by an extreme weather event force DSCC to lose communications with RTUs 2 and 3 since the two RTUs are isolated from the remaining cyber subsystem; DSCC operators therefore fail to estimate any changes in the status of buses 2 and 3 and line 2-3; even though line 2-3 could have already been on the outage, operators might still consider it in service based on the available measurements.

Although a failure of cyber subsystem components would not lead to an immediate failure in the physical subsystem, it would have a significant impact on the performance of the power distribution system by hampering human operators' situational awareness. Situational awareness, which supplies information on any anomalies in the current operating condition of the physical subsystem, lays the foundation for operators to make effective and timely decisions in emergency situations. When the cyber subsystem function is trustworthy, situational awareness allows operators to acknowledge the actual operating conditions

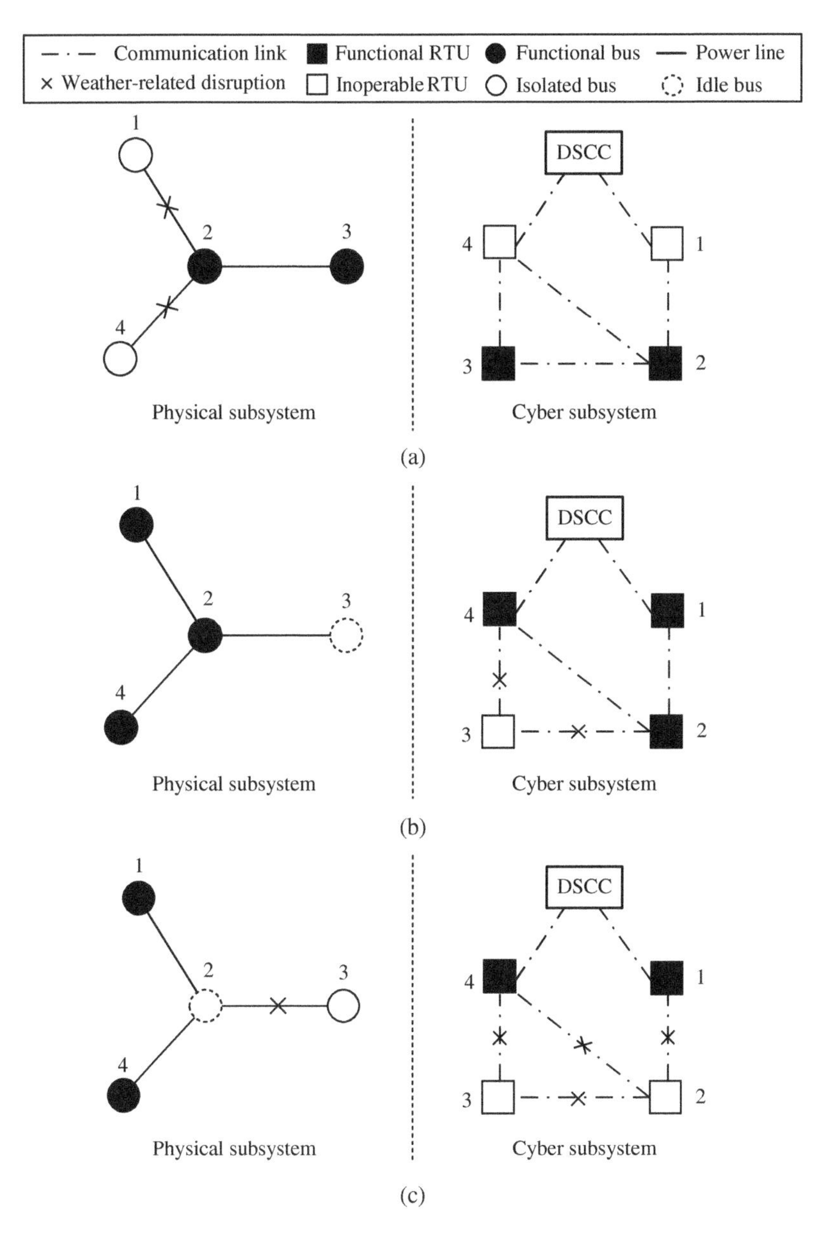

Figure 3.3 Disruptions due to cyber-physical interdependencies. (a) RTUs losing power supply. (b) Power bus becoming unobservable and uncontrollable. (c) Hidden failure of a power line.

of the physical subsystem before the system condition deteriorates. In fact, lack of situational awareness due to cyber subsystem failures has already been identified as one of the contributing factors to several major power outages [7, 8]. Hence, situation awareness is critical for strengthening the resilience of a power distribution system in extreme weather events [9].

3.3 Resilience of Power Distribution Systems

Resilience is an intrinsically complex property of major infrastructures that depend on system topology, operation strategies, and evolving conditions of disruptive events. This section presents several indices for quantifying the resilience of power distribution systems in extreme events and provides a generic framework for comprehensive resilience analysis in abnormal cases that are considered unpredictable and chaotic.

3.3.1 Evolution of Operation Performance

The performance of a power distribution system can be represented by a quantifiable index, such as generation adequacy [10], transmission security [11], and power quality [12], which allows operators to analyze system conditions in various operating circumstances. System disruptions in the wake of an extreme event can trigger a decline in operating conditions, whereas restoration efforts (i.e. repair or replacement) invested in recovering failed components could improve the system's performance after the extreme event.

Figure 3.4 shows the operation performance of a power distribution system during and after the occurrence of an extreme weather event, which is quantified in terms of the supplied power demands. In this figure, the actual operation performance is expressed as a time-varying function $Q'(t)$, and the targeted performance without the influence of the extreme weather event is expressed as $Q'_0(t)$. The actual operation performance evolves over time as follows: it retains its targeted level between t_0 (when the event unfolds) and t_1 (when the event starts to incur severe disruptions), undergoes a rapid decline until t_2 (when the event passes through), starts to improve after t_3 (when the restoration is put into place), returns to the targeted level at t_4 (when power demands are fully supplied), and remains at the targeted level through t_5 (when all failed components are recovered).

The operation performance evolution normally undergoes the following five stages:

- Prevention Stage (from t_0 to t_1): The distribution system is robust to power disruptions resulting from the extreme weather event until the onset of component failures that degrades the system condition. In fact, a power distribution system is commonly

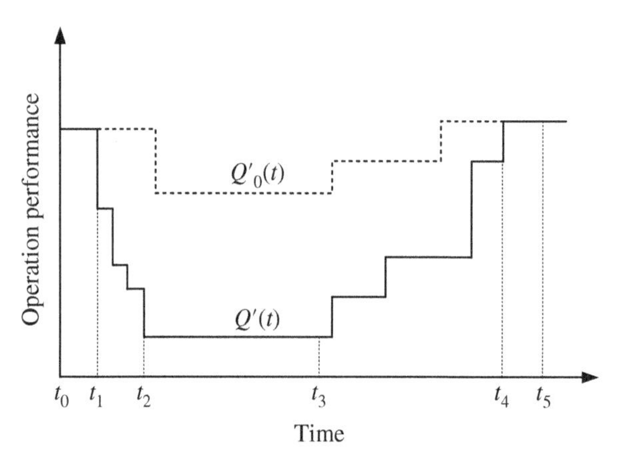

Figure 3.4 Evolution of operation performance following an extreme weather event.

designed with infrastructural redundancy so that it is capable of absorbing the impact of single component outages (e.g. $N-1$ contingencies) without hampering the operation performance.

- Accommodation Stage (from t_1 to t_2): At this stage, the power distribution system suffers degradation in its performance when the extreme weather event incurs continuous disruptions which cannot be fully absorbed. Since operators strive to adjust strategies for mitigating the implications of emergent disruptions, the performance degradation may be neither sudden nor steep.
- Investigation Stage (from t_2 to t_3): At this stage, operators investigate the extent of damages on the power distribution system following the termination of the extreme weather event. After gathering the critical information for initiating restoration, operators allocate available resources, including equipment and manpower, to recover failed components.
- Recovery Stage (from t_3 to t_4): At this stage, available resources are optimally utilized for restoring electricity services in a timely manner. Operators concentrate their efforts on improving the operation performance while assigning priorities to restoration activities. The operation performance gradually rebounds to the targeted level at the end of this stage. In particular, due to infrastructural redundancy, certain failed components may not be required for immediate recovery at this stage.
- Reinforcement Stage (from t_4 to t_5): At this stage, restoration efforts are further invested for recovering the remaining components which failed during the extreme weather event but are less critical. Meanwhile, hardening measures are also deployed in the power distribution system for reinforcing the infrastructure against future extreme events.

3.3.2 Power System Resilience Indices

The common statistics-based reliability indices, such as the system average interruption frequency index (SAIFI) and the system average interruption duration index (SAIDI), fail to capture the context-dependent nature of resilience. Accordingly, new indices specific to resilience are essential for analyzing the resilience of a power distribution system.

A power distribution system is considered resilient to an extreme weather event if it can ensure the adequacy and continuity of electricity services during and after the occurrence of the event. In particular, resilience is directly associated with the evolution of operation performance, which demonstrates how a power distribution system responds to and recovers from intensive disruptions resulting from an extreme weather event. For comparing the resilience in different circumstances (e.g. types of extreme events, configurations of system infrastructure, and strategies of energy management), the evolution of operation performance over time is normalized between 0 (total failure) and 1 (targeted level) as:

$$Q(t) = \frac{Q'(t)}{Q'_0(t)}, \quad \forall t \tag{3.1}$$

where $Q(t)$ is the normalized operation performance level at time t.

Figure 3.5 shows the operation performance evolution (depicted in Figure 3.4) after normalization, where the targeted operation performance function $Q'_0(t)$ is simplified as Q_0 with a constant value of 1, and Q_2 representing the minimum operation performance level is achieved at t_2.

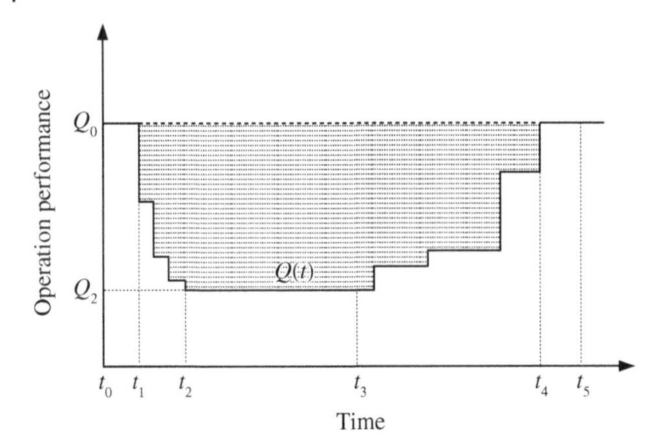

Figure 3.5 Evolution of normalized operation performance.

Given the evolution of normalized operation performance under a specific extreme weather event, we list the following indices for characterizing the resilience of the power distribution system considering four different perspectives.

- Worst Performance (R_α): This index represents the minimum level of normalized operation performance, which reflects the survivability of electricity services in the aftermath of the event. This index ranges between 0 (failure to survive) and 1 (fully preventable disruption), with a larger value being more desirable:

$$R_\alpha = \min \frac{Q(t)}{Q_0} = Q_2 \tag{3.2}$$

- Accumulated Loss (R_β): This index represents the system's temporal behavior over the reduced performance period (i.e. between t_1 and t_4), which indicates the responsiveness of operation strategies. The numerator (i.e. $\int_{t_1}^{t_4}[1 - Q(t)]\mathrm{d}t$), which corresponds to the shaded area in Figure 3.5, temporally integrates the difference between actual and targeted operation performance levels, and the denominator denotes the temporal integration of the targeted operation performance level. This index ranges between 0 (i.e. total system failure) and 1 (i.e. no influence on the operation performance), in which a smaller value is more desirable:

$$R_\beta = \frac{\int_{t_1}^{t_4} [Q_0 - Q(t)]\mathrm{d}t}{\int_{t_1}^{t_4} Q_0 \,\mathrm{d}t} = \frac{1}{t_4 - t_1} \int_{t_1}^{t_4} [1 - Q(t)]\,\mathrm{d}t \tag{3.3}$$

- Resistive Capability (R_γ): This index is calculated as the average speed of performance degradation (e.g. minutes per kilowatt of load curtailment) over the event propagation period (i.e. between t_0 and t_2), which measures the overall resistance against disruptions in the wake of the event. This index ranges between 0 (i.e. system collapses immediately after the event unfolds) and infinity (i.e. system resists the event with no degradation in operation performance), and a larger value is more desirable:

$$R_\gamma = \frac{t_2 - t_0}{Q_0 - \min Q(t)} = \frac{t_2 - t_0}{1 - Q_2} \tag{3.4}$$

- Recovery Rapidity (R_δ): This index is directly related to the average performance improvement rate (e.g. kilowatt of load restoration per minute) over the component recovery period (i.e. between t_3 and t_5), which quantifies the swiftness of recovering failed components. This index ranges between 0 (i.e. failed components require an infinitely long time to recover) and infinity (i.e. failed components are recovered instantaneously), and a larger value is more desirable:

$$R_\delta = \frac{Q_0 - \min Q(t)}{t_5 - t_3} = \frac{1 - Q_2}{t_5 - t_3} \tag{3.5}$$

Considering that resilience is a multifaceted concept regarding how a power distribution system mitigates the consequences of an extreme weather event, one-sided indices are deemed insufficient for getting a full knowledge of resilience. Accordingly, the Power System Resilience Index (PSRI) is proposed here for comprehending the resilience of a power distribution system, which is stated as:

$$R = \left[R_\alpha \cdot (1 - R_\beta) \cdot \exp\left(-\varepsilon \cdot \frac{1}{R_\delta \cdot R_\gamma}\right)\right] \times 100\% \tag{3.6}$$

where ε is a user-specified parameter that adjusts the effect of exponential term on the index, neither too large to dominate nor too small to neglect.

As indicated by Eq. (3.6), PSRI is an integrated index that combines the four indices (3.2)–(3.5) for incorporating multiple essential dimensions of resilience (e.g. survivability, responsiveness, resistance, and recovery rapidity). Accordingly, it is reasonable to use PSRI in place of multiple one-sided indices for resilience quantification. PSRI ranges between 0% (i.e. total loss of resilience) and 100% (i.e. adequacy in resilience), in which a larger PSRI denotes a higher level of resilience. If a power distribution system has an increased level of worst performance (i.e. R_α), resistive capability (i.e. R_γ), and recovery rapidity (i.e. R_δ) along with a reduced level of accumulated loss (i.e. R_β) under an extreme weather event, then the system is more resilient to the specific event. PSRI, which is dimensionless, also acts as a robust indicator for comparing the resilience quantitatively under various circumstances. For instance, it can be used to compare the resilience of various power distribution systems to the same extreme weather event, and the resilience of a power distribution system to various extreme weather events.

3.3.3 Comprehensive Resilience Analysis Framework

In order to dissect and enhance the resilience of a power distribution system to potential extreme weather events, we propose a comprehensive resilience analysis framework (shown in Figure 3.6) by capturing spatiotemporal characteristics of extreme weather events as well as synergies between infrastructure configurations and operation strategies.

The execution of the proposed framework is composed of the following seven steps.

- Step 1 (define a resilience goal): The type of extreme weather events should be first specified in every system resilience study since the resilience of a power distribution system varies with ambient weather conditions. In fact, the resilience to a specific extreme weather event could make the system vulnerable to some other types of events. For instance, a power distribution system utilizing underground cables in

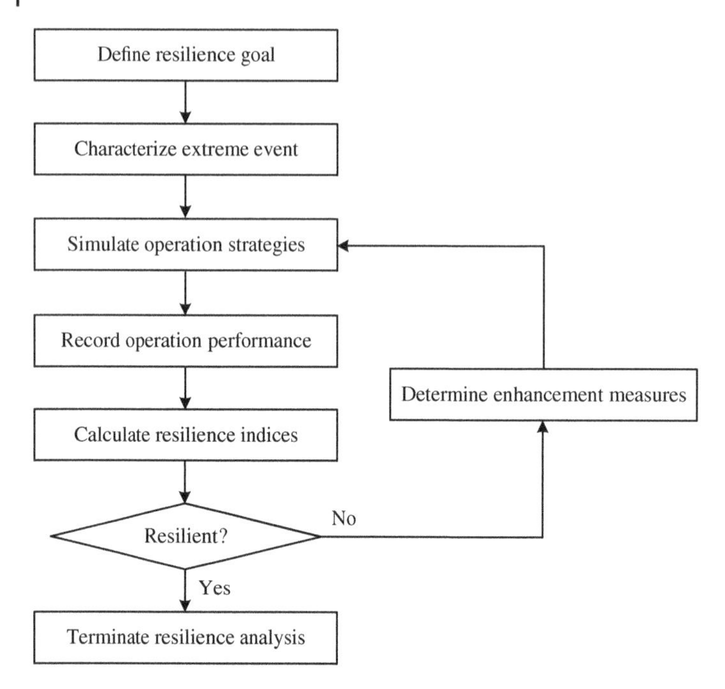

Figure 3.6 Framework for resilience analysis.

place of overhead lines tends to be resilient to hurricanes but may be vulnerable to flooding. In principle, the resilience to multiple types of extreme weather events is possible with the help of tactical infrastructure configurations and responsive operation strategies.

- Step 2 (characterize an extreme weather event): Spatiotemporal dynamics (e.g. geographical scope and damaging impact) of a specific extreme weather event that moves across a power distribution system could refer to the statistical information on the progression of that type of extreme weather (e.g. database of National Oceanic and Atmospheric Administration) [13].
- Step 3 (simulate operation strategies): As a specific extreme weather event progresses, exposed power components would stage a continuous threat of electricity service interruptions. Such interruptions would persist until adequate efforts are put in place to recover the component functionality. Meanwhile, operators strive to maintain or improve the operation performance by adjusting operation strategies (e.g. power redispatch and service restoration) that mitigate the consequences of extreme weather in a timely and effective manner. It is essential to simulate operators' decision-making process for responding to and recovering from unfolding disruptions in order to model the temporal operation performance of a power distribution system in the aftermath of an extreme weather event.
- Step 4 (monitor the operation performance): The operation performance of a power distribution system is principally a temporal function that depends on both the severity of an extreme weather event and the effectiveness of operators' mitigation strategies. The evolution of operation performance of the power distribution system is monitored and

recorded starting from the occurrence of specific extreme weather until the full system restoration.

- Step 5 (calculate resilience indices): Once the evolution curve of the operation performance is available, the power distribution system's resilience to a specific extreme weather event is measured by calculating and analyzing indices concerned with performance degradation.
- Step 6 (determine resilience enhancement measures): If the power distribution system is not sufficiently resilient to a specific extreme weather event, several reinforcement measures would be considered (e.g. deploying redundant components and enabling demand-side management) to improve the operation performance. Once a resilience enhancement plan is determined, re-execute steps 3–5 for examining the effectiveness of the resilience plan and make further decisions on deploying additional resilience enhancement measures at step 6 if necessary. Hence, steps 3–6 form a cycle of evaluation and improvement in an effort to ensure resilience to a specific extreme weather event.
- Step 7 (terminate the resilience analysis): If the power distribution system is already resilient to a specific extreme weather event, terminate the resilience analysis, present pertinent data on time-varying operation performance and prescribe enhancement measures.

The proposed framework provides dynamic interactions between the progression of an extreme weather event and the operation of a power distribution system. However, various uncertainties are to be considered for implementing the fundamental building blocks of the proposed framework, including the likelihood and the extent of disruptions following an extreme weather event, damaging impact of disruptions on a power distribution system, as well as corrective actions taken by operators in response to disruptions. Accordingly, discrete-time methods like sequential Monte Carlo simulation should be employed to characterize the representation, quantification, and propagation of such uncertainties in a systematic manner.

3.4 Mathematical Model for Resilience Analysis

This section presents the mathematical model for implementing the proposed resilience analysis framework, including the Markov chain for deriving a sequence of weather conditions, proportional hazard model for estimating the component survival probability in adverse weather conditions, second-order cone programming for optimizing the power dispatch and evaluating the operation performance, as well as greedy heuristics for restoring interdependent cyber-power subsystems. Due to page limitations, the detailed modeling of resilience enhancement measures is omitted. Table 3.1 lists the notations used in this section.

3.4.1 Markov Chain for Analyzing the Evolution of Weather Condition

The operation performance of a power distribution system is sensitive to weather conditions, mainly because of potential failures of exposed components that are fragile to

Table 3.1 Nomenclature.

	Symbol	Definition
Index	t	Time step (i.e. hour)
	i	Component in the power distribution system
	k_c, m_c, n_c	Node (i.e. RTU or DSCC) in the cyber subsystem
	k_p, m_p, n_p	Bus in the physical subsystem, and the root bus corresponds to index 0
	w	Island with the power supply capability in the physical subsystem, and the major island containing the root bus corresponds to index 0
	u	Power source in the physical subsystem
Set	$\boldsymbol{\Omega}_{\mathrm{C}}$ $(\boldsymbol{\Psi}_{\mathrm{C}})$	Set of nodes (links) in the cyber subsystem
	$\boldsymbol{\Omega}_{\mathrm{C}}^{\mathrm{O}}$ $(\boldsymbol{\Psi}_{\mathrm{C}}^{\mathrm{O}})$	Set of nodes (links) losing communication capability
	$\boldsymbol{\Omega}_{\mathrm{P}}$ $(\boldsymbol{\Psi}_{\mathrm{P}})$	Set of buses (lines) in the physical subsystem
	$\boldsymbol{\Omega}_{\mathrm{P}}^{\mathrm{G}}$	Set of buses with a power source
	$\boldsymbol{\Omega}_{\mathrm{P}}^{\mathrm{S}_w}$	Set of buses in the wth island
	$\boldsymbol{\Omega}_{\mathrm{P}}^{\mathrm{C}}$ $(\boldsymbol{\Omega}_{\mathrm{P}}^{\mathrm{U}})$	Set of buses with (without) a functional RTU
	$\boldsymbol{\Omega}_{\mathrm{P}}^{\mathrm{O}}$ $(\boldsymbol{\Psi}_{\mathrm{P}}^{\oplus})$	Set of buses (lines) actually on outage
	$\boldsymbol{\Omega}_{\mathrm{P}}^{\oplus}$ $(\boldsymbol{\Psi}_{\mathrm{P}}^{\oplus})$	Set of buses (lines) on outage as estimated by operators
	$\Delta_{m_c}\left(\nabla_{m_c}\right)$	Set of neighbors of the mth node in the backward (forward) direction
	$\Delta_{m_p}\left(\nabla_{m_p}\right)$	Set of neighbors of the mth bus in the backward (forward) direction
Parameter	T_i^r	Most recent maintenance time of the ith component
	$c_{m_p}^g\left(c_{m_p}^d\right)$	Cost of power injection (load curtailment) at the mth bus
	ρ	Penalty factor which is equal to a large positive constant
	$r_{k_p m_p}\left(x_{k_p m_p}\right)$	Resistance (reactance) of the line between the kth and mth buses
	$S_{k_p m_p}$	Capacity of the line between the kth and mth buses
	$\overline{P}_{m_p}^g$	Max active power output of the power source at the mth bus
	$S_{m_p}^g$	Generation capacity of the power source at the mth bus
	$D_{m_p,t}^p\left(D_{m_p,t}^q\right)$	Original active (reactive) power demand at the mth bus at time t
	$\underline{v}_{m_p}\left(\overline{v}_{m_p}\right)$	Min (max) square value of the voltage magnitude of the mth bus
	$v_{m_p,t}^{\dagger}$	Nominal value of the voltage magnitude of the mth bus
	$\hat{P}_{m_p,t}^g$	Latest available measurement of active power injection at the mth bus up to time t
	$\hat{D}_{m_p,t}^p\left(\hat{D}_{m_p,t}^q\right)$	Latest available measurement of active (reactive) power demand at the mth bus up to time t
	$\hat{v}_{m_p,t}$	Latest available measurement of voltage magnitude square of the mth bus up to time t
	$\tilde{P}_{m_p}^g$	Optimal injection amount of active power at the mth bus at time t
	$\tilde{L}_{m_p,t}^p\left(\tilde{L}_{m_p,t}^q\right)$	Optimal curtailment amount of active (reactive) power demand at the mth bus at time t
	$\tilde{v}_{m_p,t}$	Optimal setting of voltage magnitude square of the mth bus at time t

(Continued)

Table 3.1 (Continued)

	Symbol	Definition
Variable	X_t	Weather condition at time t
	T_i^f	Failure time of the ith component
	$F^g_{m_c,t}\ \left(F^g_{m_p,t}\right)$	Commodity injection into the mth node (bus) at time t
	$F^d_{m_c,t}\ \left(F^d_{m_p,t}\right)$	Commodity withdrawal from the mth node (bus) at time t
	$\alpha_w(\beta_w)$	Binary indicator of generation (load) curtailment in the wth island, where 0 means no curtailment and 1 means curtailment
	$\xi^g_w\ (\xi^d_w)$	Generation (load) curtailment ratio in the wth island
	$f_{m_cn_c,t}\ \left(f_{m_pn_p,t}\right)$	Commodity flow between the mth and nth nodes (buses) at time t
	$\varepsilon_{\text{DSCC},\,t}$	Commodity quantity injected at DSCC at time t
	$\varepsilon_{u,\,t}$	Commodity quantity injected at the uth generation source at time t
	$\tilde{\varepsilon}_{m_p,t}(\tilde{\varepsilon}_{m_c,t})$	Max commodity quantity consumed at the mth bus (node) at time t
	$P^g_{m_p,t}\ \left(Q^g_{m_p,t}\right)$	Active (reactive) power output of the generation source at the mth bus at time t
	$P^d_{m_p,t}\ \left(Q^d_{m_p,t}\right)$	Actual active (reactive) power demand at the mth bus after curtailment at time t
	$L^p_{m_p,t}\ \left(L^q_{m_p,t}\right)$	Curtailment amount of active (reactive) power at the mth bus at time t
	$p_{m_pn_p,t}\ \left(q_{m_pn_p,t}\right)$	Active (reactive) power flow between the mth and nth buses at time t
	$V_{m_p,t}$	Magnitude of the voltage of the mth bus at time t
	$I_{k_pm_p,t}$	Magnitude of the current between the kth and mth buses at time t
	$v_{m_p,t}$	Square of the voltage magnitude $V_{m_p,t}$
	$l_{k_pm_p,t}$	Square of the current magnitude $I_{k_pm_p,t}$
Function	$\Pi(\cdot)$	Time-dependent distribution over the state space
	$\lvert\cdot\rvert$	Cardinality of a set
	$\odot\left(\cdot,\boldsymbol{\Omega}^{\text{G}}_{\text{P}}\right)$	Ordinality of an element in the set $\boldsymbol{\Omega}^{\text{G}}_{\text{P}}$, and the root bus corresponds to ordinality 0
	$h_i\ (\cdot)$	Hazard function of the ith component
	$h_{i,0}\ (\cdot)$	Baseline hazard function of the ith component
	$\phi_i(\cdot)$	Link function to quantify the covariate's impact on the ith component
	$s_i\ (\cdot)$	Survival probability of the ith component
	$\Pr\ (\cdot)$	Probability function
	$\exp\ (\cdot)$	Exponential function
	$\mathcal{N}\ (\cdot)$	Probability density function of the normal distribution

adverse weather. Given that a power distribution system typically covers a far smaller geographical area than a power transmission system, it is reasonable to assume all distribution system components reside in the same weather condition at any given time. In other words, the progression of an extreme weather event in a power distribution system is regarded as a standstill whose damaging impact depends merely on the elapsed time, corresponding to a sequence of time-dependent and location-independent weather conditions.

The development of weather conditions is commonly considered to have the Markov property which indicates the upcoming weather condition is merely dependent on the most recent weather condition. Mathematically, the Markov property of weather condition is expressed as:

$$\Pr(X_t \mid X_{t-1}, \ldots, X_{t_0}) = \Pr(X_t \mid X_{t-1}) \tag{3.7}$$

where $\Pr(X_t \mid X_{t-1}, \ldots, X_{t_0})$ is the conditional probability of upcoming weather condition X_t given all weather previous conditions $X_{t-1},\ldots, X_{t_0}$, and $\Pr(X_t \mid X_{t-1})$ is the conditional probability of X_t given the most recent weather condition X_{t-1}.

The goal of generating a series of weather conditions conforming to the evolution of an extreme weather event is accomplished by the Markov chain. Markov chain, which is essentially a discrete-time stochastic process, enables sequential transitions among a pre-defined set of states (representing weather conditions) between adjacent time steps. Without loss of generality, the state space formed by weather conditions is discretized into four mutual-exclusive states (i.e. high-, medium- and low-impact, and normal conditions). The four weather states are expected to cover all possible weather conditions during an extreme weather event; so, the Markov chain is a powerful tool to model the evolution of an extreme weather event (e.g. improvement or deterioration of weather conditions) in a power distribution system. Besides, the Markov chain is a memoryless process such that there is no need to record previous weather conditions for predicting the future development of weather conditions.

Figure 3.7 is a directed graph representing the four-state Markov chain, where the states are interlinked by directed edges annotated with nonnegative transition probabilities λ's. The normal condition is an absorbing state in which the weather condition would remain unchanged indefinitely, so as to model the termination of the extreme weather event. For ease of presentation, the four weather states are labeled as: {L = low-impact condition, M = medium-impact condition, H = high-impact condition, N = normal condition}. The transition relationships in Figure 3.7 are represented by the transition probability matrix (3.8). In the matrix, each row corresponds to the weather condition at the current time step, and each column corresponds to the weather condition at the next time step, when the sum

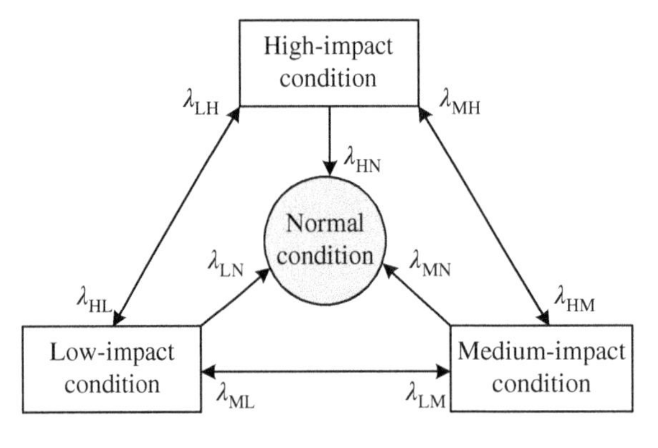

Figure 3.7 Markov chain for simulating the extreme weather event.

of entries in each row is 1, representing the transition probability distribution conditioned on the current weather state:

$$\mathbf{T} = \begin{array}{c} \\ \text{L} \\ \text{M} \\ \text{H} \\ \text{N} \end{array} \begin{array}{c} \begin{array}{cccc} \text{L} & \text{M} & \text{H} & \text{N} \end{array} \\ \begin{bmatrix} 1-\lambda_{\text{LM}}-\lambda_{\text{LH}}-\lambda_{\text{LN}} & \lambda_{\text{LM}} & \lambda_{\text{LH}} & \lambda_{\text{LN}} \\ \lambda_{\text{ML}} & 1-\lambda_{\text{ML}}-\lambda_{\text{MH}}-\lambda_{\text{MN}} & \lambda_{\text{MH}} & \lambda_{\text{MN}} \\ \lambda_{\text{HL}} & \lambda_{\text{HM}} & 1-\lambda_{\text{HL}}-\lambda_{\text{HM}}-\lambda_{\text{HN}} & \lambda_{\text{HH}} \\ 0 & 0 & 0 & 1 \end{bmatrix} \end{array} \tag{3.8}$$

The likelihood of transitions from one state to another is therefore determined by referencing entries in the transition probability matrix. For example, if the weather condition is currently in state L, the probabilities of moving to states M, H, and N subsequently are denoted by λ_{LM}, λ_{LH}, and λ_{LN}, respectively. The dynamics of Markov chain are captured by changes in a row vector of four stochastic variables specifying the probability distribution over the state space. We assume all transition probabilities that are pre-specified by expert judgment are time-independent during the Markov chain process and the transition probability matrix is constant in nature. Accordingly, the probability distribution over the four weather conditions at each time step is given as:

$$\mathbf{\Pi}(t) = \mathbf{\Pi}(t-1) \cdot \mathbf{T} = \mathbf{\Pi}(t-2) \cdot \mathbf{T}^2 = \cdots = \mathbf{\Pi}(t_0) \cdot \mathbf{T}^{t-t_0} \tag{3.9}$$

which means the probability distribution at time t can be accurately predicted by recursively multiplying the initial distribution at time t_0 with the constant transition probability matrix $\mathbf{T}$.

3.4.2 Proportional Hazard Model for Component Failure Estimation

Adverse weather is commonly regarded as a hazard with a potential to cause component failures and service disruptions in a power distribution system. Since components are continuously exposed to adverse weather and suffer a sharp increase in their failure probability, the leading cause of component failures during an extreme weather event is the ambient weather rather than physical conditions (i.e. aging). Compared with buses and RTUs which are deployed at substations with adequate hardening resources, communication links and power lines which spread over long distances are much more vulnerable to adverse weather. Accordingly, it is reasonable to assume only power lines and communication links are the components potentially disrupted by an extreme weather event.

Here, we adopt the proportional hazard model [14] to estimate the weather-dependent survival probability instead of using a constant survival rate for each component during and after the occurrence of an extreme weather event. In this model, the aging effect and weather impact are considered as two multiplicative factors, and the instantaneous failure rate (i.e. forced outage rate) of a component at t is calculated as:

$$h_i(t, X_t) = \lim_{dt \to 0} \frac{\Pr\left\{t \le T_i^f < t + \mathrm{d}t\right\}}{\mathrm{d}t} = h_{i,0}(t) \cdot \phi_i(X_t), \quad \forall i \tag{3.10}$$

where the underlying baseline hazard $h_{i,0}(t)$ only depends on the accumulated service time after last maintenance; the link function $\phi_i(X_t)$ only considers the effect of the time-dependent weather condition which is incorporated as a covariate X_t. Common mathematical forms of $h_{i,0}(t)$ and $\phi_i(X_t)$ are given as below:

$$h_{i,0}(t) = \frac{\varphi_i}{\tau_i}\left(\frac{t - T_i^r}{\tau_i}\right)^{\varphi_i - 1}, \quad \phi_i(X_t) = \exp\{\gamma_i \cdot X_t\}, \quad \forall i \tag{3.11}$$

where $h_{i,0}(t)$ is represented by the Weibull hazard rate function, and φ_i, τ_i are pre-defined constants which jointly determine the shape and scale of the Weibull distribution; $\phi_i(X_t)$ is specified as an exponential function with a pre-defined constant γ_i; the covariate X_t takes one of four values (i.e. 0, 1, 2, and 3) at each time step, where 0 denotes the normal condition, and 2, 3, and 4 denote low-, medium-, and high-impact conditions, respectively.

The survival probability function, which represents the probability of a component surviving up to time t, is then expressed as below:

$$s_i(t) = \Pr\left\{T_i^f \geq t\right\} = \exp\left\{-\int_{T_i^r}^{t} h_i(t, X_t)\mathrm{d}t\right\}, \quad \forall i \tag{3.12}$$

where the integral $\int_{T_i^r}^{t} h_i(t, X_t)\mathrm{d}t$ is the cumulative failure rate from the last maintenance time to time t. Equivalently, (3.12) is represented as below:

$$\begin{aligned} s_i(t) &= \exp\left\{-\int_{t-1}^{t} h_i(t, X_t)\mathrm{d}t\right\} \cdot \exp\left\{-\int_{T_i^r}^{t-1} h_i(t, X_t)\mathrm{d}t\right\} \\ &= s_i(t \mid t-1) \cdot s_i(t-1), \quad \forall i \end{aligned} \tag{3.13}$$

where $s_i(t \mid t-1)$ denotes the conditional survival probability at t given the survival probability at time $t - 1$. Since the weather condition is assumed to change at distinct time steps, the link function in the gap between time steps $t - 1$ and t remains a constant value $\phi_i(X_{t-1})$. Accordingly, $s_i(t \mid t-1)$ is explicitly calculated as:

$$\begin{aligned} s_i(t \mid t-1) &= \exp\left\{-\int_{t-1}^{t} h_{i,0}(t) \cdot \varphi_i(X_t)\mathrm{d}t\right\} = \exp\left\{-\varphi_i(X_{t-1}) \cdot \int_{t-1}^{t} h_{i,0}(t)\mathrm{d}t\right\} \\ &= \exp\left\{-\varphi_i(X_{t-1}) \cdot \left[\left(\frac{t - T_i^r}{\tau_i}\right)^{\varphi_i} - \left(\frac{t - 1 - T_i^r}{\tau_i}\right)^{\varphi_i}\right]\right\} \end{aligned} \tag{3.14}$$

Besides, $s_i(t - 1)$ is given by the following recursive process:

$$s_i(t-1) = s_i(t-1 \mid t-2) \cdot s_i(t-2 \mid t-3) \cdot \cdots \cdot s_i(t_0 + 1 \mid t_0) \cdot s_i(t_0) \tag{3.15}$$

where the calculation of conditional probabilities is similar to (3.14); the survival probability $s_i(t_0)$ at t_0 when the extreme weather event unfolds is only dependent on the aging process, as calculated below:

$$s_i(t_0) = \exp\left\{-\int_{T_i^r}^{t_0} h_{i,0}(t)\mathrm{d}t\right\} = \exp\left\{-\left(\frac{t_0 - T_i^r}{\tau_i}\right)^{\varphi_i}\right\} \tag{3.16}$$

The probability of a component failure between time steps $t - 1$ and t is accurately calculated in the following equation, through utilization of the calculation results of (3.14) and (3.15):

$$\Pr\left\{t - 1 \leq T_i^f < t\right\} = s_i(t-1) - s_i(t) = s_i(t-1) \cdot [1 - s_i(t \mid t-1)] \tag{3.17}$$

The estimation of component failures is realized in the process of sequential Monte Carlo simulation, when the set of evolving weather conditions is available from the four-state

Markov chain. At each time step of the sequential simulation process, a number uniformly distributed in [0, 1] is randomly generated for each component, which is then compared with the failure probability calculated by Eq. (3.17). For each component, if the randomly generated number is larger than the failure probability at the current time step, the component would remain functional; otherwise, it would fail due to the extreme weather event at the current time step. If a component fails to function at previous time steps, it would remain to be on the outage at the current time step unless it has been restored. Hence, the functional status of each component can be determined at each time step.

3.4.3 Linear Programming for Analyzing System Connectivity

Cyber-physical interdependencies would expand and exacerbate disruptions caused by an extreme weather event, triggering a cascade of component failures within and across cyber and physical subsystems [15]. Therefore, operators need a consistent perception of power system infrastructural integrity in an extreme weather event, especially when component failures impact the connectivity in either cyber or physical subsystem. In particular, when an RTU is disconnected from DSCC, it fails to communicate the data for controlling the associated bus; when a bus is disconnected from a power source, it fails to supply to the local load and the associated RTU.

Here, we present two linear programming (LP) problems for checking the connectivity of cyber and physical subsystems. First, we propose an optimization problem based on the single-commodity flow for identifying whether or not an RTU is disconnected from DSCC. The optimization problem resembles the functionality of the Ping program which is used for diagnosing Internet connectivity. The optimization problem is repeated for testing the statuses of all RTUs. Denoting RTU M_c as the RTU of concern, the proposed optimization problem is stated as follows:

$$\tilde{\varepsilon}_{M_c,t} = \text{Max}\, \varepsilon_{\text{DSCC},t} \tag{3.18}$$

$$\text{s.t.}\ \ F^g_{m_c,t} - F^d_{m_c,t} = \sum_{n_c \in \nabla_{m_c}} f_{m_c n_c,t} - \sum_{k_c \in \Delta_{m_c}} f_{k_c m_c,t},\ \ \forall m_c \in \boldsymbol{\Omega}_{\text{C}} \tag{3.19}$$

$$f_{m_c n_c,t} = 0,\ \ \forall (m_c, n_c) \in \boldsymbol{\Psi}^{\text{O}}_{\text{C}} \tag{3.20}$$

$$-1 \le f_{m_c n_c,t} \le 1,\ \ \forall (m_c, n_c) \in \boldsymbol{\Psi}_{\text{C}} \backslash \boldsymbol{\Psi}^{\text{O}}_{\text{C}} \tag{3.21}$$

$$f_{m_c n_c,t} = 0,\ \ \forall (m_c, n_c) \in \boldsymbol{\Psi}_{\text{C}} \backslash \boldsymbol{\Psi}^{\text{O}}_{\text{C}},\ \ \forall m_c \in \boldsymbol{\Omega}^{\text{O}}_{\text{C}} \tag{3.22}$$

$$f_{k_c m_c,t} = 0,\ \ \forall (k_c, m_c) \in \boldsymbol{\Psi}_{\text{C}} \backslash \boldsymbol{\Psi}^{\text{O}}_{\text{C}},\ \ \forall m_c \in \boldsymbol{\Omega}^{\text{O}}_{\text{C}} \tag{3.23}$$

$$F^g_{m_c,t} = \varepsilon_{\text{DSCC},t},\ \ \forall m_c \in \{\text{DSCC}\} \tag{3.24}$$

$$F^g_{m_c,t} = 0,\ \ \forall m_c \in \boldsymbol{\Omega}_{\text{C}} \backslash \{\text{DSCC}\} \tag{3.25}$$

$$F^d_{m_c,t} = \varepsilon_{\text{DSCC},t},\ \ \forall m_c \in \{M_c\} \tag{3.26}$$

$$F^d_{m_c,t} = 0,\ \ \forall m_c \in \boldsymbol{\Omega}_{\text{C}} \backslash \{M_c\} \tag{3.27}$$

$$0 \le \varepsilon_{\text{DSCC},t} \le 1 \tag{3.28}$$

In this model, the objective function (3.18) is to find the maximum possible commodity flow that originated from DSCC and terminated at RTU M_c, while satisfying the following constraints: Eq. (3.19) represents the conservation of commodity flows in and out of each communication node; Eq. (3.20) forces each disrupted communication link to lose the capability of conveying a commodity flow; Eq. (3.21) restricts the commodity flow volume on each communication link; Eqs. (3.22) and (3.23) state that a commodity flow either in or out of a communication node would be null when the communication node loses the communication capability due to the associated bus outage; Eqs. (3.24) and (3.25) assign commodity injections to DSCC and all RTUs, respectively; Eqs. (3.26) and (3.27) assign commodity consumptions to RTU M_c and other communication nodes, respectively; Eq. (3.28) defines the range of commodity injection by DSCC. The optimal objective $\tilde{\varepsilon}_{M_c,t}$ takes the value of either 1 or 0, and we can determine the connectivity status of RTU M_c based on the following criterion: if $\tilde{\varepsilon}_{M_c,t}$ is equal to 1, RTU M_c is connected with DSCC and the bus deployed with RTU M_c is controllable and observable; otherwise, RTU M_c is disconnected with DSCC and the functionality of the cyber subsystem is no longer intact.

Then, we propose an optimization problem based on the multi-commodity flow for identifying the bus status (i.e. whether or not disconnected from power sources). The optimization problem is repeated for checking the power supply to all buses. Denoting bus M_p as the bus of concern, the proposed optimization problem is stated as:

$$\tilde{\varepsilon}_{M_p,t} = \text{Max} \sum_{u} \varepsilon_{u,t} \tag{3.29}$$

$$\text{s.t. } F^{g}_{m_p,t} - F^{d}_{m_p,t} = \sum_{n_p \in \nabla_{m_p}} f_{m_p n_p,t} - \sum_{k_p \in \Delta_{m_p}} f_{k_p m_p,t}, \ \forall m_p \in \mathbf{\Omega}_{\mathrm{P}} \tag{3.30}$$

$$f_{m_p n_p,t} = 0, \quad \forall (m_p, n_p) \in \mathbf{\Psi}_{\mathrm{P}}^{\mathrm{O}} \tag{3.31}$$

$$-2^{|\mathbf{\Omega}_{\mathrm{P}}^{\mathrm{G}}|} + 1 \le f_{m_p n_p,t} \le 2^{|\mathbf{\Omega}_{\mathrm{P}}^{\mathrm{G}}|} - 1, \ \forall (m_p, n_p) \in \mathbf{\Psi}_{\mathrm{P}} \backslash \mathbf{\Psi}_{\mathrm{P}}^{\mathrm{O}} \tag{3.32}$$

$$F^{g}_{m_p,t} = \varepsilon_{u,t}, u = \odot\left(m_p, \mathbf{\Omega}_{\mathrm{P}}^{\mathrm{G}}\right), \ \forall m_p \in \mathbf{\Omega}_{\mathrm{P}}^{\mathrm{G}} \tag{3.33}$$

$$F^{g}_{m_p,t} = 0, \ \forall m_p \in \mathbf{\Omega}_{\mathrm{P}} \backslash \mathbf{\Omega}_{\mathrm{P}}^{\mathrm{G}} \tag{3.34}$$

$$F^{d}_{m_p,t} = \sum_{u} \varepsilon_{u,t}, \ \forall m_p \in \{M_p\} \tag{3.35}$$

$$F^{d}_{m_p,t} = 0, \ \forall m_p \in \mathbf{\Omega}_{\mathrm{P}} \backslash \{M_p\} \tag{3.36}$$

$$0 \le \varepsilon_{u,t} \le 2^{u}, \ \forall u \tag{3.37}$$

In this model, the objective function (3.29) is to find the maximum possible sum of multiple commodity flows originated from all power sources and terminated at bus M_p, while satisfying the following constraints: Eq. (3.30) represents the conservation of commodity flows in and out of each bus; Eq. (3.31) forces the commodity flow on each failed line to be zero, while (3.32) restricts the commodity flow on the available power line; Eqs. (3.33) and (3.34) assign commodity injection to buses with a power source and other buses, respectively; Eqs. (3.35) and (3.36) assign commodity consumption to bus M_p and other buses, respectively; Eq. (3.37) defines and differentiates the range of commodity injection by each power source. The optimal objective $\tilde{\varepsilon}_{M_p,t}$ can take any discrete value between 0 and $2^{|\boldsymbol{\Omega}_\mathrm{P}^\mathrm{G}|}-1$, and we identify power sources that are connected with bus M_p by factoring $\tilde{\varepsilon}_{M_p,t}$ into a unique subset of elements in the geometric series with a factor of 2. For instance, if $\tilde{\varepsilon}_{M_p,t}$ is equal to 11, we treat it as the sum of 1, 2, and 8 (i.e. 2^0, 2^1, and 2^3) and then ascertain bus M_p is connected to the root bus as well as power sources 1 and 3; if $\tilde{\varepsilon}_{M_p,t}$ is $2^{|\boldsymbol{\Omega}_\mathrm{P}^\mathrm{G}|}-1$, then the connections between bus M_p and power sources are well maintained; if $\tilde{\varepsilon}_{M_p,t}$ is an even positive number, then bus M_p loses the connection with the root bus and resides in an island with the power supply from any other power source; if $\tilde{\varepsilon}_{M_p,t}$ is equal to 0, then bus M_p is on the outage and the associated RTU loses communication capability.

3.4.4 Second-Order Cone Programming for Power Dispatch Optimization

During and after the occurrence of an extreme weather event, operators continuously optimize power dispatch for maintaining the balance between power injections and consumptions with the lowest operation costs. Power dispatch in a distribution system is typically modeled as an optimal power flow problem based on the full set of alternative-current power flow equations. However, this model pertains to a nonconvex optimization problem with a significantly high computation complexity, which often fails to converge within a short time as reported by several empirical studies (e.g. [16]). The resulting slow solution would make operators fail to adjust power dispatch in a timely way, exacerbating adverse effects of an extreme weather event on operation performance. In order for operators not to delay their response to the extreme weather event, we convexify the conventionally nonconvex power flow equations and take advantage of second-order cone programming for modeling the power dispatch problem in a much more computationally efficient form.

The physical subsystem of a power distribution system features a radial tree-like topology, where the bus interfaced with the upstream power transmission system is denoted as the root bus. Figure 3.8 illustrates a typical portion of the physical subsystem, in which power flow equations are described by the following branch flow model [17]:

$$P^g_{m_p,t} - P^d_{m_p,t} = \sum_{n_p \in \nabla_{m_p}} p_{m_p n_p,t} - \sum_{k_p \in \Delta_{m_p}} \left(p_{k_p m_p,t} - r_{k_p m_p} I^2_{k_p m_p,t} \right), \quad \forall m_p \in \boldsymbol{\Omega}_\mathrm{P} \tag{3.38}$$

$$Q^g_{m_p,t} - Q^d_{m_p,t} = \sum_{n_p \in \nabla_{m_p}} q_{m_p n_p,t} - \sum_{k_p \in \Delta_{m_p}} \left(q_{k_p m_p,t} - x_{k_p m_p} I^2_{k_p m_p,t} \right), \quad \forall m_p \in \boldsymbol{\Omega}_\mathrm{P} \tag{3.39}$$

$$V^2_{k_p,t} - V^2_{m_p,t} = 2(r_{k_p m_p} p_{k_p m_p,t} + x_{k_p m_p} q_{k_p m_p,t}) - \left(r^2_{k_p m_p} + x^2_{k_p m_p} \right) I^2_{k_p m_p,t}, \quad \forall (k_p, m_p) \in \boldsymbol{\Psi}_\mathrm{P} \tag{3.40}$$

$$I^2_{k_p m_p,t} = \frac{p^2_{k_p m_p,t} + q^2_{k_p m_p,t}}{V^2_{k_p,t}}, \quad \forall (k_p, m_p) \in \boldsymbol{\Psi}_\mathrm{P} \tag{3.41}$$

where (3.38) and (3.39) represent the balance of active and reactive power at each bus, respectively; Eq. (3.40) quantifies the voltage magnitude difference at two ends of each bus; Eq. (3.41) calculates the current magnitude on each line.

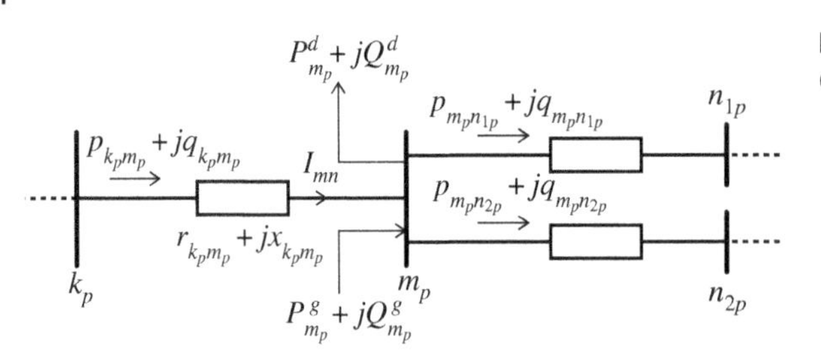

Figure 3.8 One-line diagram of a distribution feeder.

In order to leverage the computation superiority of second-order cone programming, we relax the non-convex (3.41) as the following rotated second-order cone inequality:

$$l_{k_p m_p,t} v_{k_p,t} \geq p_{k_p m_p,t}^2 + q_{k_p m_p,t}^2, \quad \forall (k_p, m_p) \in \boldsymbol{\Psi}_{\mathrm{P}} \tag{3.42}$$

where $l_{km,t} = I_{km,t}^2$ and $v_{k,t} = V_{k,t}^2$. Such a second-order cone which is based convex relaxation has been proved to be exact in relatively mild conditions [18]. Even if the relaxation is found inexact after solving the convex optimization problem, the difference-of-convex programming approach [16] can be employed to recover a feasible solution (which usually appears to be globally optimal) with the ensured relaxation exactness.

Operators make power dispatch decisions according to the estimated operating condition of the physical subsystem. An accurate operating condition observed by operators relies on their situational awareness which is in turn affected by the functionality of the cyber subsystem. However, failures occurring in the cyber subsystem render the pertinent components in the physical subsystem neither observable nor controllable. The emergence of unobservable components hampers operators' situational awareness, when, as a compromise, operators utilize the latest available measurements collected from these components for getting an estimate of the operating condition of the physical subsystem. Accordingly, the estimated operating condition might diverge from the actual condition under an extreme weather event. Uncontrollable components further complicate operators' decision-making due to the limitation of available control means. Hence, the disrupted cyber subsystem makes it more difficult for operators to curb the decline (or expedite the improvement) in the operation performance during (or after) the occurrence of an extreme weather event.

Accordingly, we formulate the following second-order cone programming-based optimization problem for the power dispatch of a power distribution system disrupted by an extreme weather event ($\forall t$):

$$\min \quad \sum_{m_p \in \Omega_{\mathrm{P}}^{\mathrm{G}}} c_{m_p}^g P_{m_p,t}^g + \sum_{m_p \in \Omega_{\mathrm{P}}^{\mathrm{C}}} c_{m_p}^d L_{m_p,t}^p + \rho \sum_{m_p \in \Omega_{\mathrm{P}}^{\mathrm{U}}} L_{m_p,t}^p \tag{3.43}$$

$$\text{s.t. } P_{m_p,t}^g - P_{m_p,t}^d = \sum_{n_p \in \nabla_{m_p}} p_{m_p n_p,t} - \sum_{k_p \in \Delta_{m_p}} (p_{k_p m_p,t} - r_{k_p m_p} l_{k_p m_p,t}), \quad \forall m_p \in \boldsymbol{\Omega}_{\mathrm{P}} \tag{3.44}$$

$$Q_{m_p,t}^g - Q_{m_p,t}^d = \sum_{n_p \in \nabla_{m_p}} q_{m_p n_p,t} - \sum_{k_p \in \Delta_{m_p}} (q_{k_p m_p,t} - x_{k_p m_p} l_{k_p m_p,t}), \quad \forall m_p \in \boldsymbol{\Omega}_{\mathrm{P}} \tag{3.45}$$

$$l_{k_p m_p,t} v_{k_p,t} \geq p_{k_p m_p,t}^2 + q_{k_p m_p,t}^2, \quad \forall (k_p, m_p) \in \boldsymbol{\Psi}_{\mathrm{P}} \tag{3.46}$$

$$S_{k_p m_p}^2 \geq p_{k_p m_p,t}^2 + q_{k_p m_p,t}^2, \quad \forall (k_p, m_p) \in \boldsymbol{\Psi}_{\mathrm{P}} \tag{3.47}$$

$$v_{k_p,t} - v_{m_p,t} = 2(r_{k_pm_p}p_{k_pm_p,t} + x_{k_pm_p}q_{k_pm_p,t}) - \left(r_{k_pm_p}^2 + x_{k_pm_p}^2\right) l_{k_pm_p,t}, \quad \forall (k_p, m_p) \in \boldsymbol{\Psi}_\mathrm{P} \backslash \boldsymbol{\Psi}_\mathrm{P}^{\oplus} \tag{3.48}$$

$$p_{k_pm_p,t} = q_{k_pm_p,t} = 0, \quad \forall (k_p, m_p) \in \boldsymbol{\Psi}_\mathrm{P}^{\oplus} \tag{3.49}$$

$$P_{m_p,t}^g = Q_{m_p,t}^g = 0, \quad \forall m_p \in \boldsymbol{\Omega}_\mathrm{P} \backslash \ \boldsymbol{\Omega}_\mathrm{P}^\mathrm{G} \tag{3.50}$$

$$0 \le P_{m_p,t}^g \le \overline{P}_{m_p}^g, \quad \forall m_p \in \boldsymbol{\Omega}_\mathrm{P}^\mathrm{G} \cap \boldsymbol{\Omega}_\mathrm{P}^\mathrm{C} \tag{3.51}$$

$$0 \le P_{m_p,t}^g \le \hat{P}_{m_p,t}^g, \quad \forall m_p \in \boldsymbol{\Omega}_\mathrm{P}^\mathrm{G} \cap \boldsymbol{\Omega}_\mathrm{P}^\mathrm{U} \tag{3.52}$$

$$\left(S_{m_p}^g\right)^2 \ge \left(P_{m_p,t}^g\right)^2 + \left(Q_{m_p,t}^g\right)^2, \quad \forall m_p \in \boldsymbol{\Omega}_\mathrm{P}^\mathrm{G} \tag{3.53}$$

$$P_{m_p,t}^d + L_{m_p,t}^p = D_{m_p,t}^p, \quad \forall m_p \in \boldsymbol{\Omega}_\mathrm{P}^\mathrm{C} \tag{3.54}$$

$$Q_{m_p,t}^d + L_{m_p,t}^q = D_{m_p,t}^q, \quad \forall m_p \in \boldsymbol{\Omega}_\mathrm{P}^\mathrm{C} \tag{3.55}$$

$$P_{m_p,t}^d + L_{m_p,t}^p = \hat{D}_{m_p,t}^p, \quad \forall m_p \in \boldsymbol{\Omega}_\mathrm{P}^\mathrm{U} \tag{3.56}$$

$$Q_{m_p,t}^d + L_{m_p,t}^q = \hat{D}_{m_p,t}^q, \quad \forall m_p \in \boldsymbol{\Omega}_\mathrm{P}^\mathrm{U} \tag{3.57}$$

$$0 \le L_{m_p,t}^p \le D_{m_p,t}^p, \quad \forall m_p \in \boldsymbol{\Omega}_\mathrm{P}^\mathrm{C} \tag{3.58}$$

$$0 \le L_{m_p,t}^p \le \hat{D}_{m_p,t}^p, \quad \forall m_p \in \boldsymbol{\Omega}_\mathrm{P}^\mathrm{U} \tag{3.59}$$

$$D_{m_p,t}^p L_{m_p,t}^q = D_{m_p,t}^q L_{m_p,t}^p, \quad \forall m_p \in \boldsymbol{\Omega}_\mathrm{P}^\mathrm{C} \tag{3.60}$$

$$\hat{D}_{m_p,t}^p L_{m_p,t}^q = \hat{D}_{m_p,t}^q L_{m_p,t}^p, \quad \forall m_p \in \boldsymbol{\Omega}_\mathrm{P}^\mathrm{U} \tag{3.61}$$

$$P_{m_p,t}^d = Q_{m_p,t}^d = 0, \quad \forall m_p \in \boldsymbol{\Omega}_\mathrm{P}^{\oplus} \tag{3.62}$$

$$v_{m_p,t} = 0, \quad \forall m_p \in \boldsymbol{\Omega}_\mathrm{P}^{\oplus} \tag{3.63}$$

$$\underline{v}_{m_p} \le v_{m_p,t} \le \overline{v}_{m_p}, \quad \forall m_p \in \boldsymbol{\Omega}_\mathrm{P} \backslash \boldsymbol{\Omega}_\mathrm{P}^{\oplus} \tag{3.64}$$

$$v_{m_p,t} = v_{m_p,t}^{\dagger}, \quad \forall m_p \in \{0\} \tag{3.65}$$

$$v_{m_p,t} = \hat{v}_{m_p,t}, \quad \forall m_p \in \ \boldsymbol{\Omega}_\mathrm{P}^\mathrm{G} \cap \boldsymbol{\Omega}_\mathrm{P}^\mathrm{U} \tag{3.66}$$

In this second-order cone programming problem, the objective function (3.43) is to minimize the total operation costs which consist of power injection costs (including energy import costs and production costs of distributed generation units), and load curtailment costs; power demands at buses with a functional RTU are curtailable at a reasonably high cost; power demands at buses without a functional RTU cannot be curtailed; accordingly, cost relations are represented by $\rho \gg c^d_{m_p} > c^g_{m_p} (\forall m_p)$. Meanwhile, the optimization problem includes the following constraints: Eqs. (3.44) and (3.45) are equivalent to (3.38) and (3.39) after substituting with $l_{k_p m_p,t} = I^2_{k_p m_p,t}$; Eq. (3.46) is the same as (3.42); Eq. (3.47) defines the apparent flow limit on each power line in a second-order cone inequality; for power lines that are not estimated on outage, (3.48) is equivalent to (3.40) after substituting with $l_{k_p m_p,t} = I^2_{k_p m_p,t}$, $v_{k_p,t} = V^2_{k_p,t}$, and $v_{m_p,t} = V^2_{m_p,t}$, but for power lines that are estimated to be on outage, (3.49) indicates neither active nor reactive power flow on such lines; Eq. (3.50) means that buses without a power source cannot inject active or reactive power into distribution lines; Eqs. (3.51) and (3.52) define the active power range for power sources located at buses with and without a functional RTU, respectively, when the upstream transmission system acts as a bulk power source associated with the root bus; Eq. (3.53) defines the apparent power limit for each source in a second-order cone inequality; Eqs. (3.54)–(3.56) and (3.57) represent both active and reactive power demands at buses with and without a functional RTU, respectively; Eqs. (3.58) and (3.59) define the curtailment ranges for active power demands at buses with and without a functional RTU, respectively; Eqs. (3.60) and (3.61) ensure the curtailed bus power demand has the same power factor as that of the original power demand; Eq. (3.62) indicates no power demands can be supplied at buses that are estimated to be on outage; Eq. (3.63) indicates the voltage magnitude of a bus on outage is zero; Eq. (3.64) restricts the voltage magnitude for buses that are not estimated to be on outage; Eq. (3.65) indicates the voltage magnitude of the root bus is regulated as a constant by the upstream power transmission system; Eq. (3.66) indicates the bus voltage magnitude of a power source remains the same as the corresponding latest available measurement when the associated RTU is not functional.

Before solving the proposed optimization problem, operators check the connectivity of cyber subsystem for getting the sets of $\mathbf{\Omega}^{\mathrm{C}}_{\mathrm{P}}$ and $\mathbf{\Omega}^{\mathrm{U}}_{\mathrm{P}}$, and then determine the sets of $\mathbf{\Omega}^{\oplus}_{\mathrm{P}}$ and $\mathbf{\Psi}^{\oplus}_{\mathrm{P}}$ (when the physical subsystem might have been sectionalized into a group of islands by checking the connectivity of the physical subsystem) based on the latest available information on power line statuses. It is important to stress that the proposed optimization problem does not state power dispatch in each island explicitly but alternatively integrates all the island-based decision-making models from the holistic view of operators in DSCC. Although supervisory control commands cannot be actually implemented at buses in the set of $\mathbf{\Omega}^{\mathrm{U}}_{\mathrm{P}}$, the proposed optimization problem still considers to virtually adjust several bus settings (i.e. $v_{m_p,t}$, $P^g_{m_p,t}$, $L^p_{m_p,t}$, $L^q_{m_p,t}$) for ensuring the solution feasibility in cases of power imbalance in certain islands, but also for facilitating the subsequent automatic execution of under-/over-frequency curtailment (as will be discussed in Section 3.4.6).

3.4.5 Greedy Heuristics for Combined Restoration of Cyber-Physical Systems

Extensive cyber-physical interdependencies not only make a power distribution system more vulnerable but also intensify the efforts in restoring electricity services after extreme weather events. Although cyber and physical subsystems have different requirements on restoration resources (e.g. maintenance duration and cost, crew, equipment), their restoration efforts cannot be carried out independently. A recovered component in the cyber subsystem could increase the controllability and observability of the physical subsystem, whereas a recovered component in the physical subsystem could enhance the functionality of the cyber subsystem. Thus, activities for recovering failed components in cyber and physical subsystems should be scheduled in tandem so that scarce restoration resources can be allocated optimally and in synergy with cyber and physical subsystems.

Restoration activities are normally initiated within a short time after the extreme weather. During this period, hidden failures of power lines are identified so that $\boldsymbol{\Omega}_{\mathrm{P}}^{\oplus}$ and $\boldsymbol{\Psi}_{\mathrm{P}}^{\oplus}$ are identical to $\boldsymbol{\Omega}_{\mathrm{P}}^{\mathrm{O}}$ and $\boldsymbol{\Psi}_{\mathrm{P}}^{\mathrm{O}}$ when operators start to make decisions on restoring electricity services. However, the optimization of cyber-physical restoration is intrinsically complex primarily due to a large number of failed components and long restoration periods.

Here, we propose a greedy heuristic algorithm for operators to determine a fast and reasonable cyber-physical restoration process. Considering that cyber and physical subsystems require different types of resources for restoring functionalities, restoration activities are conducted in parallel. Figure 3.9 shows the proposed flowchart where decision-making processes on restoring cyber and physical subsystems are interactive. Decisions on recovering failed components are made alternately between two subsystems. When one subsystem uses up all available resources for restoration, the decision-making process switches to the other subsystem. The recovery sequence of disrupted components in either cyber or physical subsystem is determined greedily by ranking their influences on reducing the total operation cost at a given time step. More specifically, the power dispatch problem (3.43)–(3.66) is solved for each disrupted component, where the disrupted component is assumed recovered and the connectivity of either the cyber or the physical subsystem is updated (i.e. sets of $\boldsymbol{\Omega}_{\mathrm{P}}^{\oplus}$ and $\boldsymbol{\Psi}_{\mathrm{P}}^{\oplus}$ are updated for recovered power lines, and sets of $\boldsymbol{\Omega}_{\mathrm{P}}^{\mathrm{U}}$

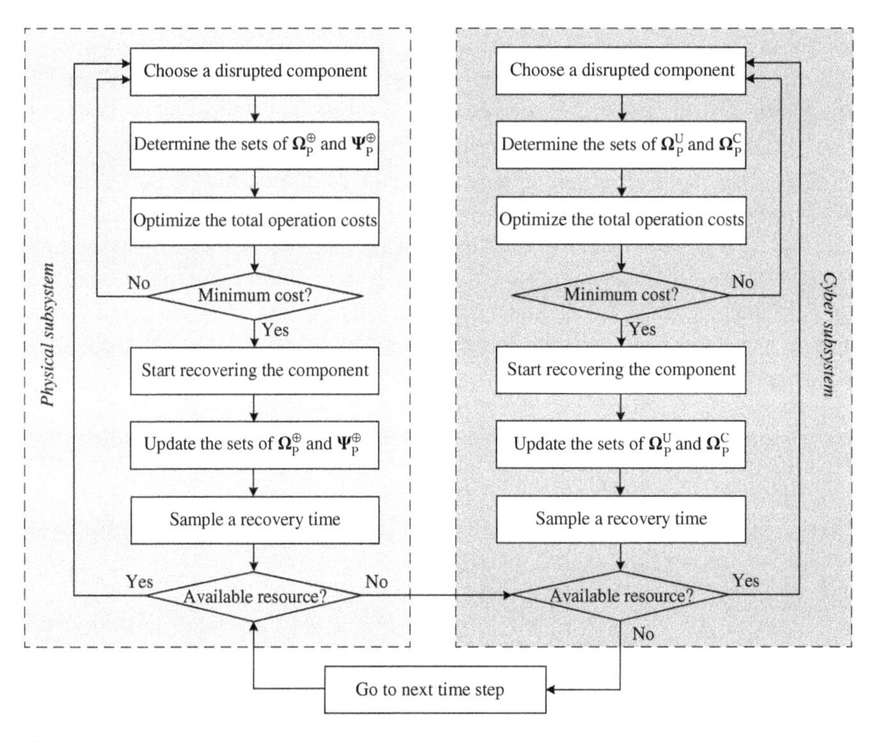

Figure 3.9 Flowchart for a cyber-physical restoration.

and $\boldsymbol{\Omega}_{\mathrm{P}}^{\mathrm{C}}$ are updated for recovered communication links). The power dispatch problem is also formulated in a look-ahead manner in which all components subjected to recovery are assumed functional in the problem. The total operation cost is then compared and a disrupted component corresponding to a less costly operation is recovered with a higher priority. The recovery time for a disrupted component is assumed random (i.e. depends on the system operating conditions) which is sampled from an empirical probability distribution function.

3.4.6 Integrated Optimization for Quantifying the Operation Performance

An extreme weather event can drive a physical subsystem to a series of islands. Notably, an island without a power source would undoubtedly suffer severe power outages. So each empowered island possesses at least one power source. The island containing the root bus is labeled as the major island (considering that a power distribution system commonly has a single bus interfaced with the upstream power transmission system) and others are considered as minor islands. In order to get an accurate evaluation of time-varying operation performance of the power distribution system, we perform deterministic power flow analysis in all islands to derive a complete power flow solution for simulating the operating condition of the physical subsystem.

Without loss of generality, we assume the reactive power balance is always maintained in each island. Power source buses except the root bus are regarded as PV buses whose voltage magnitude and active power injection are maintained at scheduled values, and power demands are considered at scheduled values. The scheduled values are periodically provided by DSCC according to the latest power dispatch decisions. In the cases where a bus is uncontrollable due to the failure of the associated RTU, the scheduled values are still set at the latest available DSCC supervisory control commands. Since the main island is directly connected to the upstream power transmission system, any active power deficit or surplus in this island can be easily compensated via the root bus which is regarded as a slack bus. Each minor island maintains its active power balance according to the following rules: if the total active power of local sources is larger than power demand plus local line losses, power sources automatically apply over-frequency generation curtailment until the active power is balanced; otherwise, all buses automatically execute under-frequency load curtailment with a uniform ratio until the active power is balanced.

Here, we propose an integrated optimization problem that provides power flow solutions in all the empowered islands ($\forall t$):

$$\min \sum_{m_p \in \boldsymbol{\Omega}_{\mathrm{P}}^{\mathrm{G}}} c_{m_p}^{g} P_{m_p,t}^{g} + \sum_{m_p \in \boldsymbol{\Omega}_{\mathrm{P}}} c_{m_p}^{d} L_{m_p,t}^{p} \tag{3.67}$$

s.t. Eqs. (3.44)–(3.46) and (3.50)

$$P_{m_p,t}^{d} + L_{m_p,t}^{p} = D_{m_p,t}^{p}, \quad \forall m_p \in \boldsymbol{\Omega}_{\mathrm{P}} \tag{3.68}$$

$$Q_{m_p,t}^{d} + L_{m_p,t}^{q} = D_{m_p,t}^{q}, \quad \forall m_p \in \boldsymbol{\Omega}_{\mathrm{P}} \tag{3.69}$$

$$v_{k_p,t} - v_{m_p,t} = 2(r_{k_p m_p} p_{k_p m_p,t} + x_{k_p m_p} q_{k_p m_p,t}) - \left(r_{k_p m_p}^2 + x_{k_p m_p}^2\right) l_{k_p m_p,t}, \quad \forall (k_p, m_p) \in \boldsymbol{\Psi}_{\mathrm{P}} \backslash \boldsymbol{\Psi}_{\mathrm{P}}^{\mathrm{O}} \tag{3.70}$$

$$p_{k_p m_p,t} = q_{k_p m_p,t} = 0, \quad \forall (k_p, m_p) \in \boldsymbol{\Psi}_{\mathrm{P}}^{\mathrm{O}} \tag{3.71}$$

$$P^d_{m_p,t} = Q^d_{m_p,t} = 0, \quad \forall m_p \in \boldsymbol{\Omega}_{\mathrm{P}}^{\mathrm{O}} \tag{3.72}$$

$$v_{m_p,t} = 0, \quad \forall m_p \in \boldsymbol{\Omega}_{\mathrm{P}}^{\mathrm{O}} \tag{3.73}$$

$$v_{m_p,t} = \tilde{v}_{m_p,t}, \quad \forall m_p \in \boldsymbol{\Omega}_{\mathrm{P}}^{\mathrm{G}} \tag{3.74}$$

$$P^g_{m_p,t} = \tilde{P}^g_{m_p,t}, \quad \forall m_p \in \left(\boldsymbol{\Omega}_{\mathrm{P}}^{\mathrm{G}} \setminus \{0\}\right) \cap \boldsymbol{\Omega}_{\mathrm{P}}^{\mathrm{S}_0} \tag{3.75}$$

$$P^d_{m_p,t} = D^p_{m_p,t} - \tilde{L}^p_{m_p,t}, \quad \forall m_p \in \boldsymbol{\Omega}_{\mathrm{P}}^{\mathrm{S}_0} \tag{3.76}$$

$$Q^d_{m_p,t} = D^q_{m_p,t} - \tilde{L}^q_{m_p,t}, \quad \forall m_p \in \boldsymbol{\Omega}_{\mathrm{P}}^{\mathrm{S}_0} \tag{3.77}$$

$$P^g_{m_p,t} = \tilde{P}^g_{m_p}\left(1 - \xi^g_w\right), \quad \forall m_p \in \boldsymbol{\Omega}_{\mathrm{P}}^{\mathrm{G}} \cap \boldsymbol{\Omega}_{\mathrm{P}}^{\mathrm{S}_w}, \forall w \neq 0 \tag{3.78}$$

$$P^d_{m_p,t} = \left(D^p_{m_p,t} - \tilde{L}^p_{m_p,t}\right)\left(1 - \xi^d_w\right), \quad \forall m_p \in \boldsymbol{\Omega}_{\mathrm{P}}^{\mathrm{S}_w}, \forall w \neq 0 \tag{3.79}$$

$$Q^d_{m_p,t} = \left(D^q_{m_p,t} - \tilde{L}^q_{m_p,t}\right)\left(1 - \xi^d_w\right), \quad \forall m_p \in \boldsymbol{\Omega}_{\mathrm{P}}^{\mathrm{S}_w}, \forall w \neq 0 \tag{3.80}$$

$$0 \le \xi^g_w \le \alpha_w, \quad \forall w \neq 0 \tag{3.81}$$

$$0 \le \xi^d_w \le \beta_w, \quad \forall w \neq 0 \tag{3.82}$$

$$\alpha_w + \beta_w \le 1, \quad \forall w \neq 0 \tag{3.83}$$

This is a mixed-integer second-order cone programming problem in which the objective function (3.67) is to minimize the total operation cost including those of bus power injection and load curtailment; power demands at all buses are curtailable, regardless of the functional status of associated RTUs. The objective function will guarantee the exactness of second-order cone relaxation in (3.46); even if the relaxation is inexact, we can apply the difference-of-convex programming approach [16] to recover the exactness. The optimization problem includes the following constraints in addition to (3.44)–(3.46), and (3.50). Here, (3.68) and (3.69) represent active and reactive power demands at all buses, respectively; Eq. (3.70) specifies power flow characteristics on operational lines, and (3.71) indicates that failed lines cannot convey any flows; Eq. (3.72) indicates that failed buses cannot supply any power demands, and (3.73) enforces such bus voltage magnitudes to be zero; Eq. (3.74) assumes buses with a power source have an adequate reactive power capacity to maintain voltage magnitudes at scheduled values; Eq. (3.75) indicates power-source buses in the major island except the root bus maintain active power injections at scheduled values; Eqs. (3.76) and (3.77) represent active and reactive bus power demands in the major island following scheduled load curtailments; Eq. (3.78) indicates buses in minor islands (without sufficient power sources) may suffer inevitable generation curtailments for over-frequency concerns, in which the generation curtailment ratio is unified among power-source buses in the minor island; Eqs. (3.79) and (3.80) indicate active and reactive power demands in a minor island may suffer additional load curtailments (without changing the power factor) for under-frequency concerns, in

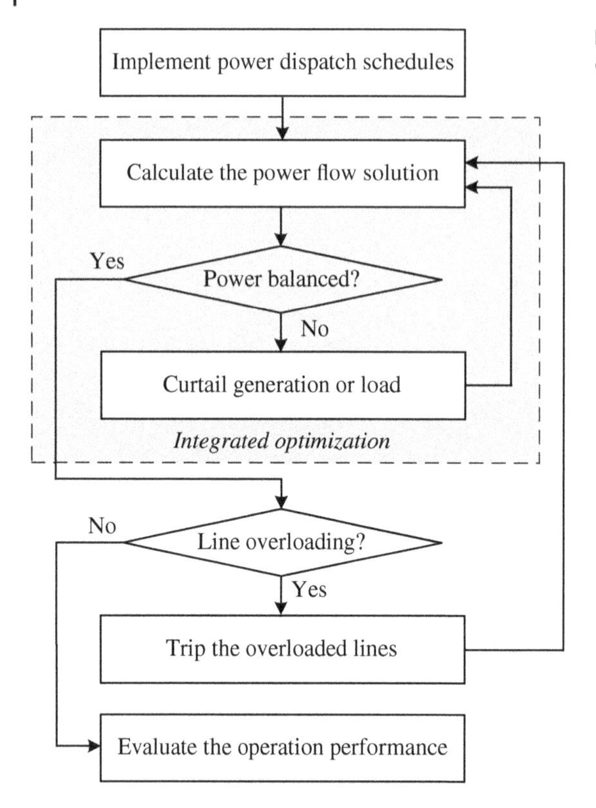

Figure 3.10 Flowchart for the evaluation of operation performance.

which the ratio of additional load curtailment is unified among buses in a minor island; Eqs. (3.81) and (3.82) restrict the ranges of generation and load curtailment ratios by two binary indicators in each minor island; Eq. (3.83) indicates that two binary indicators will not be equal to 1 simultaneously as over-frequency generation curtailment and under-frequency load curtailment would not occur concurrently in a minor island.

It has been shown that the power flow solution with feasible voltage magnitudes always exists and is unique in a radial power distribution system [19]. Accordingly, the proposed integrated optimization problem produces an accurate power flow solution once the second-order cone relaxation is exact. When a line flow solution exceeds line capacity, the line would be tripped resulting in a potentially more fragmented physical subsystem whose power flow solution needs to be recalculated. Accordingly, the power flow solution would not be final as long as line outages could form new islands. Eventually, the operation performance will be quantified according to the operating condition represented by the power flow solution. Figure 3.10 shows the corresponding flowchart for the evaluation.

3.5 Simulation Results

This section presents the simulation results for applying the proposed framework to analyze the resilience of a test cyber-physical power distribution system in an extreme weather event.

3.5.1 Configuration of the Power Distribution System

In order to form a cyber-physical power distribution system for our case studies, we integrate the 33-bus radial power distribution grid [17] (i.e. physical subsystem) with a synthetic communication network (i.e. cyber subsystem). In the 33-bus physical subsystem, bus 1 is interfaced with the upstream power transmission system, whose voltage magnitude is regulated at 1.0 per unit. Tables 3.2 and 3.3 list the characteristics of buses and lines, respectively. Table 3.4 lists the characteristics of two distributed generation units located at buses 12 and 28, when the cost of importing energy via bus 1 is 50 $/MWh and load curtailment costs are 500 $/MWh.

Given that most practical communication networks are estimated to be scale-invariant networks whose node degree (i.e. number of communication links incident to a communication node) conforms to the power-law probability distribution, the cyber subsystem is formed by using the following method [20]: Initially, DSCC is connected, via three dedicated communication links, with RTUs 1, 12, and 28 installed at buses with a power source. Subsequently, nodes which are randomly permutated are added sequentially to the communication network, when each newly added node is linked to an existing node with a linkage probability that is proportional to the existing node degree. Eventually, we obtain a randomized network with 34 nodes (including DSCC and 33 RTUs) and 33 links. Table 3.5 lists the detailed topology information of the cyber subsystem. DSCC is equipped with uninterruptible power supply apparatus for avoiding the occurrence of the single point of failure. The hourly sequential Monte Carlo simulation is performed since an hourly resolution is usually deemed adequate for modeling the weather condition evolution.

Table 3.6 lists other parameters for the proposed resilience analysis. An extreme weather event occurs at $t_0 = 0$. The most recent maintenance time for all components is assumed to be 1000 hours prior to the occurrence of the event (i.e. $T_i^r = -1000, \forall i$). After the

Table 3.2 Buses in physical subsystem.

	Load			Load			Load	
Bus	**Active (kW)**	**Reactive (kVar)**	**Bus**	**Active (kW)**	**Reactive (kVar)**	**Bus**	**Active (kW)**	**Reactive (kVar)**
1	0	0	12	60	35	23	90	50
2	100	60	13	60	35	24	420	200
3	90	40	14	120	80	25	420	200
4	120	80	15	60	10	26	60	25
5	60	30	16	60	20	27	60	25
6	60	20	17	60	20	28	60	20
7	200	100	18	90	40	29	120	70
8	200	100	19	90	40	30	200	600
9	60	20	20	90	40	31	150	70
10	60	20	21	90	40	32	210	100
11	45	30	22	90	40	33	60	40

Table 3.3 Power lines in physical subsystem.

Start bus	End bus	Impedance/Ω	Limit/MVA	Start bus	End bus	Impedance/Ω	Limit/MVA
Bus 1	Bus 2	$0.0922+j0.047$	5	Bus 17	Bus 18	$0.732+j0.574$	0.5
Bus 2	Bus 3	$0.986+j0.5022$	2.5	Bus 2	Bus 19	$0.164+j0.1565$	1
Bus 2	Bus 3	$0.986+j0.5022$	2.5	Bus 19	Bus 20	$1.5042+j1.3554$	0.5
Bus 3	Bus 4	$0.366+j0.1864$	2.5	Bus 20	Bus 21	$0.4095+j0.4784$	0.5
Bus 4	Bus 5	$0.3811+j0.1941$	2.5	Bus 21	Bus 22	$0.7089+j0.9373$	0.5
Bus 5	Bus 6	$0.819+j0.707$	2.5	Bus 3	Bus 23	$0.4512+j0.3083$	1.5
Bus 6	Bus 7	$0.1872+j0.6188$	1.5	Bus 23	Bus 24	$0.898+j0.7091$	1.5
Bus 7	Bus 8	$0.7114+j0.2351$	1	Bus 24	Bus 25	$0.896+j0.7011$	1
Bus 8	Bus 9	$1.03+j0.74$	1	Bus 6	Bus 26	$0.203+j0.1034$	1.5
Bus 9	Bus 10	$1.044+j0.74$	1	Bus 26	Bus 27	$0.2842+j0.1447$	1.5
Bus 10	Bus 11	$0.1966+j0.065$	1	Bus 27	Bus 28	$1.059+j0.9337$	1.5
Bus 11	Bus 12	$0.3744+j0.1238$	1	Bus 28	Bus 29	$0.8042+j0.7006$	1.5
Bus 12	Bus 13	$1.468+j1.155$	1	Bus 29	Bus 30	$0.5075+j0.2585$	1
Bus 13	Bus 14	$0.5416+j0.7129$	1	Bus 30	Bus 31	$0.9744+j0.963$	1
Bus 14	Bus 15	$0.591+j0.526$	1	Bus 31	Bus 32	$0.3105+j0.3619$	0.5
Bus 15	Bus 16	$0.7463+j0.545$	0.5	Bus 32	Bus 33	$0.341+j0.5302$	0.5
Bus 16	Bus 17	$1.289+j1.721$	0.5				

Table 3.4 Characteristics of distributed generation units.

Unit	Location	Cost/MWh	Capacity/MVA	Active power output/MW
1	Bus 12	80	2	0–1
2	Bus 28	100	2	0–1

termination of the extreme weather event, the duration of the investigation stage is sampled from a normal distribution $\mathcal{N}(\mu_1, \sigma_1)$, and the recovery time of a failed component is sampled from another normal distribution $\mathcal{N}(\mu_2, \sigma_2)$. Power demands are altered uniformly in the physical subsystem. Figure 3.11 shows the power demand after an extreme weather event in which the load ratio at a time step is considered the corresponding power demand over the power demand immediately before the extreme weather event.

3.5.2 Impact of Extreme Weather on Component Failures

Figure 3.12 illustrates the four-state Markov chain for predicting the sequence of weather condition during an extreme event, where each directed edge is annotated with the corresponding transition probability. The initial weather condition at t_0 has a low impact.

Table 3.5 Topology of cyber subsystem.

Start node	End node	Start node	End node	Start node	End node
RTU 1	DSCC	RTU 12	DSCC	RTU 23	RTU 13
RTU 2	RTU 8	RTU 13	RTU 12	RTU 24	DSCC
RTU 3	RTU 15	RTU 14	DSCC	RTU 25	RTU 13
RTU 4	RTU 13	RTU 15	RTU 14	RTU 26	RTU 1
RTU 5	DSCC	RTU 16	RTU 13	RTU 27	RTU 1
RTU 6	DSCC	RTU 17	RTU 28	RTU 28	DSCC
RTU 7	RTU 20	RTU 18	RTU 13	RTU 29	RTU 31
RTU 8	RTU 28	RTU 19	RTU 1	RTU 30	RTU 28
RTU 9	RTU 23	RTU 20	RTU 12	RTU 31	RTU 28
RTU 10	RTU 13	RTU 21	DSCC	RTU 32	RTU 1
RTU 11	RTU 13	RTU 22	RTU 32	RTU 33	RTU 24

Table 3.6 Parameters for resilience analysis.

Parameter	ϵ	t_0	τ	φ	γ	μ_1	σ_1	μ_2	σ_2
Value	0.01	0	15,000	1.5	1	4	1	6	1.5

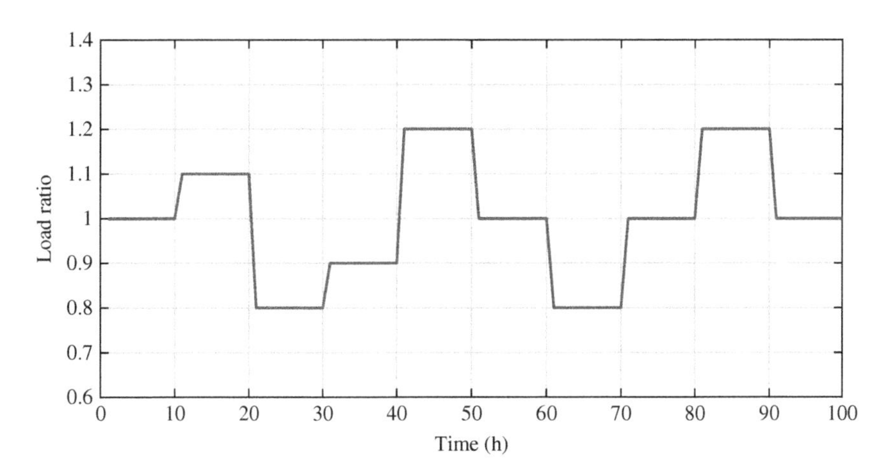

Figure 3.11 Power demand evolution.

Accordingly, we calculate the transition probability matrix and the initial distribution over the state space as:

$$\mathbf{\Pi}(0) = [1 \quad 0 \quad 0 \quad 0], \quad \mathbf{T} = \begin{bmatrix} 0.3 & 0.3 & 0.2 & 0.2 \\ 0.2 & 0.35 & 0.3 & 0.15 \\ 0.2 & 0.4 & 0.35 & 0.05 \\ 0 & 0 & 0 & 1 \end{bmatrix} \tag{3.84}$$

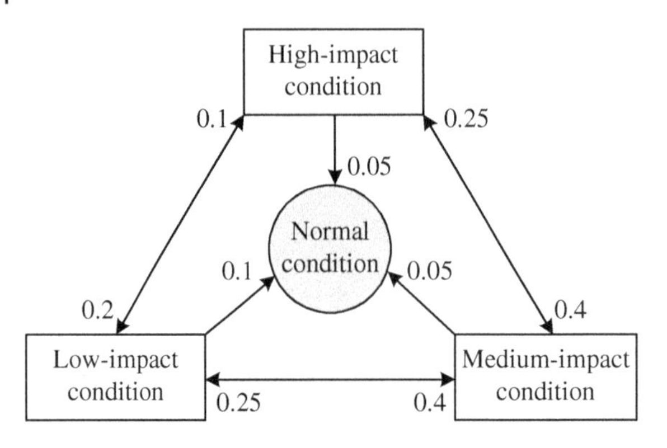

Figure 3.12 Markov chain for the extreme weather event.

Table 3.7 Probability distribution over weather conditions.

	$t = 10$		$t = 20$		$t = 50$	
Probability	**Markov model**	**Monte Carlo**	**Markov model**	**Monte Carlo**	**Markov model**	**Monte Carlo**
X_1	0.1587	0.1472	0.0818	0.0751	0.0108	0.0099
X_2	0.2391	0.2197	0.1206	0.1120	0.0147	0.0148
X_3	0.1277	0.1191	0.0690	0.0607	0.0083	0.0080
X_4	0.4745	0.5140	0.7286	0.7522	0.9662	0.9672

which can be utilized in (3.9) to derive the distribution over the four weather conditions at any time after the extreme weather event.

Table 3.7 compares the probability distribution obtained by the direct calculation of (3.9) with that averaged by 10,000 scenarios for the sequential Monte Carlo simulation which undergoes step-by-step transitions. It is obvious in Table 3.7 that the two distributions are very close at various time steps and the weather condition becomes normal with an extremely high probability after 50 time steps. Accordingly, we can employ the four-state Markov chain to estimate the probability of the termination of the extreme weather event.

After obtaining a sequence of weather conditions by recording the weather condition at each time step in the four-state Markov chain stochastic process, we can estimate component failures in both cyber and physical subsystems based on the proposed proportional hazard model. In a particular case, the weather condition returns to the normal state at the 10th time step. Figure 3.13 shows the effect of adverse weather on component survival probability when all components are assumed to possess an identical proportional hazard model in the extreme weather event. As illustrated in this figure, the survival probability of a component is sensitive to adverse weather conditions, which witnesses an obvious decline due to the damaging impact of the extreme weather event. Table 3.8 gives the detailed list of component failures during the extreme weather event which accounts for the failures of seven power lines and eight communication links.

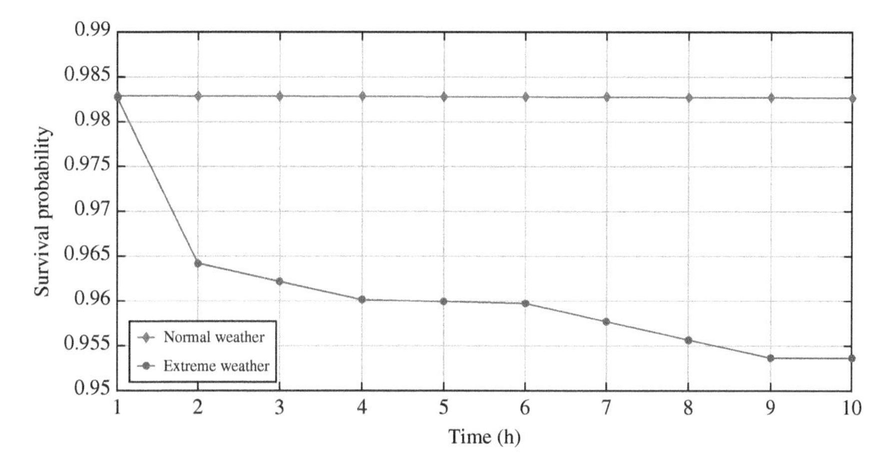

Figure 3.13 Effects of adverse weather on the component survival probability.

Table 3.8 Component failures in extreme weather condition.

		Component failure	
Time step	**Weather condition**	**Physical subsystem**	**Cyber subsystem**
1	Low-impact	Lines 4-5	/
2	High-impact	Lines 2-3	Links 2-8, 12-D, and 16-13
3	Medium-impact	Lines 8-9	/
4	Medium-impact	Lines 15-16	/
5	Low-impact	Lines 12-13 and 26-27	Links 1-D, 7-20, and 26-1
6	Low-impact	Lines 6-26	/
7	Medium-impact	/	/
8	Medium-impact	/	/
9	Medium-impact	/	Links 24-D and 25-13
10	Normal	/	/

3.5.3 Power Dispatch in Extreme Weather Condition

Our optimization problems, including LP, second-order cone programming, and mixed-integer programming problems, are solved by CPLEX 12.6 [21]. In order to investigate the exactness, the difference between the two sides of (3.46) is defined as the relaxation error. Figure 3.14 shows the relaxation errors for all lines, when power dispatch at t_0 is optimized by solving the second-order cone programming problem proposed in Section 3.4.4. In this figure, the relaxation error is negligible and the second-order cone relaxation is exact and efficient for lowering the computation complexity of optimal dispatch in a power distribution system. We compare the power flow solution obtained

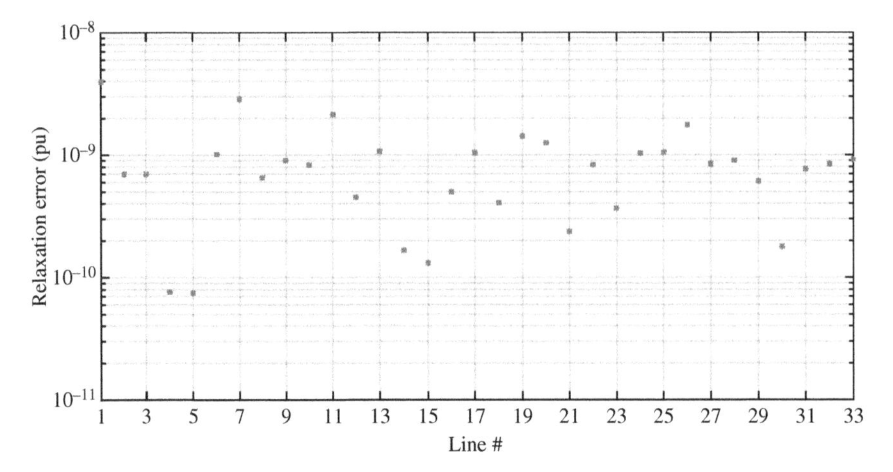

Figure 3.14 Second-order cone relaxation error.

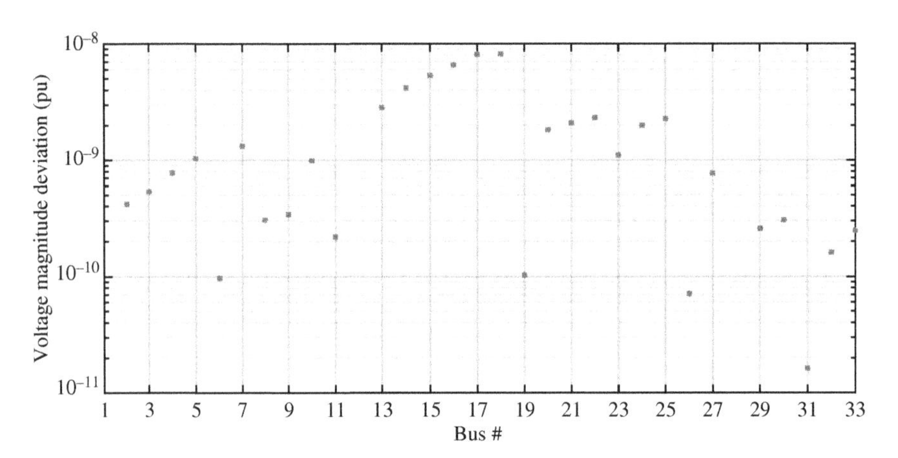

Figure 3.15 Voltage magnitude deviation.

by using MATPOWER toolbox [22] with that produced by the integrated optimization problem proposed in Section 3.4.6. Figure 3.15 shows the comparison of two power flow solutions at t_0 in terms of voltage magnitude deviation, where extremely small errors prove that the integrated optimization problem is capable of generating an exact power flow solution.

After the extreme weather event unfolds, operators continuously adjust power dispatch decisions based on the estimated operating condition of the physical subsystem in order to mitigate the consequences of the event. However, insufficient situational awareness resulting from component failures fails to provide operators with an accurate perception of the operating condition. Figure 3.16 presents the operating conditions of cyber and physical subsystems immediately after the extreme weather event terminates. In this figure, the cyber subsystem suffers severe disruptions caused either directly by the extreme weather

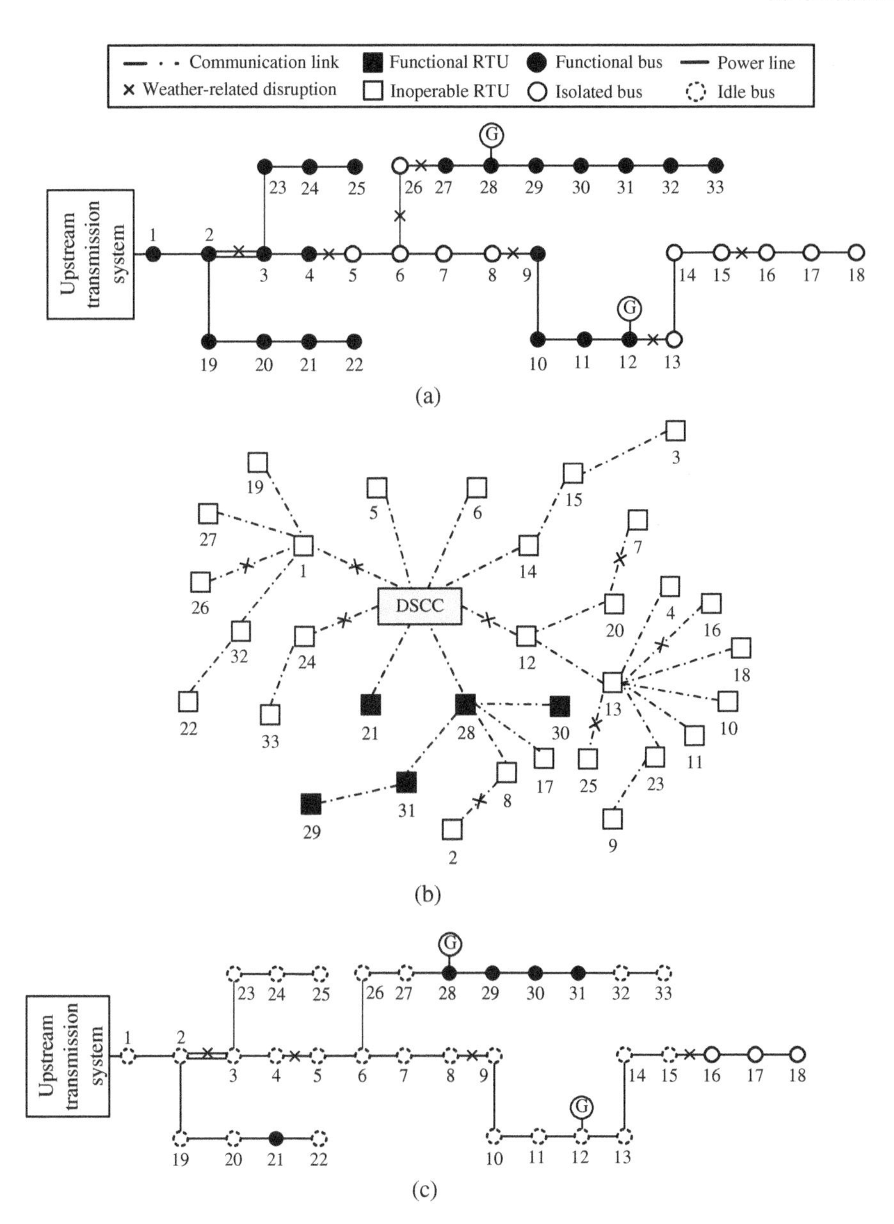

Figure 3.16 Operating condition after the extreme weather event. (a) Actual operating condition of the physical subsystem. (b) Operating condition of the cyber subsystem. (c) Estimated operating condition of the physical subsystem.

event or indirectly by cyber-physical interdependencies so that the physical subsystem's operating condition estimated by operators is vastly different from the actual condition.

The synthetic cyber subsystem is so fragile in the extreme condition that its configuration would need to be reinforced urgently (e.g. by adding redundant communication links) for assuming its responsibility of monitoring and controlling the physical subsystem in an

Table 3.9 Power output mismatch due to cyber subsystem failures.

	Scheduled dispatch (MW + *j* MVar)			Actual output (MW + *j* MVar)		
Time (Hour)	**Import**	**Unit 1**	**Unit 2**	**Import**	**Unit 1**	**Unit 2**
1	$1.62 + j0.8$	$1 + j0.41$	$1 + j0.73$	$1.62 + j0.8$	$1 + j0.41$	$1 + j0.73$
2	$1.62 + j0.8$	$1 + j0.28$	$1 + j0.86$	$1.62 + j0.8$	$1 + j0.28$	$1 + j0.86$
3	$1.62 + j0.8$	$0.68 + j0.31$	$1 + j0.47$	$1.62 + j0.8$	$0.68 + j0.31$	$1 + j0.47$
4	$1.62 + j0.8$	$0.47 + j0.23$	$1 + j0.47$	$1.62 + j0.8$	$0.47 + j0.23$	$1 + j0.47$
5	$1.62 + j0.8$	$0.47 + j0.23$	$1 + j0.47$	$1.62 + j0.8$	$0.23 + j0.11$	$0.42 + j0.19$
6	$1.62 + j0.8$	$0.47 + j0.23$	$1 + j0.47$	$1.62 + j0.8$	$0.23 + j0.11$	$0.42 + j0.19$
7	$1.62 + j0.8$	$0.47 + j0.23$	$1 + j0.47$	$1.62 + j0.8$	$0.23 + j0.11$	$0.42 + j0.19$
8	$1.62 + j0.8$	$0.47 + j0.23$	$1 + j0.47$	$1.62 + j0.8$	$0.23 + j0.11$	$0.42 + j0.19$
9	$1.62 + j0.8$	$0.47 + j0.23$	$1 + j0.48$	$1.62 + j0.8$	$0.23 + j0.11$	$0.42 + j0.20$
10	$1.62 + j0.8$	$0.47 + j0.23$	$1 + j0.48$	$1.62 + j0.8$	$0.23 + j0.11$	$0.42 + j0.20$

extreme event. Accordingly, operators are misguided by faulty operating conditions to make proper power dispatch decisions. Furthermore, failures in the cyber subsystem cause certain components to be uncontrollable in the physical subsystem. Consequently, operators fail to maintain the power balance in certain islands where automatic frequency adjustment is executed for balancing the active power. Table 3.9 compares the scheduled and the actual outputs of three power sources during the extreme weather event. Obviously, the uncontrollability of units 2 and 3 from the fifth time step deteriorates the operation performance of the power distribution system by incurring additional under-frequency load curtailment in the pertinent islands.

3.5.4 Restoration of Power Distribution System

The cyber-physical interdependencies in the power distribution system may necessitate additional efforts for the restoration of cyber and physical subsystems. Assume available restoration resources are so limited that at most one failed component is being recovered in each subsystem within any time period. The restoration schedule of disrupted power lines and communication links is illustrated by a Gantt chart in Table 3.10, where the restoration period of each component is represented between the start and finish times. In this table, the restoration is initiated at the 15th time step and completed at the 44th time step, where cyber and physical subsystems are restored in parallel.

3.5.5 Quantification of Resilience in Extreme Weather Condition

When all failed components are recovered, we can quantify the resilience of the power distribution system to the extreme weather event according to the recorded evolution of the normalized operation performance depicted in Figure 3.17. Accordingly, we divide the

Table 3.10 Restoration schedule (in time steps) for the power distribution system.

Component		t = 15–20						t = 21–25					t = 26–30					t = 31–35					t = 36–40					t = 41–44			
Power line	4-5	■	■	■	■	■																									
	12-13						■	■	■																						
	15-16									■	■	■																			
	6-26												■	■	■	■	■	■													
	26-27																		■	■											
	8-9																				■	■	■	■							
	2-3																								■	■	■				
Communication link	2-8	■																													
	7-20		■	■	■	■																									
	16-13						■	■																							
	1-D								■	■	■																				
	12-D											■	■	■	■	■	■	■													
	24-D																		■	■	■	■	■	■							
	25-13																								■	■	■	■			
	26-1																												■	■	■

Shades denote the duration of each restoration activity.

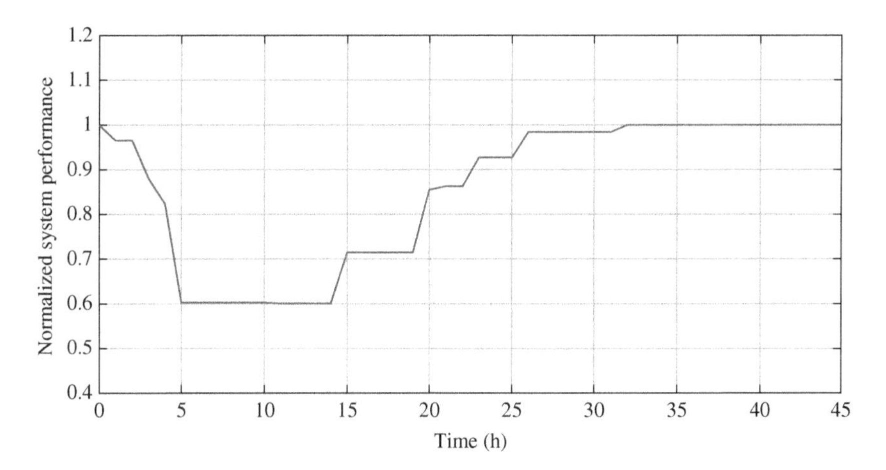

Figure 3.17 Evolution of the normalized operation performance.

evolution curve into five stages as discussed in Section 3.3.1 and calculate several resilience indices as listed in Table 3.11.

We perform multiple sequential Monte Carlo simulations by considering the same sequence of weather conditions generated by the four-state Markov chain in order to mitigate the implications of probabilistic parameters (e.g. component failure rate,

Table 3.11 Resilience evaluation.

Time	t_1	t_2	t_3	t_4	t_5
Value	1	11	15	32	45
Index	R_α	R_β	R_γ	R_δ	R
Value	0.6005	0.1447	27.5334	0.0222	0.4998

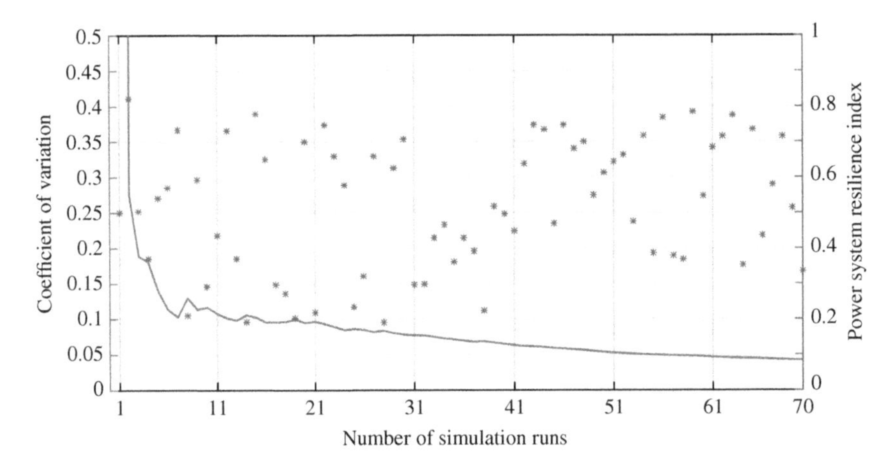

Figure 3.18 Convergence of Monte Carlo simulation.

component recovery time) on the simulation results for resilience analysis. The variability among these sequential Monte Carlo simulations is measured by the coefficient of variation [14] defined as the ratio of standard deviation to the mean for multiple simulations. We set 0.05 as the threshold for the coefficient of variation so as to determine whether or not we should continue with simulations. If the present coefficient of variation is less than the threshold, the simulation results have converged and we can terminate the simulation. The simulation results represent the resilience of the power distribution system in a specific extreme event. Otherwise, more simulations would be required for convergence. Figure 3.18 shows the Monte Carlo simulation converges in 55 runs, given that the coefficient of variation at the 55th run is 0.0495. Correspondingly, the PSRI R in the power distribution system is 0.5213 (which is the mean over the first 55 simulations).

3.6 Conclusions

This chapter presents a comprehensive framework that enables the resilience analysis of a power distribution system, regardless of the system's design, operation, and control settings. Although extreme weather events are chosen as an example to illustrate the proposed implementation process, the resilience analysis framework is generic enough to be adopted for any rare and disastrous events such as massive physical failures and terrorist attacks, thereby contributing to the overall resilience of a power distribution system. Although the occurrence of such catastrophic events can hardly be predicted nor prevented, the proposed resilience analysis framework lays the theoretical foundation for modernizing power distribution systems to withstand various disruptions without compromising electricity services.

In fact, power distribution systems are conventionally designed with centralized management of generation resources and long-distance delivery of electricity. As observed in the simulation results of such a power distribution system, the failure of a single component could potentially lead to extensive power outages to the extent that end consumers would be faced with a high possibility of losing electricity services in the aftermath of an extreme weather event. Therefore, power distribution systems require ongoing resilience-oriented efforts pertaining to the configuration and the management of cyber and physical subsystems, especially when there is an ever-growing number of extreme weather events prompted by the changing climate.

In addition, microgrids, which are increasingly regarded as building blocks of future power systems, open the door to embedding resilience as an inherent capability of a power distribution system. The implementation of microgrids not only promotes localized electricity services by enabling on-site generation and demand response, but also lessens the burden of communicating with DSCC for optimal energy management [23]. When electricity services in an adjoining utility grid are unexpectedly interrupted, microgrids will have the capability of islanding themselves and sustaining electricity services through the strategic management of locally available resources [24]. Microgrids with a sufficient generation capacity can also act as virtual feeders for facilitating restoration services in adjoining utility grids [4]. It is therefore beneficial to reconfigure a power distribution system as a set of networked microgrids so as to achieve the goal of resilience in providing electricity services for end customers.

References

1 NOAA National Centers for Environmental Information (2017). U.S. Billion-Dollar Weather and Climate Disasters. https://www.ncdc.noaa.gov/billions/.

2 Arghandeh, R., Brown, M., Rosso, A.D. et al. (2014). The local team: leveraging distributed resources to improve resilience. *IEEE Power and Energy Magazine* 12 (5): 76–83.

3 Li, Z., Shahidehpour, M., Aminifar, F. et al. (2017). Networked microgrids for enhancing the power system resilience. *Proceedings of the IEEE* 105 (7): 1289–1310.

4 Liu, C.C. (2015). Distribution systems: reliable but not resilient? [In My View]. *IEEE Power and Energy Magazine* 13 (3): 93–96.

5 Havlin, S., Parshani, R., Paul, G. et al. (2010). Catastrophic cascade of failures in interdependent networks. *Nature* 464 (7291): 1025–1028.

6 Li, Z., Shahidehpour, M., and Aminifar, F. (2017). Cybersecurity in distributed power systems. *Proceedings of the IEEE* 105 (7): 1367–1388.
7 Panteli, M., Crossley, P., Kirschen, D.S., and Sobajic, D.J. (2013). Assessing the impact of insufficient situation awareness on power system operation. *IEEE Transactions on Power Systems* 28 (3): 2967–2977.
8 Panteli, M. and Kirschen, D.S. (2015). Situation awareness in power systems: theory, challenges and applications. *Electric Power Systems Research* 122: 140–151.
9 Ton, D.T. and Wang, W.T.P. (2015). A more resilient grid: the U.S. Department of Energy joins with stakeholders in an R&D plan. *IEEE Power and Energy Magazine* 13 (3): 26–34.
10 Amjady, N. (2006). Generation adequacy assessment of power systems by time series and fuzzy neural network. *IEEE Transactions on Power Systems* 21 (3): 1340–1349.
11 Shahidehpour, M., Tinney, W.F., and Fu, Y. (2005). Impact of security on power systems operation. *Proceedings of the IEEE* 93 (11): 2013–2025.
12 Santoso, S., Grady, W.M., Powers, E.J. et al. (2000). Characterization of distribution power quality events with Fourier and wavelet transforms. *IEEE Transactions on Power Delivery* 15 (1): 247–254.
13 NODC-National Oceanic and Atmospheric Administration (2017). http://www.nodc.noaa.gov/.
14 Wang, Y., Li, Z., and Shahidehpour, M. (2016). Stochastic co-optimization of midterm and short-term maintenance outage scheduling considering covariates in power systems. *IEEE Transactions on Power Systems* 31 (6): 4795–4805.
15 Billinton, R. and Wenyuan, L. (1991). A novel method for incorporating weather effects in composite system adequacy evaluation. *IEEE Transactions on Power Systems* 6 (3): 1154–1160.
16 Wei, W., Wang, J., Li, N., and Mei, S. (2017). Optimal power flow of radial networks and its variations: a sequential convex optimization approach. *IEEE Transactions on Smart Grid* 99: 1.
17 Baran, M.E. and Wu, F.F. (1989). Network reconfiguration in distribution systems for loss reduction and load balancing. *IEEE Transactions on Power Delivery* 4 (2): 1401–1407.
18 Gan, L., Li, N., Topcu, U., and Low, S.H. (2015). Exact convex relaxation for optimal power flow in distribution networks. *IEEE Transactions on Automatic Control* 60 (1): 351–352.
19 Chiang, H.-D. and Baran, M.E. (1990). On the existence and uniqueness of load flow solution for radial distribution power networks. *IEEE Transactions on Circuits and Systems* 37 (3): 410–416.
20 Barabási, A.-L. and Albert, R. (1999). Emergence of scaling in random networks. *Science* 286: 509–512.
21 Cplex, I.B.M. (2014). *ILOG CPLEX 12.6 Optimization Studio*. New York, NY, USA: IBM.
22 Zimmerman, R.D., Murillo-Sánchez, C.E., and Thomas, R.J. (2011). MATPOWER: steady-state operations, planning, and analysis tools for power systems research and education. *IEEE Transactions on Power Systems* 26 (1): 12–19.

23 Khodayar, M.E., Barati, M., and Shahidehpour, M. (2012). Integration of high reliability distribution system in microgrid operation. *IEEE Transactions on Smart Grid* 3 (4): 1997–2006.

24 Gholami, A., Aminifar, F., and Shahidehpour, M. (2016). Front lines against the darkness. *IEEE Electrification Magazine* 4 (1): 18–24.

4

Measuring and Enabling Resiliency for Microgrid Systems Against Cyber-attacks

Venkatesh Venkataramanan[1], *Adam Hahn*[2], *and Anurag K. Srivastava*[3]

[1] *National Renewable Energy Laboratory, Golden, CO, USA*
[2] *Washington State University, Pullman, WA, USA*
[3] *Lane Department of Computer Science and Electrical Engineering, West Virginia University, Morgantown, WV, USA*

4.1 Introduction

The microgrid concept has been evolving and gaining prominence over the last decade, especially with more investments [1]. In the evolving digital grid, massive amounts of data need to be transferred from the field devices to the control centers. As more optimal algorithms are deployed in the grid to produce optimal control at faster rate, the communication infrastructure becomes critical for successful implementation. At the same time, increased number of "smart" devices in the grid also increase the attack surface for potential cyber-attacks. Given these developments, cyber-physical system-based security analysis for the power grid is very critical.

The increased possibility of cyber-attacks on the smart grid has been a growing concern, and the recent attacks on the Ukraine power grid [2] has increased these concerns. In the face of the growing threat of cyber-attacks, it is important to realize that the threat of cyber-attacks can never be completely neutralized because the threats are constantly evolving and changing. Hence, it is important for the critical infrastructure to be resilient in face of failures. The definition of resiliency for critical infrastructure becomes important and also ways of measuring and computing the resiliency.

Research on the resiliency of microgrids is growing, with work such as the Energy Surety Microgrid from Sandia National Laboratories [3]. Given the ambiguity, various definitions of resiliency are explored. The NERC Report on "Severe Impact Resilience: Considerations and Recommendations" of 2012 defines that during severe events, it will not be possible to meet all consumers, and entities will prioritize their work for rapid restoration of service [4]. Hence, we can understand that resilience becomes an important property especially in face of contingencies and the need for operation in this state. In some microgrids, it is necessary that some critical loads are kept operational even in the presence of contingencies. While reliability-based problems have been studied for the power grid [5], resiliency is a relatively new area of study. Resilience in this work is defined as the ability of the microgrid to supply the critical load in the case of contingencies. When considering the microgrid's resiliency, it is important to consider the microgrid as a cyber-physical system and also consider the

Resiliency of Power Distribution Systems, First Edition.
Edited by Anurag K. Srivastava, Chen-Ching Liu, and Sayonsom Chanda.

communication architecture and associated services. Presidential Policy Directive 21 states that resilience is "the ability to prepare for and adapt to changing conditions and withstand and recover rapidly from disruptions." Resilience in this work is defined as the ability of the microgrid to supply the critical load even in the case of contingencies. Before enabling resilience, it is important to determine the resilience of a system numerically to enable the operator to quickly understand the changing nature of the system. A report from the National Academy of Sciences titled "Disaster Resilience – A national imperative" states that "without some numerical basis for assessing resilience, it would be impossible to monitor changes or show that community resilience has improved."

Two tools are introduced which enable the operator to determine the resilience of the microgrid by considering the holistic cyber-physical system. The first tool, CyPhyR (Cyber-Physical Resilience) tool is used to assess the impact of vulnerabilities on the resiliency [6]. The tool has two stages – a planning stage where the vulnerabilities are evaluated in terms of their position in the microgrid, and an operation phase where the real time status of the vulnerability is monitored to determine the impact on the resiliency. The second tool CP-SAM (Cyber-physical Security Assessment Metric) considers various factors affecting the resiliency and combines it into a single cyber-physical metric that can be used to monitor the overall resiliency [7]. These metrics are generated using data-driven techniques and also using the cyber-physical microgrid testbed at Washington State University (WSU).

4.2 Testbed Description for Validating Resilience Tools

The developed testbed can be configured in multiple ways depending on the simulation need [8]. Here, we demonstrate the microgrid cyber-physical security analysis capability using the following components:

- RTDS for simulating the power system,
- CORE for emulating the communication network,
- A resiliency-based reconfiguration algorithm.

The testbed is shown in Figure 4.1 and further details about the interfacing of these components can be found in [9].

4.3 Test System for Validating Cyber-Physical Resiliency

The CERTS microgrid has been used in this work to model the networked microgrid. For the purposes of making the microgrid suitable for reconfiguration, a few modifications have been made as detailed below:

1. There is no substation transformer modeled specifically between main grid and the microgrid.
2. Zone-3 and Zone-4 are three-wire system and without neutral.
3. There is no isolation transformer in Zone-5.

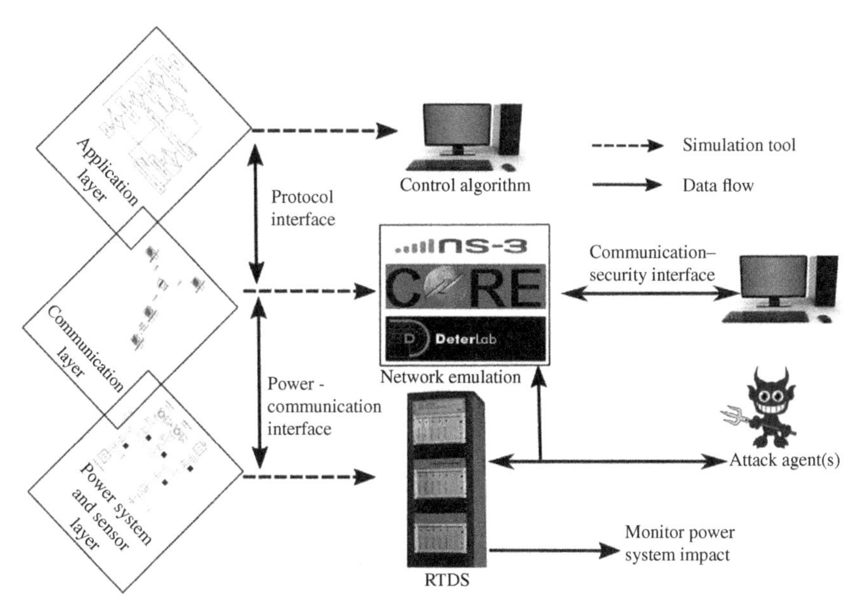

Figure 4.1 Cyber-physical testbed for validating resiliency tools. Sources: Sweetym/Getty Images; RTDS image courtesy RTDS Technologies Inc.

4. The PV panel supplies both priority loads and nonpriority loads in Zone-3, Zone-4, and Zone-5.
5. Additional distributed energy resources (DER) units are added to increase the number of feasible paths.

The modified model of the microgrid is shown in Figure 4.2. In order to provide more options for reconfiguration and also to have a more complex feasible network architecture, two CERTS microgrids have been connected together. The microgrids connect with main

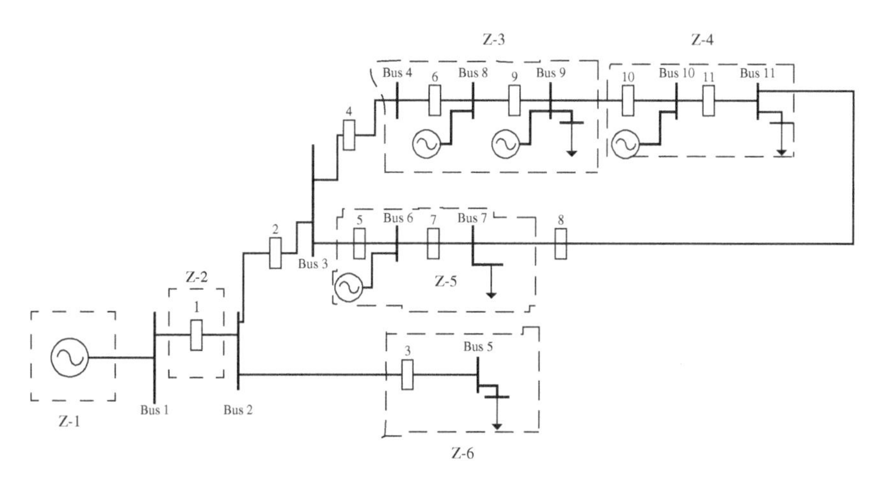

Figure 4.2 Modified CERTS microgrid system.

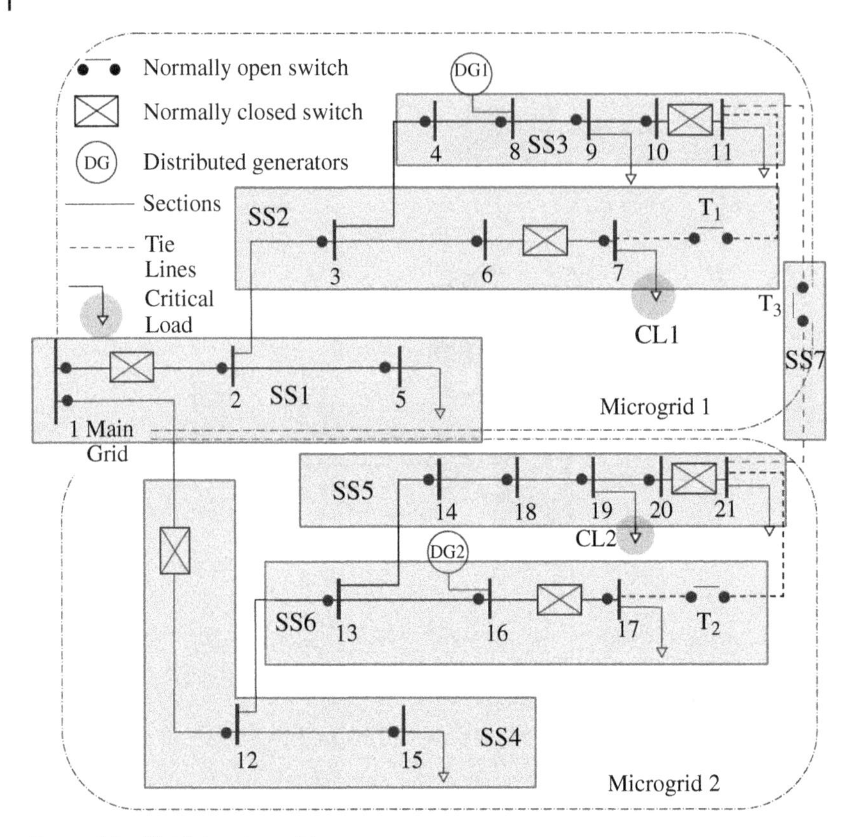

Figure 4.3 Multiple microgrid system.

grid at the same substation, but have different feeders. In addition, there is a tie line switch between the two individual microgrids, which enable the microgrids to exchange power in cases where the connection through the main grid is unavailable. The overall power system topology is shown in Figure 4.3.

For the communication model, each node of the microgrid is modeled as a node in CORE. The DMS (Distribution Management System) is at the control center, and this is where the reconfiguration algorithm is implemented. The communication system can be configured in various topologies such as point to point, aggregated, or mesh topology. For point-to-point network topology, there are no aggregators in the communication model, and each node is assumed to be directly connected to the control center. The mesh and aggregated networks require an aggregator to be present between the nodes and the control center. In this case, each of the aggregator can be considered as a substation, while the reconfiguration is implemented at the control center with DMS. The key difference between the mesh and the aggregated communication architectures is that the mesh network also has connections between the various nodes in the system, while in the aggregated architecture, the nodes are only connected to the aggregator. In the mesh architecture, more redundancy can also be built in, such as a connection between the node and the DMS directly similar to the point-to-point configuration for some critical nodes. The planning engineer needs to study the cost-reliability ratio and determine the redundancy required for the system. Here, we

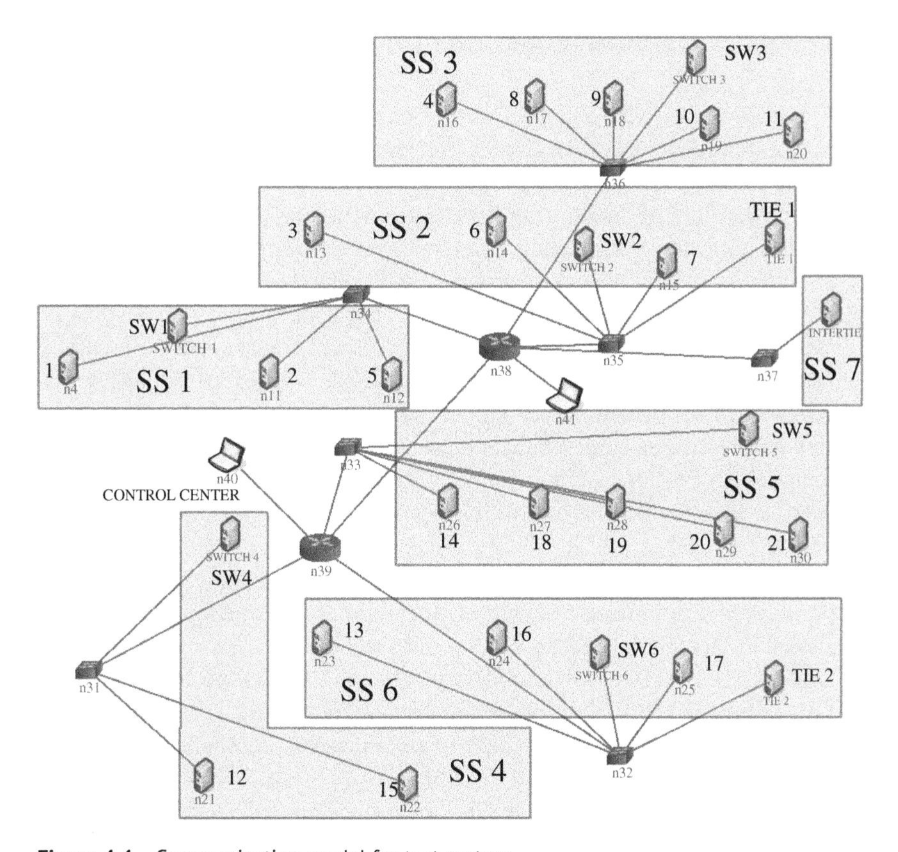

Figure 4.4 Communication model for test system.

consider an aggregated system topology that aggregates the nodes to substations, which are again connected to the DMS at the control center. The associated communication topology for the test system is shown in Figure 4.4.

In the communication model, each substation is associated with several breakers, and this architecture is reflected in the communication model developed. The CORE model shows several hosts in the figure, which represents the different relays that control the breakers in the simulation, and switches which can be controlled remotely. Each of these devices are connected to an aggregator or a substation gateway in the substation, which is modeled as a switch in the model. There are six such substations or switches in the communication model. Finally, there is also the control center in which the control application is run. It is typical to have a backup control center in case the primary one fails. This is modeled in the network.

We assume there are also Intrusion Detection Systems (IDSs) running at a network level which will look for attack patterns and monitor vulnerabilities to see if they are exploited. This feeds into the Cyber Impact Severity (CIS) score. In addition, firewalls restrict access between different devices so that the attacker does not have access to the entire network after exploiting one vulnerability.

4.4 Dependencies Between Cyber and Physical Systems

The dependencies between the cyber and physical models need to be explored to understand the impact of vulnerabilities on the cyber-physical system. Traditional approaches focus on an analysis of the physical system or the communication system separately. Physical system analysis includes power flow, optimal power flow, continuation power flow, state estimation, and such. Communication system analysis includes performance quantification using metrics such as latency, bandwidth, packets dropped, and such others.

In combined cyber-physical analysis, it is important to consider the effect of one system on the other, meaning consider the effect of what is perceived as a power system or communication event and consider the impact on the system as a whole. While some of these effects are apparent, such as latency in communication on the control action, it is also important to consider other factors such as vulnerabilities present in the various devices used in the smart grid. Vulnerabilities exist in regular computer systems, such as a substation computer running Windows, or even the field devices which can be compromised by sending crafted packets.

Vulnerability information can be found by techniques, such as penetration testing, and other testing procedures. The information about vulnerabilities can be found in resources such as the National Vulnerability Database (NVD), or by information sharing among utilities and stakeholders. The next step after discovering vulnerabilities is to evaluate what the impact of the vulnerability is. This step is important because most times not every vulnerability can be fixed immediately, and there needs to be an evaluation procedure to determine the priority in which the vulnerabilities need to be fixed.

Common Vulnerability Scoring System (CVSS) is a scoring system that classifies vulnerabilities into high, medium, or low categories based on various criteria. CVSS rates a base score, a temporal score, and an environmental score. The base score rates the vulnerability based on an exploitability and impact sub-score, which are again based on factors, such as attack vectors, and permissions required. The temporal score is based on factors such as report confidence and remediation level required. The environmental score considers the effect of the vulnerability on the confidentiality, integrity, and availability (CIA) of the application.

While the CVSS is very useful and is based on expert's opinion, it helps the system operator to have a quantitative score based on system topology, and is tailored to provide increased visibility and helps the operator make control decisions quickly. In the next sections, we introduce two tools one of which allows the operator to evaluate the impact of vulnerabilities, and the other that includes vulnerabilities when generating an overall resiliency score.

4.5 Cyber-Attack Implementations

To understand the microgrid cyber-physical resiliency, it is important to understand how various cyber-attacks are implemented and what factors that contribute to the resiliency are affected. Depending on the type of communication simulator used, various cyber-attacks can be executed with differing levels of accuracy. For the power grid, the integrity and availability impacts are given greater weight usually as they can have a direct physical impact. These attacks and their implementations are described below.

4.5.1 Availability Attack

The DOS attack is one of the most popular cyber-attacks that can cause a severe impact on the smart grid. DOS attack affects the availability in the CIA requirements. The theory behind the DOS attack is simple – the objective is to disable a device in the network. However, implementing the DOS attack can be tricky based on the simulator used. As previously discussed, some implementations of the DOS attack need to also model the computing device on which the DOS is performed. Consider the transmission control protocol (TCp) SYN flood attack. The attack involves the attacker sending multiple SYN packets to the compromised node until it becomes nonresponsive. However, if only the network stack is simulated, it is not possible to simulate the node running out of computation space. In these cases, the DOS attack has to be implemented by using the bandwidth restriction on the link. The attacker has to send enough traffic that the node is unable to send any information through that particular channel. In cases where the actual implementation of the attack itself is not important, rather the point is to study the power system impact, it may be sufficient to simply disable the node in question and then monitor the power system impact. However, to test defense mechanisms, it is important to choose the right simulator and model to accurately simulate the DOS attack.

4.5.2 Integrity Attack

MiTM is also a popular attack to study the performance of the smart grid. The MiTM attack is typically used to study the response of the smart grid to integrity-based attacks. The theory behind the attack is that the attacker gains access to the compromised node and then proceeds to modify the system in a malicious manner. This may involve changing a measurement from the node, changing the control signal sent by the node, manipulating the data stored in a node, or triggering false control actions. Depending on the level of detail needed, the user can either choose to simulate the process of actually gaining control of the node or choose to manipulate the node directly to study the power system impacts. In order to simulate the process of gaining access to the node, the user has to go through multiple steps such as modeling the ARP traffic, modeling vulnerabilities which allow elevation of privilege, and modeling intelligent attack agents capable of manipulating the node without detection. If the end goal is to study the power system impact and not the defense mechanisms, the process is simpler.

4.6 Cyber-Physical Resiliency Metrics and Tools – CyPhyR and CP-SAM

In this section, two tools that have been developed to quantify the resiliency based on the cyber-physical model of the microgrid are introduced.

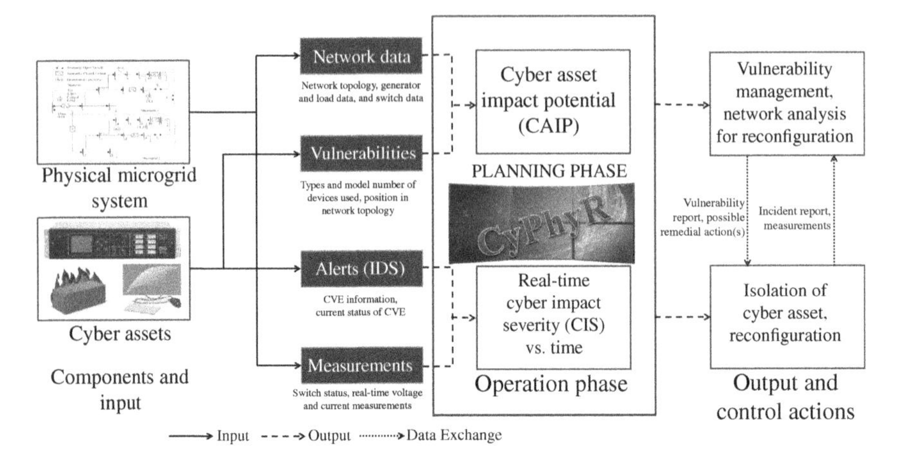

Figure 4.5 Cyber-physical resilience impact and analysis. Source: Schweitzer Engineering Laboratories, Inc.

4.6.1 CyPhyR

The CyPhyR tool is aimed at improving the microgrid's resiliency. The overview of the CyPhyR tool is shown in Figure 4.5.

We consider two different stages: a planning phase and an operation phase. In the planning phase, the tool is to be used by the planning engineers to design a resilient microgrid and study the impact of different components used in the microgrid on the microgrid's resiliency. Similar to the transmission grid where the planning engineer is responsible for contingency management, in the microgrid, we consider that the planning engineer is responsible for the evaluation of new vulnerabilities found during operation. The planning engineer evaluates these vulnerabilities, provides threat reports, and suggests possible remedial actions in terms of reconfiguration. In planning phase, a metric CAIP (Cyber Asset Impact Potential) is defined using which the planning engineer can study the impact of each cyber asset used in the microgrid considering their criticality to the microgrid. This criticality is determined using the centrality of the cyber asset in a graph representation, which is further explained in the next section. The overall algorithm for the planning phase is in Figure 4.6. In the operation phase, the tool is to be used by the microgrid operator. A metric CIS is defined, using which the operator can determine in real time the resiliency of the microgrid. The metric CIS is based on the CVSS, and the definition is explained in the next section. Using the CIS metric, the operator can decide on the control actions to improve the resiliency of the microgrid, such as isolation of a cyber-asset, and reconfiguring of the microgrid. The algorithm for the operation phase is in Figure 4.6.

The cyber-physical resiliency is developed across two stages. Initially, the physical power system resiliency is considered by analyzing the power system using a graph theory-based algorithm considering power system constraints. From the obtained resiliency values and reconfiguration paths, the cyber-physical resiliency is determined, by computing different metrics for planning and operation phase. A graph-based algorithm which also considers

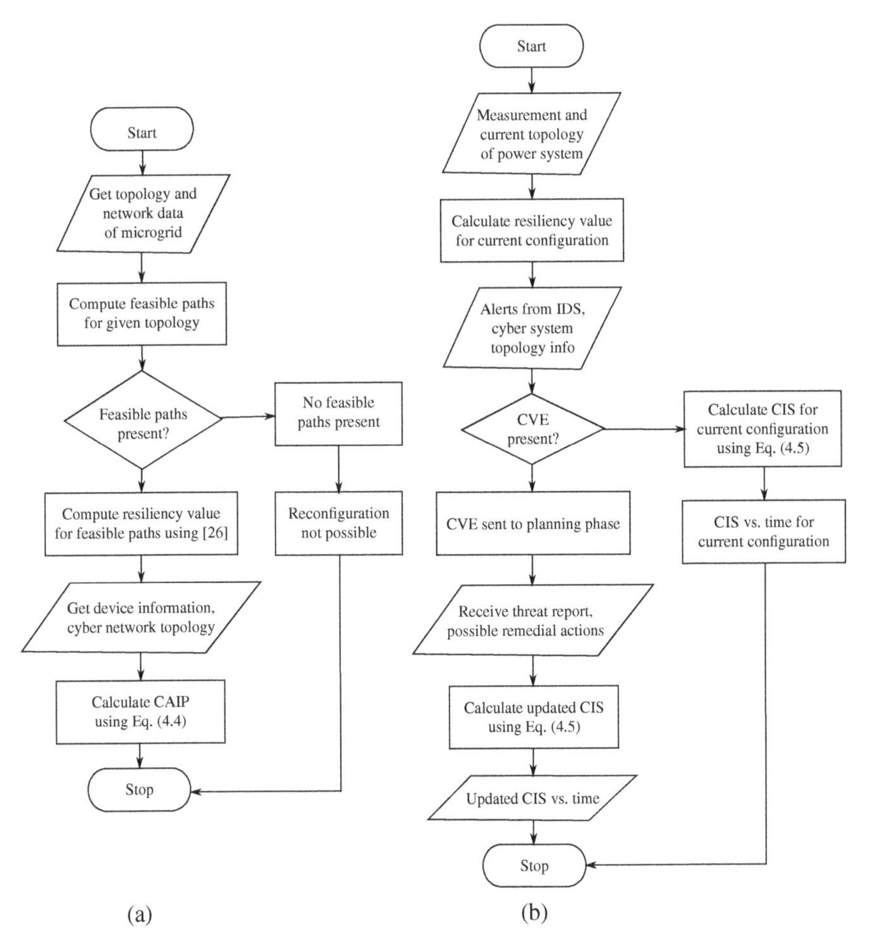

Figure 4.6 Algorithm for (a) planning phase and (b) operation phase.

power flow constraints is used to determine the feasible paths for a given microgrid. The steps for calculating physical resiliency value are the following:

1. Find all possible paths from the critical load to the sources.
2. Determine if critical load is islanded, else continue.
3. Ensure that there are no loops present in the path, else eliminate the loops.
4. Eliminate similar paths based on switch configurations.
5. Run forward–backward powerflow for all the paths.
6. Create a list of feasible paths based on powerflow.
7. Compute Choquet Integral resiliency value for all feasible paths.

An overview of this process is shown in Figure 4.7.

4.6.1.1 Cyber-Physical Resiliency

In order to understand the complexity of the cyber attack, the CVSS has been used. CVSS is a vulnerability evaluation and scoring system developed by "FIRST" [10, 11].

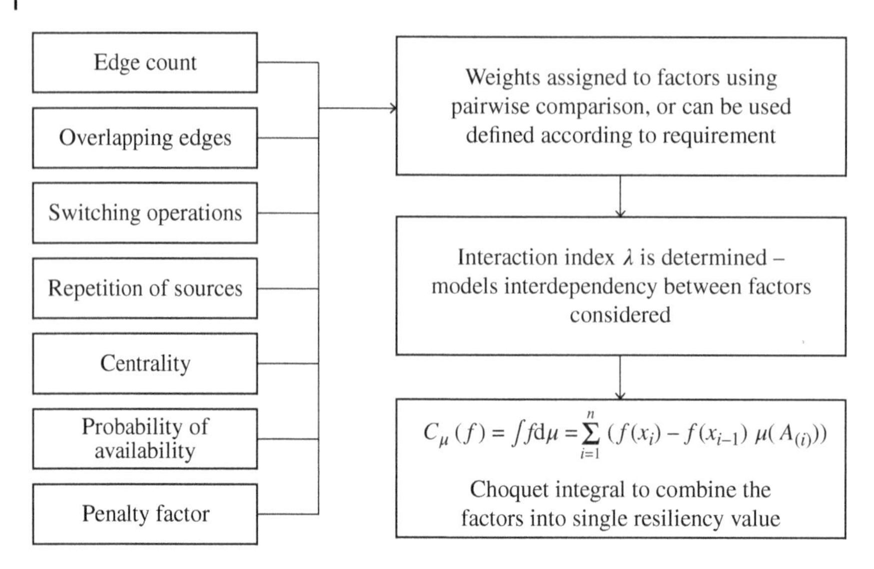

Figure 4.7 Overview of physical resiliency quantification.

The CVSS system considers known vulnerabilities in devices (known as Common Vulnerabilities and Exposures or CVEs) and tries to assess the impact of the vulnerability by looking at various parameters such as complexity of attack vector, CIA impacts. The study of vulnerabilities and CVSS play an important role in the management of cyber-physical systems. The NVD provides statistics of how many vulnerabilities it adds each year [12].

An example of the CVSS score for a particular vulnerability is discussed here. The CVSS base score consists of two parts, an exploitability metric and an impact metric. The exploitability metric considers the factors of Access Vector (if the exploit needs physical, local, or network access), Access Complexity (low, medium, or high), and Authentication (single, multiple, or none). The impact metrics considers the impact on the confidentiality, integrity, and availability (low, medium, or high).

Three scores are computed – an impact score, an exploitability score, and an overall base score. The equations for calculating these scores are shown as follows:

$$\begin{aligned}\text{Impact} = 10.41 &* (1 - (1 - \text{ConfidentialityImpact}) \\ &* (1 - \text{IntegrityImpact}) \\ &* (1 - \text{AvailabilityImpact}))\end{aligned} \tag{4.1}$$

$$\begin{aligned}\text{Exploitability} = 20 &* \text{AccessComplexity} * \text{AccessVector} \\ &* \text{Authentication}\end{aligned} \tag{4.2}$$

$$\begin{aligned}\text{BaseScore} &= (0.6 * \text{Impact} + 0.4 * \text{Exploitability} - 1.5) \\ &\quad * f(\text{Impact}), \\ \text{where } f(\text{Impact}) &= 0 \ \text{ if Impact is 0} \\ &= 1.176 \text{ otherwise}\end{aligned} \tag{4.3}$$

In the CVSS, the following factors contribute to the environmental score:

1. Confidentiality requirement
2. Integrity requirement
3. Availability requirement.

Of these, the CIA requirements (items 1, 2, and 3) are used to assess how valuable the CIA properties are for that particular device in that environment. While computing the environmental CVSS score could give a qualitative understanding of the impact of exploiting a vulnerability for a specific device, the power system architecture is unique in the sense that the impact can be physically and numerically defined. In this work, the impact of resiliency is considered to measure the impact of the environmental score.

In the planning phase, it is assumed that the planning engineer ensures that there are no vulnerabilities in the design. However, the planning engineer needs to take into consideration the potential effect of future vulnerabilities on the system. Hence, the tool evaluates the CAIP for each device in the microgrid. CAIP is used to study the maximum impact each device can cause rather than the impact of a specific vulnerability. This is done by assuming maximum values for impact and exploitability for each device in the microgrid, and then using its position in the network to calculate its impact potential. The criticality of the node and edge in the network is calculated by using its central point dominance otherwise called betweeness centrality. The equation to calculate this value is give as $C_{\mathrm{B}}(d) = \sum_{j \neq d \neq k} \frac{\sigma_{jk}(d)}{\sigma_{jk}}$. In the equation, $\sigma_{jk}(d)$ is the number of paths that pass through node d, and σ_{jk} is the total number of shortest paths from node j to k. By assuming maximum values for exploitability and impact, we effectively have a CVSS value of 10, which is the maximum impact that the device can have. Now, we use the central point of dominance $C_{\mathrm{B}}(d)$ to study the effect of position of the device on the resiliency of the microgrid. For nodes, C_{B} is used, while for the switches E_{B}, the edge centrality is used. The edge centrality is used to represent the controllability that is lost when a particular switch is compromised. E_{B} is given by a similar formula as C_{B}, which is $E_{\mathrm{B}}(d) = \sum_{j \neq d \neq k} \frac{\sigma_{jk}(d)}{\sigma_{jk}}$, where d is every edge on the paths.

The environmental score of the CVSS depends on the CIA requirements, as discussed earlier. For CAIP, we use the quantitative resiliency impact score in place of environmental impact. This score is given by the change in the resiliency value from the base configuration to the new configuration. The final component is the controllability score, which is explained in next section. The controllability score looks at the number of affected nodes for availability and integrity-based attacks and determines what portion of the system is available for reconfiguration.

$$\begin{aligned} \text{CAIP} = {} & \text{CVSS} * C_{\mathrm{B}}(d) * \\ & \text{Environmental Impact} * \text{Controllability} \end{aligned} \tag{4.4}$$

4.6.2 CIS

For the operation phase, CIS is used. This phase assumes that an IDS is present in the system and relies on alerts from the IDS. This metric uses the IDS to monitor the cyber network during operation and provides the operator with a cyber severity score. The IDS is used to define the f(Impact) of any vulnerability in real time. The f(Impact) is adapted from CVSS formulation. The f(Impact) score can vary between three values based on the state of the vulnerability. This is shown in Eq. (4.5). The CIS uses the modified CVSS score and refines it for situational awareness. Hence, the impact component can have three states: vulnerable, exploited, and physical impact. The impact is considered to be 0.2 if the operator knows that the vulnerability exists, but the IDS shows that it has not been exploited. The impact score is considered to be 0.5 if the vulnerability is exploited but has no physical impact as yet. In this case, the operator knows that the attacker has successfully exploited the CVE but has not caused any problems to the system's resilient operation. Finally, if the attacker causes physical impact on the system, the impact component of the CIS is considered to be 1.

$$\begin{aligned} \text{CIS} &= \text{CAIP} * f(\text{Impact}) \\ \text{where } f(\text{Impact}) &= 0.2 \quad \text{if CVE is discovered} \\ &= 0.5 \quad \text{if exploited but no impact} \\ &= 1 \text{ for physical Impact} \\ &\quad \text{due to CVE} \end{aligned} \tag{4.5}$$

The CIS also uses measurements from the system to determine the resiliency impact, which is the change in resiliency value from the previous configuration. The resiliency impact is defined as the normalized change in resiliency for the new configuration. CIS detects the new configuration and its impact from the measurements such as breakers statuses, voltage, and current values.

In using CyPhyR, it is important to understand the relationship between the nodes, switches, substations, and the control center. It is important because to choose the appropriate reconfiguration path for a given contingency, it is essential to understand the impact of each device on the overall system. Assuming that the device at a particular node is being controlled by the attacker, we determine the most resilient path possible by isolating that node. The relationship can be formalized by using partially order sets (POSET). A Hasse diagram is a visual representation of a POSET, which can be used to easily understand these relationships. The Hasse diagram simplifies the binary relationships between the elements and makes them easier to visualize and understand. The diagram for the test system of two proximal CERTS microgrids is shown in Figure 4.8.

In the diagram, we have the nodes at the bottom of the graph. These nodes represent the relays in the system at various nodes. The control algorithm at the control center does not have control over all these nodes for reconfiguration. Only certain switches are

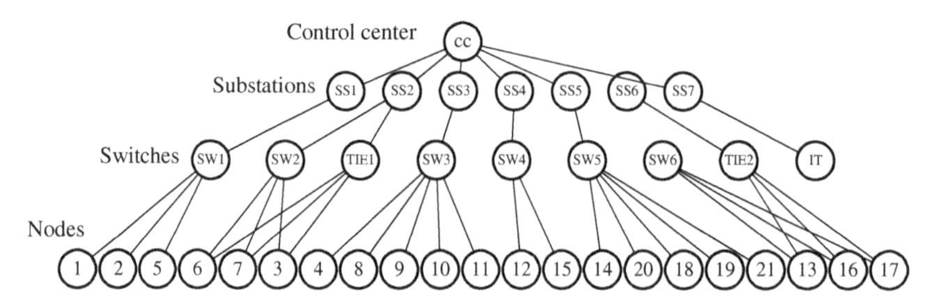

Figure 4.8 Hasse diagram of POSET for test system.

SCADA-enabled and can be controlled from the control center. These switches are shown in the power system model in Figure 4.3 and also include the various tie-line switches. In case of a cyber attack or other contingencies, the operator can isolate part of the power system by using these switches. The relation between the nodes and the corresponding switches that can be used to isolate the affected part of the network. It can be seen that certain nodes are connected to the various switches SW1 through SW7 and the tie-line switches. Some nodes are also connected to two switches depending on the position in the network. In the level above these switches, we have the substations. In the communication model, it can be assumed that these relays and SCADA switches are connected to the control center through a network switch. Typically, these network switches are present in substations which act as a point of aggregation. As seen in the communication model in Figure 4.4, the various nodes are connected to the network switches represented by substations SS1 thorough SS7. These network switches are then connected to the control center where the reconfiguration algorithm is implemented. The diagram helps us derive the controllability lost during a particular contingency. Now, the reconfiguration result is used to derive the feasible paths by looking at the switch statuses, from which the most resilient path is chosen.

4.6.3 CP-SAM

CyPhyR is a simplified tool that helps the operator decide on the right control action depending on the status of the vulnerabilities. CyPhyR has a clear delineation between the cyber and physical components, and in some cases might not accurately reflect the true cyber-physical nature of the power grid. CyPhyR works best to study the impact of vulnerabilities in the context of the physical impact that they can have. CP-SAM on the other hand is a more comprehensive tool that can be used to provide the operator with visibility based on a more complex system model. Both CyPhyR and CP-SAM provide various levels of sophistication to enable the operator to have greater visibility, and the user can choose whichever tool is suitable for their purposes. The architecture of CP-SAM is shown in Figure 4.9.

Different factors to be considered for the cyber-physical security affecting resiliency as listed in Table 4.1.

Various factors are considered for the power system resilience by considering the power system as a graph. Consider a graph G, with N nodes and E edges. The order of the graph

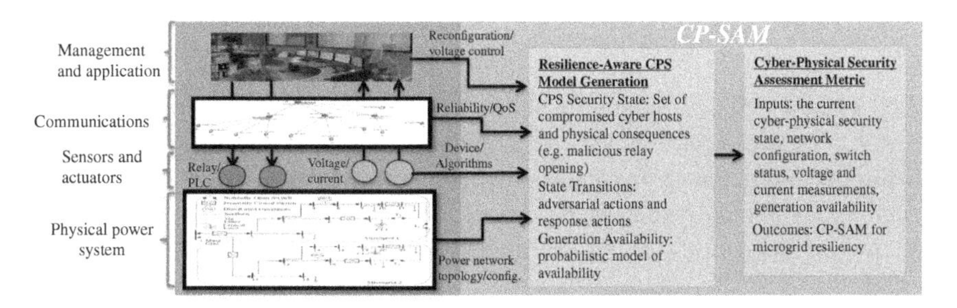

Figure 4.9 Cyber physical modeling for microgrid security assessment to enable resiliency.

Table 4.1 Factors affecting microgrid resiliency and solution approach.

Layer in microgrid system	Factors affecting resiliency	Solution approach for formulation
Power system	Switching operations	Number of switch state changes for new configuration
	Redundant paths	Depth first search (DFS)
	Redundant generations	System information
	Probability of source	Reliability models of source availability
	Average path length	Graph metric
Sensor and actuator	Visibility of system	Topological observability
	Algorithmic and device level resilience	Input/output data value validation
Communication	Security	Vulnerabilities present in system
	Reliability and robustness	Network topology
	Quality of service	Network performance metrics
Application	Feasibility of reconfiguration	Path availability
	Power flow constraints	Forward backward power flow
	Application resilience	Algorithmic resilience of reconfiguration application

G is given by $G = (N, E)$, and the graph distance d between two nodes is the number of sections in the shortest path connecting them. A path in the graph G_t can be represented as a subgraph G' of the form,

$$N(G') = [n_0, n_1, n_2, \dots, n_i] \tag{4.6}$$

$$E(G') = [(n_0, n_1), (n_0, n_2), \dots, (n_{i-1}, n_i)] \tag{4.7}$$

such that $N(G') \subset N$ and $E(G') \subset E$. $|E(G')]$ gives the length of the new path G'. The average path length for the graph is expressed by

$$l_G = \frac{2}{i(i-1)} * \sum_{a \neq b} d(n_a, n_b) \tag{4.8}$$

The average path length provides a measure of how far the load is from the source. Shorter path lengths lead to increased resiliency as there are usually multiple redundant paths from the source to the load. From the graph G, the path redundancy can also be calculated. There are various definitions of redundancy that can be used, such as scalar, vector, and matrix redundancy. The equations for the redundancy are given in Eq. (4.9):

$$R_{\mathrm{m}}(i,j) = \sum_{l=1}^{L} P(i,j,l) \tag{4.9}$$

The matrix redundancy represents the total redundancy between the nodes of the graph. The higher the number, the more paths that exist between the nodes i, and j. Finaly, the reconfiguration of the microgrid must satisfy power flow and line limits for distribution of power. Other network constraints are the possible loop formations, which must be avoided to not cause any problems with power system protection [13]. The power flow equations used here are the forward–backward sweep approach. This approach is suited for the distribution system/microgrids as they are typically operated in a radial fashion and have high R/X ratios. In this work, the distribution system software GridLAB-D is used to determine the power flow. The equations for the power flow are referred from standard textbooks and given in equations as follows:

$$V_j = [A] * [V_i] - [B] * [I_j] \tag{4.10}$$

$$V_i = [a] * [V_j] + [b] * [I_j] \tag{4.11}$$

$$I_i = [c] * [V_j] + [d] * [I_j] \tag{4.12}$$

where i, and j are the sending end and receiving end, respectively. In the above equations, Eq. (4.10) represents the forward sweep, and the other two equations represent the backward sweep step. A, B, a, b, c, and d represent the generalized matrices that are developed using characteristics of the individual components.

Other factors that affect reconfiguration such as auxiliary control actions required for stable operation, such as capacitor switching, tap changing of transformers that should be taken into account. The number of switching operations should also be minimized as more number of switching operations also increases the probability of failure. The reconfiguration depends on the information provided by the measurement devices. The devices in this level are essential to maintain visibility of the system, so that the reconfiguration algorithm knows the state of the system. Availability and accuracy of measurements are key to avoid inaccurate reconfiguration of the system. Also, the control devices should not give wrong control action (due to various reasons such as cyber attack, wrong estimate of system state) and trigger the reconfiguration when it is not necessary. When the actual reconfiguration control signal is received, the control devices should not fail [14].

The control and monitoring algorithms become useless essentially if the communication infrastructure between the control center and the deployed sensing and actuation devices on the field do not operate properly. The communication layer must be secure and should be resilient to integrity- and availability-based attacks. The communication system should also be reliable and robust and must use appropriate protocols to ensure successful transportation of control signals [15]. Availability of control signal on time is vital for a successful reconfiguration and the control signal must not be delayed due to network constraints. Quality of Service (QoS) must also be ensured, and in addition to cyber attacks, the integrity of the control signal should be impervious to noise or disturbance in the communication channel [16]. Our proposed metric evaluates how a given power grid communication infrastructure either maintains and ensures the satisfaction of the abovementioned operational requirements despite various types of incidents or violates those requirements and cannot tolerate the incidents. The tolerance level will be measured by the metric and used for the infrastructural resilience assessment.

4.6.4 Computation of CP-SAM

After identifying the factors from the various layers of the cyber-physical model, we integrate them in a single metric using fuzzy Choquet integral which is a solution method for solving Multi-criteria Decision-making (MCDM) problems. A single value for resiliency at any given time will provide an insight to the system operator about the state of the microgrid.

MCDM problems are an ideal fit as we aim to combine several ordinal and cardinal parameters to obtain a single, easily understandable resilience metric. There exist multiple solution approaches to the MCDM problem [17, 18]. The method chosen in the work is fuzzy Choquet Integral. While several solution strategies do not consider the relation between the parameters used, Choquet integral uses a parameter called the "interaction index" to model the dependencies between the various factors. This ensures that the problem solution considers the interdependence of the parameters used in the problem, making it ideal for a holistic cyber-physical system.

We consider a finite universal set $N = \{1, 2, \dots, n\}$, which can be thought as the index set of a set of criteria, attributes, players, etc. For the case of determining resiliency, these factors are detailed in Table 4.1. For each of these factors, a fuzzy measure is defined as follows: $\mu : \rho(N) \mapsto [0,1]$ such that

1. $\mu(\emptyset) = 0, \mu(N) = 1$
2. $\mu(A) \leq \mu(B)$, whenever $A \subset B \subset N$ (monotonicity property)

In addition to the properties defined above, we define an additional property for the fuzzy integral as follows:

$$g(A \cup B) = g(A) + g(B) + \lambda g(A)g(B), \text{ for some fixed } \lambda \geq -1 \tag{4.13}$$

Equation (4.13) can be used to determine the value of the fuzzy measure by solving $g(X) = 1$, which results in the following equation:

$$\lambda + 1 = \prod_{i=1}^{n} (1 + \lambda g_i) \tag{4.14}$$

The steps for computing the CP-SAM using the fuzzy Choquet Integral is in Figure 4.10. The factors that are considered for the CP-SAM are already listed in Table 4.1. These factors, as mentioned before, can be modified as required by the user. In the next stage, each factor is assigned an initial weight based on user-defined criteria. Based on the initial weights provided by the user, the fuzzy weights are determined. The fuzzy weights is the weights used for the interaction between the various factors in computing the final CP-SAM score. Various methods such as Singleton value measure, Shapely value, and Input Number Standard exist to determine the fuzzy weight from the initial set of weights. In this work, a method called the Shapely value has been used to determine the fuzzy weights. The fuzzy Choquet Integral is defined as follows:

$$\int_X h(\bullet) \circ g(\bullet) \tag{4.15}$$

The Choquet Integral is an aggregation operator that combines the various interdependent fuzzy measures using their weight and the interaction index. For the case with a finite

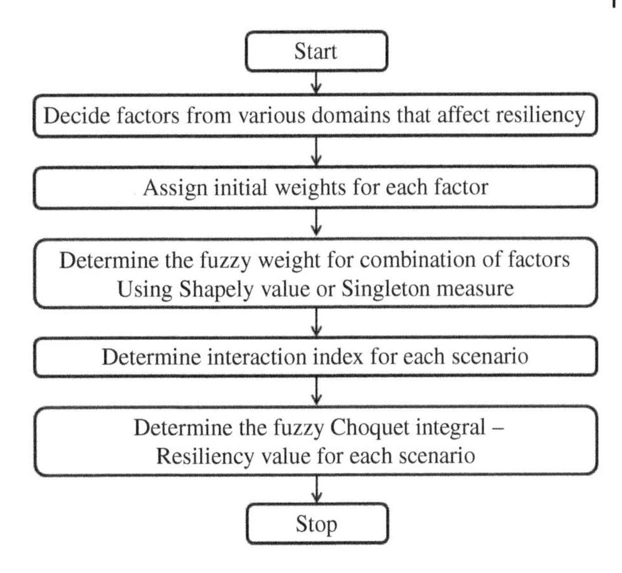

Figure 4.10 CP-SAM computation.

criteria set of X, the fuzzy Choquet integral can be defined as follows:

$$\mathrm{CI}(h) = \sum_{i=1}^{n} = [h(x_i) - h(x_{i-1})]\ g(A_i) \tag{4.16}$$

where $h(x)$ is the input weight for each of the different factors, and $g(x)$ is the fuzzy measure for the particular scenario. The output of the Choquet Integral is normalized between 0 and 100 to help the operator understand the resilience score easily.

4.7 Case Studies for Cyber-Physical Resiliency Analysis

Two tools have been proposed above to determine the cyber-physical resiliency of the microgrid. CyPhyR focuses on the effect of vulnerabilities on the microgrid during the planning and operation phase. CP-SAM combines various factors that affect resiliency into a single metric using decision-making processes and provide an easy to understand metric that evolves with changing scenario in the grid.

4.7.1 CyPhyR Results

The results are presented across two different phases: a planning phase and an operation phase. The operation phase can be considered to have two parts – a situational awareness part by which the operator can keep track of the cyber-physical resiliency and impact of the system, and a decision support part which will aggregate information from planning and awareness to provide information to the operator to support decision-making.

The CAIP for the nodes and switches are shown in Figures 4.11 and 4.12. The figure shows the various nodes affected due to an availability-based attack and integrity-based attacks. For availability-based attacks since once a single node is taken out of service, the impact is less when compared to integrity-based attacks. In integrity-based attacks, the operator has

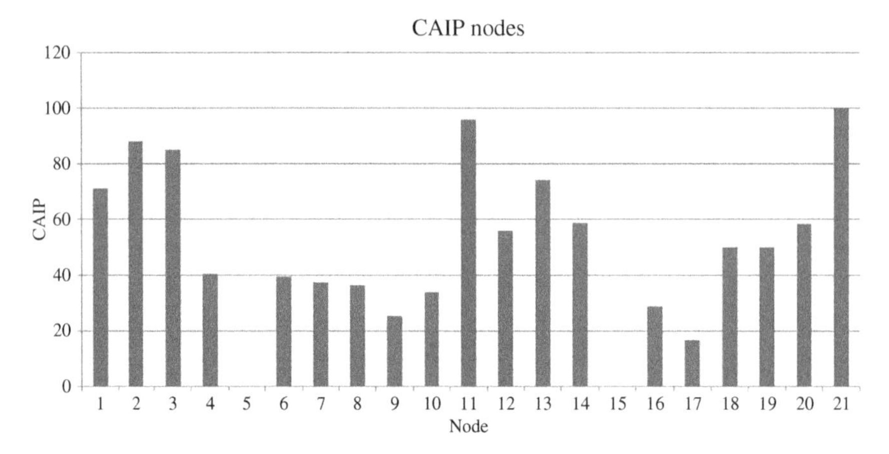

Figure 4.11 CAIP for nodes.

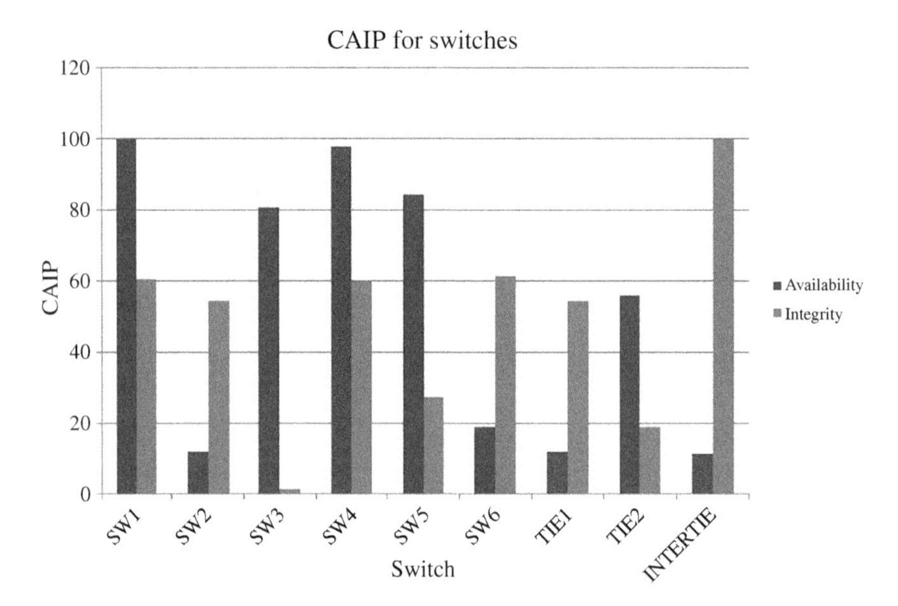

Figure 4.12 CAIP for switches.

to take some control actions to isolate that particular node which results in a broader impact. The CAIP value has been scaled between 0 and 100 for more intuitive understanding of the results.

In case of the switches, the impact is much higher for both availability- and integrity-based attacks. This is because even an availability-based attack on a switch restricts the number of feasible reconfiguration states that the operator can go to. In some cases, the availability-based attack has an higher impact than the integrity-based attack. This is because if the attacker restricts certain critical switches, the operator has no choice but to go to a less resilient configuration than the current configuration.

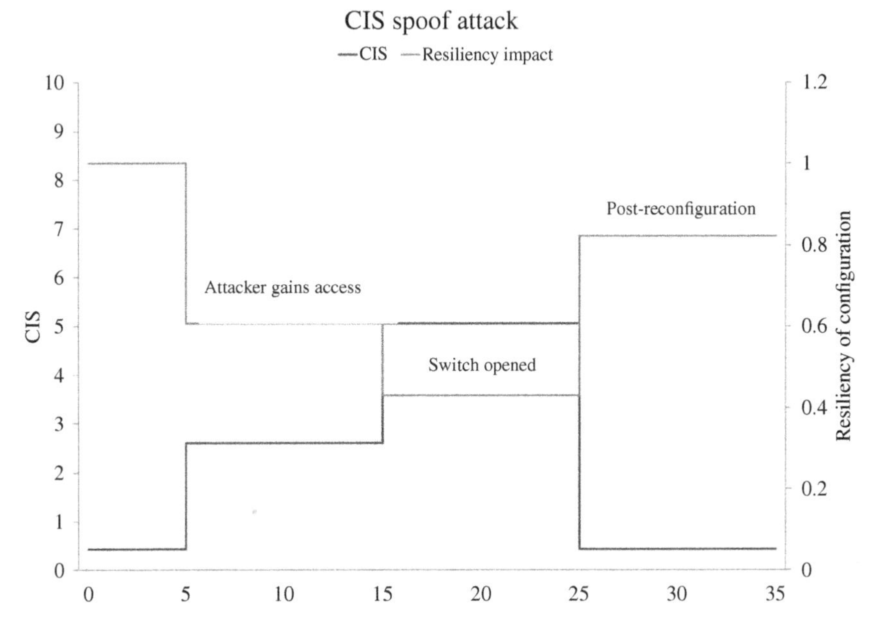

Figure 4.13 CIS vs. time.

Based on the feedback from planning phase, the operator is aware of the effect each device in the microgrid can have on the microgrid resiliency. The value of the CIS and the resiliency for every five minutes is shown in the Figure 4.13. The right axis is for normalized resiliency which ranges between 0 and 1, and the left axis is for the CIS value at that time.

The operators can use the real-time CIS and resiliency value from Figure 4.13 to understand the current state of the system. The combination of these results can be used by the operator to understand the current CIS of the microgrid and the impact of vulnerabilities in any device on the resiliency of the microgrid.

The results clearly bring out the problems with hierarchical network and control structure currently used in the power grid. Also, it provides a better understanding for network operators to decide which vulnerabilities to patch first and provides a way to quantify the physical impact of the vulnerability on the resiliency. The same methodology can be used for other CVEs to determine the impact on the microgrid.

4.7.2 CP-SAM Results

The microgrid operator can obtain a better understanding of the state of the system by using metrics such as CP-SAM. Various case studies are considered to better understand the usefulness of the proposed metric. Fault trees are developed for each these scenarios to identify the key interdependencies between the various factors for that particular scenario.

The microgrid operator can obtain a better understanding of the state of the system by using metrics such as CP-SAM. Various case studies are considered to better understand the usefulness of the proposed metric. Fault trees are developed for each these scenarios to identify the key interdependencies between the various factors for that particular scenario.

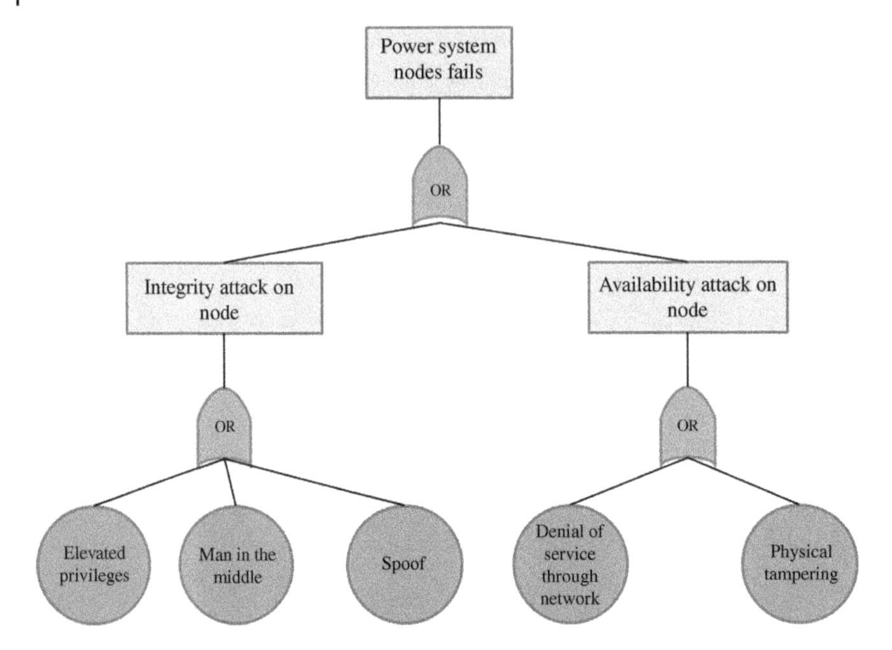

Figure 4.14 Fault tree for a power system node failure.

4.7.3 Loss of Cyber-Physical Node

In this scenario, the loss of a power system component node at node 2 from the system is considered. The loss of a power system might be due to an integrity-based attack or an availability-based attack. The fault tree in Figure 4.14 presents some of the typical cases by which a power system node can be compromised. The red line shows the sequence of events for a node failure. Initially, the system is in normal operation. When the vulnerability is detected, the resiliency score decreases proportionate to the severity of the vulnerability, and its position in the system. Now, when the attacker exploits the vulnerability, initially the latency in communication increases. This step depends on the type of vulnerability exploited and may not be the same for all attacks. The goal of the attacker is to disconnect the microgrid from the main grid, which results in the next two steps where the attacker gains elevated privileges on the node connecting the microgrid to the main grid and proceeds to disconnect the microgrid. This triggers the reconfiguration algorithm, and the microgrid is reconfigured to the most resilient state according to the predecided control action, which results in the vulnerability being isolated and the microgrid being reconfigured to the next most resilient state.

The fault tree used to decide on the pertinent factors is not comprehensive, but only includes a small subset of possible attacks on the power system node. However, the fault tree allows us to identify which factors affect the resiliency of the microgrid for this case. In this case, when the power system node fails, the reconfiguration algorithm reconfigures to the next most resilient configuration.

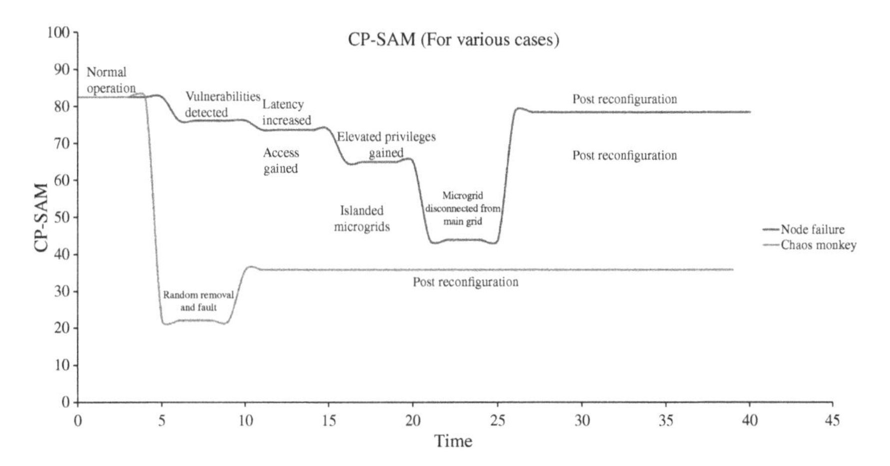

Figure 4.15 CP-SAM for various cases.

4.7.4 Chaos Monkey – Random Removal of Communication Links

In this case, the concept of Chaos Monkey is utilized. Chaos Monkey is a tool used by Netflix to test its resilience to failures. The service identifies a group of systems, and randomly terminates a system or a group of systems to test the resilience. A similar concept can be applied to microgrid resiliency. The communication system is essential to understand the state of the system and perform control actions. Hence, to test the communication resilience, a random subset of nodes and their associated communication links are removed, and the system performance is studied. The results for this study is presented in Figure 4.15.

In this case, when the communication system is compromised, the critical loads still get power because the electrical network is not damaged. To illustrate the severity of this case, a combined cyber-physical attack is considered where the system operator loses control over all the normally closed switches. In this case, the system operator has only one feasible path left for reconfiguration where the inter-tie switch between the microgrids is closed, connecting the two microgrids. This results in the lowest resilient configuration, and any further attacks will result in the critical load losing power. Other random failures will results in scenarios where the critical load loses power. This tool helps the system planning engineer design more resilient systems considering random failures and analyze the trade-offs between resiliency and cost.

4.8 Summary

In this chapter, we have explored the concept of cyber-physical resiliency for microgrids. We studied the need for defining the resiliency and explored various contributing factors toward resilience. Two tools – CyPhyR and CP-SAM – are introduced which can be used to determine the resiliency of the microgrid. CyPhyR focuses on the quantification of the impact of vulnerabilities, and CP-SAM provides a holistic microgrid resiliency score. Various scenarios with different cyber-attacks are presented, and the resulting microgrid resiliency is discussed.

References

1. Department of Energy (2016). Energy Department Announces $8 Million to Improve Resiliency of the Grid. http://energy.gov/articles/energy-department-announces-8-million-improve-resiliency-grid (accessed 15 July 2016).
2. Whitehead, D.E., Owens, K., Gammel, D., and Smith, J. (2017). Ukraine cyber-induced power outage: analysis and practical mitigation strategies. *2017 70th Annual Conference for Protective Relay Engineers (CPRE).* IEEE, pp. 1–8.
3. Sandia National Laboratories (2016). Energy Surety Microgrid (ESM). http://energy.sandia.gov/energy/ssrei/gridmod/integrated-research-and-development/esdm/ (accessed 15 July 2016).
4. North American Electric Reliability Corporation (2016). Severe Impact Resilience: Considerations and Recommendations. http://goo.gl/iTQVFB (accessed 19 October 2022).
5. Stamp, J., McIntyre, A., and Ricardson, B. (2009). Reliability impacts from cyber attack on electric power systems. *2009 Power Systems Conference and Exposition. PSCE '09. IEEE/PES*, March 2009.
6. Venkataramanan, V., Srivastava, A., and Hahn, A. (2019). CyPhyR: A cyber-physical analysis tool for measuring and enabling resiliency in microgrids. *IET Cyber-Physical Systems: Theory & Applications* 4 (4): 313–321.
7. Venkataramanan, V., Hahn, A., and Srivastava, A. (2019). CP-SAM: Cyber-physical security assessment metric for monitoring microgrid resiliency. *IEEE Transactions on Smart Grid* 11 (2): 1055–1065.
8. Venkataramanan, V., Sarker, P.S., Sajan, K.S. et al. (2020). A real-time federated cyber-transmission-distribution testbed architecture for the resiliency analysis. *IEEE Transactions on Industry Applications* 56 (6): 7121–7131.
9. Venkataramanan, V., Srivastava, A., and Hahn, A. (2016). Real-time co-simulation testbed for microgrid cyber-physical analysis. *2016 Workshop on Modeling and Simulation of Cyber-Physical Energy Systems (MSCPES).*
10. Mell, P., Scarfone, K., and Romanosky, S. (2006). Common vulnerability scoring system. *IEEE Security and Privacy* 4 (6): 85–89.
11. Schiffman, M. (2011). Common Vulnerability Scoring System (CVSS). http://www.first.org/cvss/cvss-guide.html (accessed 10 October 2022).
12. NVD (2016). CVE and CCE Statistics. https://web.nvd.nist.gov/view/vuln/statistics (accessed 10 October 2017).
13. Brown, R.E. (2008). Impact of smart grid on distribution system design. *2008 IEEE Power and Energy Society General Meeting-Conversion and Delivery of Electrical Energy in the 21st Century.* IEEE, pp. 1–4.
14. Singh, C. and Patton, A.D. (1980). Protection system reliability modeling: unreadiness probability and mean duration of undetected faults. *IEEE Transactions on Reliability* R-29 (4): 339–340.

15 Gungor, V., Sahin, D., Kocak, T. et al. (2011). Smart grid technologies: communication technologies and standards. *IEEE Transactions on Industrial Informatics* 7 (4): 529–539.

16 Li, H. and Zhang, W. (2010). QoS routing in smart grid. *2010 IEEE Global Telecommunications Conference (GLOBECOM 2010)*, December 2010.

17 Golden, B.L., Wasil, E.A., and Harker, P.T. (2003). *Analytic Hierarchy Process*, vol. 113. Springer.

18 Saaty, T.L. (2006). *The Analytic Network Process*. Springer.

5

Resilience Indicators for Electric Power Distribution Systems

Julia Phillips[1] *and Frédéric Petit*[2]

[1] *The Perduco Group, Beavercreek, OH, USA*
[2] *European Commission, Ispra, Lombardy, Italy*

5.1 Introduction

The United States faces the ongoing challenge of protecting its national critical infrastructure from significant damage caused by extreme weather events. As these natural hazards continue to increase in both frequency and intensity, the efforts of owners and operators to enhance the resilience of their systems and assets to extreme weather are more crucial than ever. Resilience of the electric grid has gained attention since the implementation of the Energy Independence and Security Act of 2007. Recent attention on the impacts of climate change, including The President's Climate Action Plan, released in June 2013, and the two installments of the Quadrennial Energy Review have underscored the importance of modernizing the electric grid.

In response to the need to modernize the grid and following the recommendations made in the two installments of the Quadrennial Energy Review [1, 2], Department of Energy (DOE) developed the DOE Grid Modernization Laboratory Consortium (GMLC) with the objective to help shape the future of the electric grid and ensure its protection and resilience [3]. Among all initiatives conducted by the GMLC, one project addresses the development and application of analysis metrics (i.e. reliability, resilience, sustainability, flexibility, affordability, and security) for assessing the evolving state of the U.S. electricity system and monitoring progress in modernizing the system [4]. Performance-based and attribute-based methods are the two main approaches used for developing resilience metrics [4].

Performance-based methods are generally used for cost-benefit analysis and planning, while multi-attribute evaluation methodology provides a basic understanding of system resiliency. This methodology leverages previous work conducted for developing the Department of Homeland Security (DHS) Infrastructure Protection (IP) Resilience Measurement Index (RMI) and a survey tool developed for DOE Office of Energy (OE) to collect information for characterizing the resilience posture of the electric distribution infrastructure.

A comprehensive understanding of the grid and of the components required to deliver electric power successfully to end users is necessary to characterize the resilience of the

Resiliency of Power Distribution Systems, First Edition.
Edited by Anurag K. Srivastava, Chen-Ching Liu, and Sayonsom Chanda.

electric power distribution systems. In alignment with the four major resilience domains, defined in the RMI, collected information will be used to assess the distribution system's ability to prepare for, mitigate, respond to, and recover from a disturbance.

Resilience is a function of both physical reliability factors as well as intangible resilience indicators. Commonly considered reliability factors include system voltage, feeder length, exposure to natural elements (i.e. overhead or underground conductor routing), sectionalizing capability, redundancy, dependencies and interdependencies, conductor type/age, and number of customers on each feeder. Intangible resilience indicators that complement standard reliability factors include planning (e.g. business continuity, emergency management, procurement management, and preventive maintenance); training and exercising of plans focused on responding to disruptions to normal operations; relationships with local emergency responders; and the ability of the facilities that depend on the electric distribution system to perform their core mission. Variability of electric distribution system characteristics related to geography, the number of customers served, and the type of extreme weather events that are encountered also must be considered in the overall assessment of resilience.

This chapter provides an overview of the multi-attribute methodology used to estimate resilience and to provide resilience comparisons for critical infrastructure sectors and subsectors. Section 5.1 explains the motivations supporting the development of resilience indicators. Section 5.2 presents the decision analysis methodologies used for developing resilience indicators. Section 5.3 explains how these approaches are applied to electric power distribution systems. Finally, Section 5.4 identifies future developments.

5.2 Motivations for Resilience Indicators

The need to understand and enhance the protection and resilience of the U.S. critical infrastructure has been a national focus since the President's Commission on Critical IP was established in 1996. In 2011 and 2013, the release of *Presidential Policy Directive (PPD)-8 on National Preparedness* and *PPD-21 on Critical Infrastructure Security and Resilience* expanded on the importance of understanding the resilience of citizens, communities, and critical infrastructure.

PPD-8 aimed at strengthening the security and resilience of the United States by directing the development of a national preparedness goal that identifies the core capabilities necessary for preparedness and a national preparedness system to guide activities that will enable the resilience enhancement of the nation [5]. PPD-21 is more specific to critical infrastructure because it establishes the roles and responsibilities of the Secretary of DHS to strengthen the security and resilience of all 16 critical infrastructure sectors. PPD-21 specifically calls for operational and strategic analysis to inform planning and operational decisions regarding critical infrastructure and to recommend resilience enhancement measures [6]. The 2013 edition of the National IP Plan follows PPD-21 and reinforces the need to "enhance critical infrastructure resilience by minimizing the adverse consequences of incidents through advance planning and mitigation efforts, and employing effective responses to save lives and ensure the rapid recovery of essential services" [7].

These strategy documents are completed by government agencies' strategic plans that operationalize the consideration of both security and resilience measures in risk management approaches. DHS has developed several of these plans. The DHS Office of IP *Strategic Plan 2012–2016* establishes the goals for the DHS IP to improve risk management activities and enhance resilience through better understanding of critical assets, systems, and networks' operations [8].

In 2015, the DHS developed the *National Critical Infrastructure Security and Resilience Research and Development Plan* (CSIR) to guide prioritization of research and development efforts within DHS. Among the tenets articulated for the CISR, is that "[m]etrics, standard methods of assessment, and baselines must continue to be developed and refined to effectively measure resilience" [9].

DHS has developed specific programs, such as the Regional Resiliency Assessment Program (RRAP), to address the objectives of its strategic plans. RRAP is an interagency and cooperative assessment of specific critical infrastructure within a designated geographic area [10]. RRAP projects are initiated to respond to stakeholders' requirements. Some RRAPs specifically analyze infrastructure systems and propose resilience enhancement options and operational alternatives to reinforce the robustness of critical infrastructure.

In addition to national-level government directives, plans, and policies, numerous private sector industry standards require the consideration of resilience measures in their risk management and business continuity practices. Some of these standards constitute the foundation of the DHS PS-Prep™ program, which provides a systemic approach to business continuity and recovery that allows organizations to unify their preparedness activities [11].

DHS is not the only U.S. agency that has developed plans and programs to specifically address resilience measures and enhancement options in all-hazards risk management processes. In particular, resilience of the electric grid has also gained attention several years since the Energy Independence and Security Act of 2007. In 2009, the Office of the President extended Congress' efforts through the American Recovery and Reinvestment Act (ARRA) of 2009. The ARRA allocated US$4.5 billion to DOE for investments in electric delivery and energy reliability in support of grid modernization. To facilitate these ventures, in June 2011, the Office of the President, National Science and Technology Council, released a policy framework focusing on making cost-effective investments, encouraging innovation, educating and enabling consumers to make smart decisions, and securing the grid from attacks.

The President's Climate Action Plan, released in June 2013, underscores the importance of modernizing the electric grid. Grid modernization not only emphasizes the integration of clean fuel for energy production, but it also highlights the need for increased reliability of electric power delivery and cost savings for consumers. DOE is also active in this area. In the two installments of the Quadrennial Energy Review, DOE specifically recognizes the need to modernize the grid and to ensure the electricity system reliability, security, and resilience [1, 2].

The electric grid is a highly complex, engineered, and interconnected network that connects the production and delivery of power to customers. Electricity enables and supports all critical infrastructure sectors, and our society's dependence on electricity only continues to increase. States and utility companies are at the front lines of assuring this system resilience [12]. In general, the primary goals for utilities are to protect the system, reduce the impact of damage sustained, reduce the area affected by damage, and improve restoration time [12]. Despite all the work underway to ensure the resilience of the grid, no coordinated industry or government initiative is in place to develop a consensus on or to implement standardized resilience metrics [2].

In response to the need to modernize the grid and following the recommendations made in the two installments of the Quadrennial Energy Review to develop resilience metrics [1, 2], DOE developed the DOE GMLC with the objective to help shape the future of the electric grid and ensure its protection and resilience [3]. Among all initiatives conducted by the GMLC, one project addresses the development and application of analysis metrics (i.e. reliability, resilience, sustainability, flexibility, affordability, and security) for assessing the evolving state of the U.S. electricity system and monitoring progress in modernizing the system [4]. The resilience metrics combine two approaches: performance-based and attribute-based [4].

Performance-based methods are generally quantitative methods that are used to interpret quantitative data that describe infrastructure outputs in the event of specified disruptions and formulate metrics of infrastructure resilience [4]. Attribute-based methods typically include categories of system properties that are generally accepted as being beneficial to resilience. Application of these methods typically requires that analysts follow a process to review their system and determine the degree to which the properties are present within the system [4].

5.3 Decision Analysis Methodologies for Resilience Indicators

Risk indicators are critically important in assisting decision-makers in evaluating their organizations' and facilities' readiness to handle crises. Although studies examining specific risks, such as economic or financial risk, have been conducted [13], studies assessing the risks related to an organization in its entirety are lacking.

Some work has been conducted to assess risk indicators for nuclear power plants, refineries, and chemical plants [14, 15]. However, these studies only propose specific indicators to assess a particular operational component, such as safety in offshore operations. Although these indicators are useful, it is difficult to use them in other types of facilities or for different operations. In addition, many of these indicators are often constructed using probabilistic risk assessment [14], which is extremely data-intensive and difficult to employ in information-poor environments.

In addition to these measures and methods, the use of decision analysis supports the development of indicators that address the performance of a system in terms of resilience to potential hazards.

5.3.1 Decision Analysis

Decision analysis is a systematic and logical set of procedures for analyzing complex, multiple-objective (multi-criteria) decision problems. It utilizes a "divide and conquer" philosophy in which hard-to-define, high-level objectives are successively divided into lower-level objectives that are more easily understood, defined, and evaluated. Decision analysis develops meaningful and useful measurement scales for objectives, examines trade-offs among conflicting objectives, and incorporates uncertainty and risk attitudes as appropriate.

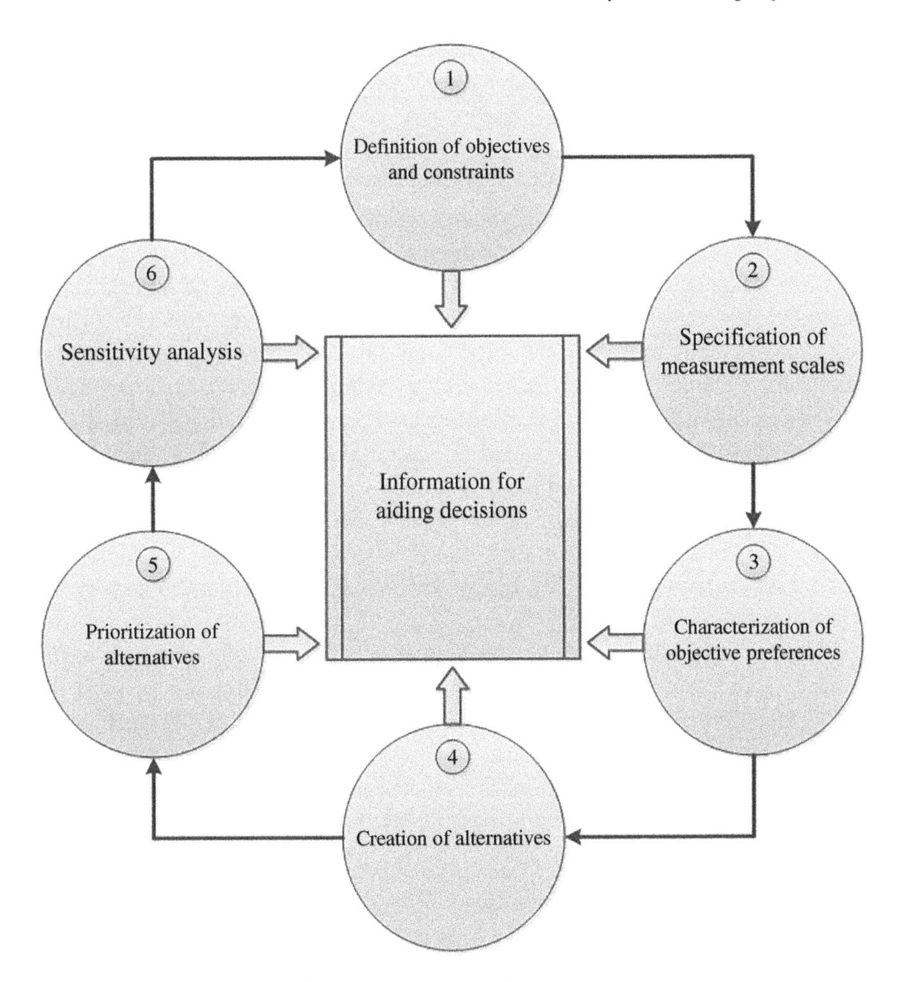

Figure 5.1 Development of system resilience index.

Among all decision analysis methods, value-focused thinking follows the axioms of multi-attribute utility theory, in which a utility function can be constructed for a set of attributes representing the preferences of an individual [16]. The value-focused technique [17] uses this concept to evaluate criteria considered when making decisions. An elicitation is conducted with subject matter experts (SMEs) to determine the relative importance of each criterion to the overall decision.

The decision analytic approach, used for developing the system resilience index (RI), consists of six iterative phases (see Figure 5.1).

5.3.1.1 Definition of Objectives and Constraints

Development of a good resilience indicator requires an established and agreed-upon set of objectives or goals. These objectives include reducing consequences and increasing preparedness, mitigation, response, and recovery mechanisms. Additional objectives can address reducing cost and time to implement the resilience mechanisms identified.

5.3.1.2 Specification of Measurement Scales

After the identification of objectives, the next step is to develop measures that identify the degree to which each objective can be achieved. A good measure not only spans the plausible range of performance by potential alternatives, but it also is operational (i.e. it discriminates among alternatives) and understandable.

Decision analytic approaches typically use three types of scales to capture data: natural, constructed, and proxy. Dollars are a direct measurement of cost and are thus a natural scale for any economic cost objective. Constructed scales are often created when no natural scale exists for the specific objective. For example, a five-point scale that rates comfort of care is a constructed scale. Finally, a proxy scale is used in place of a direct measurement scale when the latter is too difficult or costly to capture outright. For example, if an assessment tool aimed to capture the overall human health impact of air pollution, it might measure pounds of air emissions rather than the actual effects on human health (i.e. measuring pounds of emissions is used as a proxy for health effects).

As with objectives, it is important to ensure that the scale selected captures the range of values important to the decision-maker, that each level in the scale is distinct such that no overlap occurs, and that there is little ambiguity in the scale's levels (particularly when using constructed scales).

Finally, it is important to determine the *value* associated with (and distinct from) each level of the scale. Scales are not all linear, and value-focused thinking can be used to determine the shape of the curve that describes the value (to the individual) at each level. For example, it is generally recognized that the value of $2 million to an individual is less than twice the value of $1 million. Furthermore, the (negative) value of losing $1000 is much greater than the (positive) value of gaining $1000 [18].

5.3.1.3 Characterization of Objective Preferences

After the objectives and associated scales have been agreed upon, it is necessary to determine the relative importance of each objective to the decision-maker via an elicitation. Relative importance can be assessed by having decision-makers "swing-weight" each objective. Swing-weighting asks decision-makers to attribute an importance weight to each objective based on the importance of moving from that objective's worst performance to its best. These weights can be input into a utility function to help determine a "best-fit" alternative; more detail on this concept is provided below.

5.3.1.4 Creation of Alternatives

The next step consists of identifying or constructing alternatives. Each desired characteristic is represented by an objective that should be achieved to the greatest feasible extent, subject to factors such as time and cost, among others. The performance of each facility with respect to each of the objectives must be determined (i.e. characterized). Once this is accomplished, different combinations of specific resilience measures define a possible alternative.

Given the preferences within and across the objectives assessed in the previous step (see Section 2.1.3), it is possible to create alternatives that will integrate stakeholders' objectives and requirements.

5.3.1.5 Prioritization of Alternatives

After alternatives are characterized, they must be prioritized, typically using either holistic ranking or applying a utility function using swing-weights. Although both prioritization processes may appear to yield a final recommendation, often these recommendations differ; thus, they should only be used as information to aid decision-makers. As always, a review should be conducted to ensure that each recommendation makes sense and that no factors are missing from the analysis.

5.3.1.6 Sensitivity Analysis

The last step is the sensitivity analysis, which enables analysts to assess the importance of uncertainties in the criteria value judgments and scale measures.

Infrastructure facilities are growing increasingly interconnected, both with each other and with their environments. To ensure their continued performance and support the well-being of society as a whole, it is important to understand both facility- and system-level resilience. Using decision analytics aids in analyzing existing measures at facility and system levels and identifying ways to improve resilience. The indicators allow for comparison of like assets (e.g. substations or urban electricity distribution system) by providing managers a report on both the strengths and weaknesses of their resilience posture. The ultimate objective is to provide insightful information to help decision-makers make better-informed management decisions.

Argonne National Laboratory, in collaboration with DHS IP used this six-step decision analysis process to develop the Infrastructure Survey Tool (IST) and two related indices, the protective measures index (PMI) and the RMI, which, respectively, characterize the level of protection and resilience of critical infrastructure facilities.

5.3.2 The Resilience Measurement Index

The RMI is a descendent of an earlier index called the resilience index (RI). Both indices support decision-making in risk management, disaster response, and business continuity. Argonne National Laboratory developed the RI in 2010 using a comprehensive methodology of consistent and uniform data collection and analysis. This index was built using the National Infrastructure Advisory Council (NIAC) definition of critical infrastructure resilience: resilience is the "ability to reduce the magnitude and/or duration of disruptive events" [19]. The effectiveness of a resilient infrastructure or enterprise depends on its "ability to anticipate, absorb, adapt to, and rapidly recover from a potentially disruptive event, whether naturally occurring or human caused" [19].

The RI characterized the resilience of critical infrastructure in terms of robustness, resourcefulness, and recovery [20, 21]. The main benefit of the RI was to give the critical infrastructure owners/operators a performance indicator of the resilience of their facilities that could support their decisions in risk and resilience management. In early 2012, a review of the index methodology resulted in enhancements to the structure of the RI and the information collected in order to develop a more comprehensive and informative index – the RMI.

The first step in revising the RI was a literature search to determine how to incorporate additional information and provide a better indicator of infrastructure resilience. This work

Table 5.1 Major resilience components.

Preparedness	Mitigation	Response	Recovery
Anticipate	Resist Absorb	Respond Adapt	Recover
Activities taken by an entity to define the hazard environment to which it is subject	Activities taken prior to an event to reduce the severity or consequences of a hazard	Immediate and ongoing activities, tasks, programs, and systems that have been undertaken or developed to manage the adverse effects of an event	Activities and programs designed to be effective in returning conditions to a level that is acceptable to the entity

was finished in 2012 and led to the publication of a report titled "Resilience: theory and Applications" [22]. This document outlined the definition of resilience and defined the resilience pillars used for developing the RMI:

resilience is *"the ability of an entity – e.g. asset, organization, community, region – to anticipate, resist, absorb, respond to, adapt to, and recover from a disturbance"* [22].

This definition of resilience is broader than the one proposed by NIAC in 2009 by considering not only the capabilities to anticipate, absorb, adapt to, and recover from a disruptive event, but also the notions of resistance and response to the event. The RMI structures the information collected in four attribute categories (preparedness, mitigation measures, response capabilities, and recovery mechanisms) that characterize the resilience capability of a facility [23]. These attribute categories are used as proxy for the six pillars contributing to infrastructure resilience. Table 5.1 illustrates how the four attribute categories constituting the RMI are connected to the six pillars that define resilience.

The RMI combines information gathered with the IST. The questions constituting the IST have been developed on the basis of business continuity and resilience standards and specifically draw from the following:

- NFPA 1600 Standard on Disaster/Emergency Management and Business Continuity Programs [24];
- ANSI/ASIS SPC.1-2009 Standard on Organizational Resilience [25]; and
- International Organization for Standardization (ISO) 22301 Societal Security – Business Continuity Management Systems – Requirements 06-15-2012 [26].

The RMI organizes the information collected with the IST into six levels in order of increasing specificity; raw data are gathered at levels 6 and 5. They are then combined further through levels 4, 3, 2, and finally to level 1. Each of the level 1 components is defined by the aggregation of level 2 components that allow analysts to characterize a facility. The RMI is constituted from 4 level 1 components, 10 level 2 components, and 29 level 3 components, as defined by SMEs. Table 5.2 shows the level 1 and 2 components.

The RMI is defined by the aggregation of its six levels of information. For each component, an index corresponding to the weighted sum of its components is calculated. This process

Table 5.2 Major level 1 and level 2 components constituting the RMI[a].

Preparedness	Mitigation measures	Response capabilities	Recovery mechanisms
• Awareness (2) • Planning (4)	• Mitigating construction (4) • Alternate site • Resource mitigation measures (8)	• On-site capabilities (2) • Off-site capabilities (3) • Incident management and command center characteristics (2)	• Restoration agreements (2) • Recovery time (2)

a) The number of subcomponents is shown in parentheses.

results in an overall RMI that ranges from 0 (low resilience) to 100 (high resilience) for the critical infrastructure analyzed, as well as an index value for each level 1 through 5 component. This method for characterizing the resilience of a critical infrastructure makes it possible to consider the specificity of all critical infrastructure subsectors and to compare the efficiency of different measures to enhance resilience in the studied system.

The value of the RMI is 0 if the facility does not have any of the elements that contribute to the index and 100 if the facility has implemented the best option for all the elements contributing to the RMI. The RMI is an indicator of the degree to which a given facility has implemented the important elements contributing to resilience (e.g. business continuity plan, backups and alternatives, and mitigation measures). A value of 0 does not mean that the facility has no resilient features or that every type of threat will lead to its immediate shutdown. For instance, other elements, such as the capabilities of the emergency services sector that will affect the ultimate consequences to the facility, may not be captured in the RMI calculation. On the other hand, an RMI of 100 does not mean that the facility can resist, respond to, and recover from all types of events. Thus, an RMI of 50 can be interpreted as meaning that the resilience value or worth of elements present at the facility contributes resilience features that, in total, amount to half of the maximum RMI. However, a value of 50 does not mean that 50% of the elements considered in the RMI calculation are in place at the facility. Indeed, an RMI of 50 can be obtained in different ways by combining various components of resilience. If the value of the RMI increases, the resilience capabilities of the facility in terms of preparedness, mitigation, response, and recovery are improved.

DHS IP started using the RMI in 2013. At the end of 2017, more than 350 facilities were assessed with this decision analytics method only for the electric subsector facilities. Since 2015, a version of the RMI has been adapted to be used by Public Safety Canada in the Critical Infrastructure Resilience Tool [27], which is one of the main tools supporting the Canadian Regional Resilience Assessment Program [28]. Modified versions of the RMI are also currently used in several research projects, such as IMPROVER [29] and SmartResilience [30], funded through the European Commission H2020 program [31].

Even if indicators, the RMI and derivative indices, already exist to characterize the resilience of critical infrastructures, they are used at the facility level and are not totally adapted to characterize a system combining several facilities with various characteristics. However, the decision analytics method and the work conducted for developing the RMI can be leveraged to construct an index that characterizes the resilience of the electric grid.

5.4 An Application to Electric Power Distribution Systems

Similar to the RMI, the conceptual components of resilience are captured using data collection questions covering the four major domains of resilience. Table 5.3 provides a framework for data collection under the resilience domains: preparedness, mitigation, response, and recovery. The subcomponents listed include the types of topics that would fall under each of the top-level components.

As shown in Table 5.3, the index characterizing the resilience of the electric power distribution system has several similarities when compared with the RMI. The four main attribute categories (i.e. preparedness, mitigation measures, response capabilities, and recovery mechanisms) remain the same. The first noticeable differences are the types and number of the second level of elements constituting this system-level resilience index. For example, the planning section also includes procurement management and prevention of maintenance planning, which are not part of the RMI. Two other major differences are that some elements characterize the system as a whole (e.g. awareness and planning sections), while others are specifics to the assets constituting the system. For example, the system capabilities section characterizes the control centers, distribution voltage subsystem, and subtransmission voltage subsystem. The voltage subsystems are characterized in more detail by differentiating between substations and lines. Another major difference is the way in which both indices take into account infrastructure dependencies. The RMI characterizes eight types of upstream dependencies that potentially affect the performance of a specific facility when the system resilience index considers dependencies at both facility and system levels.

For characterizing the resilience of the electric power distribution system, Argonne National Laboratory used the concepts of multi-attribute utility theory [16] to construct a common metric to inform decisions. For developing the value hierarchies (i.e. identifying the resilience attributes and defining the overall structure of the index), Argonne used a combined standard approach [32]. It consisted of reviewing approved documents (i.e. resilience and business continuity standards, energy policy and regulations) and conducting facilitated discussions with SMEs representing critical infrastructure owners and operators, the DOE, standards authorities, and professional associations.

In addition to the resilience and business continuity standards that supported the development of the RMI, Argonne used several other sources to define the elements that contribute to the resilience of the electric grid distribution system:

Table 5.3 Resilience attributes for the electric power distribution system[a].

Preparedness	Mitigation measures	Response capabilities	Recovery mechanisms
• Awareness (2) • Planning (6)	• Mitigating construction (4) • Utility mitigation (3) • Resource mitigation measures (2)	• System capabilities (3) • External capabilities (4) • Incident management and command center characteristics (3)	• Restoration agreements (3) • Recovery time (2)

a) The number of subcomponents is shown in parentheses.

- North American Electric Reliability Corporation;
- Federal Energy Regulatory Commission;
- DOE, Office of Electricity Delivery and Energy Reliability;
- Quadrennial Energy Review;
- National Association of State Energy Officials;
- National Association of Regulatory Utility Commissioners;
- Electric Power Research Institute; and
- Edison Electric Institute.

The document review was completed by electric utility industry inputs through facilitated discussions.

Definition of objectives and constraints and specification of measurement scales are crucial first steps to define the index components and their measurement. Experts can assist in the development of these attributes, given the ultimate objective to be achieved. Several properties are desired for these attributes, driven by the underlying utility theory foundation, as described below:

1. Complete: all *significant* attributes necessary to meet the decision objectives are captured.
2. Practical: the necessary information can be obtained for all attributes.
3. Decomposable: difficult-to-use attributes can be broken down into more understandable components.
4. Non-redundant: double counting is avoided; the attributes should be as independent from one another as possible.
5. Minimal: selected attributes provide the decision-maker with sufficient information, yet not so much information that the decision-maker finds it is cost- or time-prohibitive to implement the collection.

Meeting the desired properties for the decision attributes leads to transparent, defensible, and repeatable results. The value-focused thinking methodology creates a strong basis for each alternative considered and is generalized so that it may be applied to alternatives with similar sets of attributes.

Within each of the broad categories outlined in Table 5.3, the new survey tool captures information on the utility characteristics (i.e. control and dispatch center, lines, and substations), the potential consequences generated by the disruptions of the system (including system reliability indicators), extreme weather exposure, existing agreements and information sharing processes, resilience planning (i.e. business continuity, emergency operations, cybersecurity, procurement management, preventive maintenance), response capabilities (e.g. deployable capabilities and incident management and command center [IMCC]), and the main dependencies supporting the system operations (e.g. electric power, communications, fuels, chemicals, and transportation). The following sections present attribute grouping defined for characterizing the resilience of the electric grid.

5.4.1 Preparedness

Preparedness refers to activities undertaken by the electric distribution system to anticipate the effects of potential natural hazards (Figure 5.2).

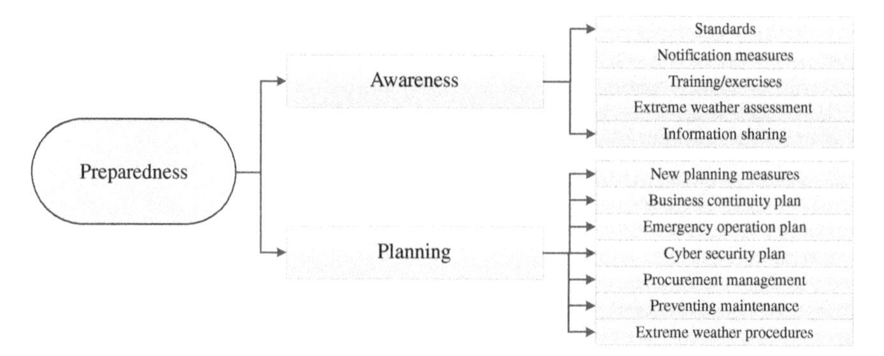

Figure 5.2 Resilience components contributing to preparedness.

The awareness category combines elements characterizing internal and external communications procedures, information mechanisms, and measures designed to anticipate potential natural hazards. The planning category combines information characterizing the types and content of plans (e.g. business continuity, emergency operations/emergency action, cybersecurity, procurement management, preventive maintenance) as well as specific extreme weather procedures implemented for the entire distribution system.

5.4.2 Mitigation Measures

Mitigation measures refer to activities undertaken prior to an event by the electric distribution system to resist potential natural hazards or to absorb the negative effects from these hazards.

The mitigating construction category combines measures taken at distribution system level to offset naturally occurring adverse events by implementing construction to mitigate impacts, specific plans/procedures for long-term and immediate mitigation measures, and deployable mitigation measures. The utility mitigation category combines measures taken at facility level to mitigate the effects of natural hazard events on control centers, substations, and distribution lines. The resource mitigation measures category combines measures taken to mitigate the effects of loss of resource supplies (e.g. energy, telecommunications, and water systems) on core operations at both facility and system levels.

5.4.3 Response Capabilities

The response capabilities category refers to activities undertaken or developed by the electric distribution system to respond and adapt to the adverse effects of an event (Figure 5.3).

The system capabilities category combines elements characterizing the system's capabilities to respond to an event without needing support from external organizations. The external capabilities category combines elements characterizing the interactions and agreements existing among the electric distribution system, emergency responders, and public work organizations. The IMCC characteristics category combines information that captures the electric distribution system's capabilities for managing response, continuity, and recovery operations if an incident occurs.

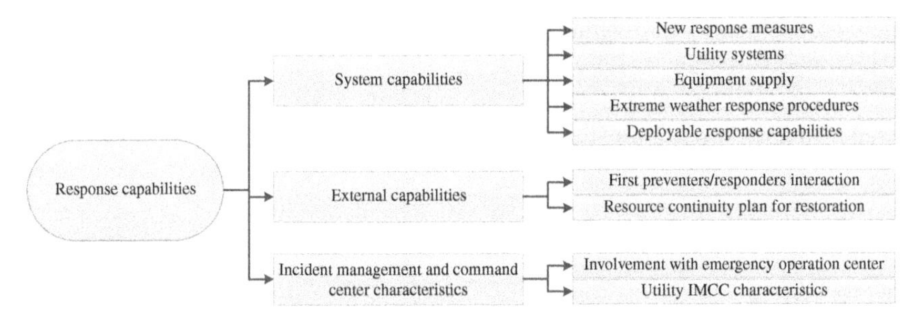

Figure 5.3 Resilience components contributing to responses capabilities.

5.4.4 Recovery Mechanisms

The recovery mechanisms section refers to activities undertaken by the electric distribution system to resume the system's operations to an acceptable level of performance (Figure 5.4).

The restoration agreements category combines information relative to existing agreements with other critical infrastructure systems, as well as procedures/equipment that will support restoration procedures. The recovery time category combines information characterizing the time necessary for the system to resume full operations after the loss of one of its significant facility assets.

Value-focused thinking [17], a decision analytic technique, will be used to determine the contribution of each component captured in the survey tool to the overall distribution system resilience. Through a formal elicitation process, various SMEs will place relative importance values on all of the individual components contributing to resilience. These relative value judgments will then be used to calculate a system resilience performance metric, or indicator, similar to the RMI. The resulting indicator will aid critical decision-making by identifying gaps in resilience measures and investment strategies for key stakeholders.

Once the methodology is completely established and the survey tool is fully created, the tool can be made available to all electric distribution owners and operators to use on a voluntary basis. The organizational framework outlined in Table 5.3 will act as a guide to the private entity owners and operators for identifying and understanding the applicable components of their system-specific resilience. Defining the characteristics and components contributing to the resilience of the system will enable stakeholders to see and/or propose possible options for enhancing the resilience of the system and lowering the consequences potentially generated by system disruption.

Similar to the RMI, the system index is a relative measure that varies from 0 to 100. Interpretation and implication of the index are important for decision-makers to understand. A low overall value does not mean that every type of hazard will lead to an immediate

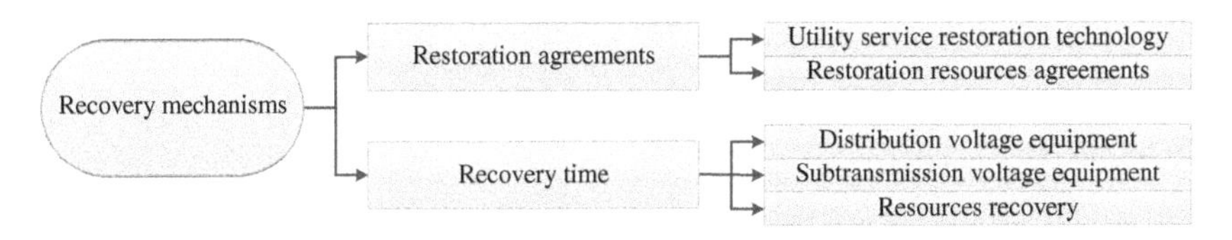

Figure 5.4 Resilience components contributing to recovery mechanisms.

system shutdown. A high overall value does not mean that a specific event will have minimal consequences. Simply stated, the index allows comparison of different levels of resilience. The scaling of the index is such that improvement from 20 to 40 is equivalent to improvement from 60 to 80. Determining the system resilience index and how different options affect this index can be used to establish the most effective ways to improve a system's overall resilience.

5.5 Future Work

The first phase of this project focused primarily on the two first steps of the decision analytics approach: defining the objectives and constraints and specifying the attributes and scales of the future electricity distribution system resilience index. This first phase contributed to the development of a system resilience survey tool that allows one to gather all information and elements that characterize the resilience of the electricity distribution system.

The next step consists of working on the third step of the decision analytic approach by characterizing the objective preferences. This second phase consists of elicitation sessions with SMEs to define the relative importance value to each attribute, component, and subcomponent defined during the two first steps of the decision analytics process. The process is threat agnostic and consists of defining, for each category, which element would be implemented first to enhance the resilience of the electric grid distribution. Once the SMEs define the ranks and relative values for a given set of information, a period of discussion allows them to exchange and explain the elements that guided their thinking. On the basis of this discussion, the SMEs can revise the ranks and relative values and even provide comments on modifying the questions, wording, answer types, and the categorization of attributes and components. If any aspect of a question or category changes, the relevant rank and relative importance assignments are revisited and updated if needed. At the end of this third step, the relative importance of each index component will be integrated in a calculation tool to create an overall resilience index score and individual scores for the four main resilience attributes.

After completion of the third step, use cases are then conducted to perform the last three steps of the decision analytics approach to validate and refine the resilience metrics. These three last steps help to refine the data collection process to ensure the uniformity and reproducibility of the data collected by developing explanations, training, and quality assurance review. Visualization capabilities can be developed to perform sensitivity analyzes on potential grid modernization options. An interactive dashboard can be used to compare existing characteristics of the grid and future developments, to see if the grid resilience has improved. The dashboard allows the user to see – in real time – the impacts of component modifications on the overall index value as well as on the specifically selected and modified components.

5.6 Conclusion

As extreme weather events continue to gain recognition as a worsening problem, and as infrastructure resilience rises to the forefront of discussions, the need to understand impacts and reactions of natural hazards to key national lifeline sectors is critical. In response to recent presidential directives and initiatives, the DOE is actively pursuing greater understanding of the resilience of the national electric grid and identifying opportunities for government support in enhancing national resilience. Assessing the resilience related to system operations is not easy, and defining specific indicators that characterize resilience is a challenge. Decision analytic principles can help to promote a proactive approach to improve preparation to disruptive events. The electric distribution resilience assessment methodology and survey tool discussed here are a step toward the acknowledgment of vital support for grid resilience. With the help of information on best practices and needs from the private sector, improvements can be made today to ensure the operation and resilience of the grid for tomorrow. Ultimately, decision-makers must decide what is acceptable in terms of resilience, but the decision analytics method should help inform and assist decision-makers through the application of a method that provides a consistent and reproducible approach.

Acknowledgments

The authors gratefully acknowledge the contributions of many people who helped bring this project to its current state of development, including the U.S. Department of Energy Grid Modernization Laboratory Consortium and the U.S. Department of Energy Office of Electricity Delivery and Energy Reliability. The authors are also particularly thankful to the Argonne National Laboratory team who has been involved in the development of the resilience indices presented in this chapter.

References

1 DOE (2015). Quadrennial energy review: energy transmission, storage, and distribution infrastructure. DOE Office of Policy, Washington, DC. https://www.energy.gov/sites/prod/files/2015/07/f24/QER%20Full%20Report_TS%26D%20April%202015_0.pdf (accessed 10 October 2022).
2 DOE (2017). Quadrennial energy review – transforming the nation's electricity system: the second installment of the QER. DOE Office of Policy, Washington, DC. https://www.energy.gov/sites/prod/files/2017/02/f34/Quadrennial%20Energy%20Review--Second%20Installment%20%28Full%20Report%29.pdf (accessed 10 October 2022).
3 DOE (2018). Grid modernization lab consortium. https://energy.gov/under-secretary-science-and-energy/grid-modernization-lab-consortium (accessed 10 October 2022).
4 Pacific Northwest National Laboratory (2017). Grid modernization metrics analysis (GMLC 1.1), reference document, Version 2.1, PNNL-26541. Grid Modernization Laboratory Consortium. https://gridmod.labworks.org/sites/default/files/resources/GMLC1%201_Reference_Manual_2%201_final_2017_06_01_v4_wPNNLNo_1.pdf (accessed 10 October 2022).

5 DHS (2011). Presidential policy directive/PPD-8: national preparedness. National Protection and Programs Directorate, Washington, DC. http://www.dhs.gov/presidential-policy-directive-8-national-preparedness (accessed 10 October 2022).

6 White House (2013). Presidential policy directive – critical infrastructure security and resilience, PPD-21. Washington, DC. http://www.whitehouse.gov/the-press-office/2013/02/12/presidential-policy-directive-critical-infrastructure-security-and-resil (accessed 10 October 2022).

7 DHS (2013). National infrastructure protection plan 2013: partnering for critical infrastructure security and resilience. Washington, DC. https://www.dhs.gov/sites/default/files/publications/national-infrastructure-protection-plan-2013-508.pdf (accessed 10 October 2022).

8 DHS (2012). Office of infrastructure protection strategic plan: 2012–2016. Washington, DC. http://www.dhs.gov/sites/default/files/publications/IP-Strategic-Plan-FINAL-508.pdf (accessed 10 October 2022).

9 DHS (2015). National critical infrastructure security and resilience research and development plan. Washington, DC. http://www.dhs.gov/sites/default/files/publications/National%20CISR%20R%26D%20Plan_Nov%202015.pdf (accessed 10 October 2022).

10 DHS (2017). Regional resiliency assessment program. Washington, DC. https://www.dhs.gov/regional-resiliency-assessment-program (accessed 10 October 2022).

11 Federal Emergency Management Agency (2017). PS-Prep™ program resources. https://www.fema.gov/program-resources (accessed 10 October 2022).

12 Finster, M., Phillips, J., and Wallace, K. (2016). Front-line resilience perspectives: the electric grid, ANL/GSS-16/2. Argonne National Laboratory, Global Security Sciences Division, Argonne, IL, USA. https://energy.gov/sites/prod/files/2017/01/f34/Front-Line%20Resilience%20Perspectives%20The%20Electric%20Grid.pdf (accessed 10 October 2022).

13 Simpson, D.M. (2008). Disaster preparedness measures: a test case development and application. *Disaster Prevention and Management* 17 (5): 645–661.

14 Charkraborty, S., Flodin, Y., and Grint, G. et al. (2003). Risk-based safety performance indicators for nuclear power plants. In: *Transactions of the 17th International Conference on Structural Mechanics in Reactor Technology (SMiRT 17)*, Prague, Czech Republic, 17–22 August 2003.

15 Oien, K. (2001). A framework for the establishment of organizational risk indicators. *Reliability Engineering and System Safety* 74: 147–167.

16 Keeney, R.L. and Raiffa, H. (1976). *Decisions with Multiple Objectives: Preferences and Value Tradeoffs*. New York: Wiley.

17 Keeney, R.L. (1992). *Value-Focused Thinking: A Path to Creative Decision Making*. Cambridge, MA, USA: Harvard University Press.

18 Kahneman, D. and Tversky, A. (1979). Prospect theory: an analysis of decision under risk. *Econometrica* 47 (2): 263–292.

19 National Infrastructure Advisory Council (2009). Critical infrastructure resilience, final report and recommendations. U.S. Department of Homeland Security, Washington, DC. http://www.dhs.gov/xlibrary/assets/niac/niac_critical_infrastructure_resilience.pdf (accessed 10 October 2022).

20 Fisher, R.E., Bassett, G.W., and Buehring, W.A. et al. (2010). Constructing a resilience index for the enhanced critical infrastructure protection program, ANL/DIS-10-9. Argonne National Laboratory, Decision and Information Sciences Division, Argonne, IL, USA. http://www.ipd.anl.gov/anlpubs/2010/09/67823.pdf (accessed 10 October 2022).

21 Petit, F., Eaton, L., Fisher, R. et al. (2012). Developing an index to assess the resilience of critical infrastructure. *International Journal of Risk Assessment and Management* 16 (1/2/3): 28–47. Inderscience Publishers, Geneva, Switzerland.

22 Carlson, L., Bassett, G., and Buehring, W. et al. (2012). Resilience theory and applications, ANL/DIS-12-1. Argonne National Laboratory, Decision and Information Sciences Division, Argonne, IL, USA. http://www.ipd.anl.gov/anlpubs/2012/02/72218.pdf (accessed 10 October 2022).

23 Petit, F.D., Bassett, G.W., and Black R. et al. (2013). Resilience measurement index: an indicator of critical infrastructure resilience, ANL/DIS-13-01. Argonne National Laboratory, Decision and Information Sciences Division, Argonne, IL, USA. http://www.ipd.anl.gov/anlpubs/2013/07/76797.pdf (accessed 10 October 2022).

24 National Fire Protection Association (2016). Standard on disaster/emergency management and business continuity/continuity of operations programs –2016 edition. Quincy, MA, USA. https://www.nfpa.org/codes-and-standards/all-codes-and-standards/list-of-codes-and-standards/detail?code=1600 (accessed 10 October 2022).

25 ASIS International (2009). Organizational resilience: security, preparedness, and continuity management systems – requirements with guidance for use, ASIS SPC.1-2009. https://www.ndsu.edu/fileadmin/emgt/ASIS_SPC.1-2009_Item_No._1842.pdf (accessed 10 October 2022).

26 International Organization for Standardization (2012). Societal security – business continuity management systems – requirements, ISO 22301:2012. http://www.iso.org/iso/catalogue_detail?csnumber=50038 (accessed 10 October 2022).

27 Parenteau, M.-P., Guziel, K., Petit, F., and Norman, M. (2015). The critical infrastructure resilience tool – a tool to evaluate Canadian critical infrastructure. The CIP Report, Center for Infrastructure Protection and Homeland Security, George Mason University School of Law, Washington, DC, USA, vol. 14, no. 8, pp. 6–10, May 2015. https://cip.gmu.edu/wp-content/uploads/2013/06/154_The-CIP-Report-May-2015_InternationalIssues.pdf (accessed 10 October 2022).

28 Public Safety Canada (2017). The regional resilience assessment program. https://www.publicsafety.gc.ca/cnt/ntnl-scrt/crtcl-nfrstrctr/crtcl-nfrstrtr-rrap-en.aspx (accessed 10 October2022).

29 IMPROVER (2018). The IMPROVER project. http://improverproject.eu/ (accessed 10 October 2022).

30 SmartResilience (2018). SmartResilience project. http://www.smartresilience.eu-vri.eu/ (accessed 10 October 2022).

31 European Commission (2018). Horizon 2020. https://ec.europa.eu/programmes/horizon2020/ (accessed 10 October 2022).

32 Parnell, G.S., Bresnick, T.A., Tani, S.N., and Johnson, E.R. (2013). *Handbook of Decision Analysis*. NJ, USA: Wiley & Sons.

6

Quantitative Model and Metrics for Distribution System Resiliency

Alexis Kwasinski

Department of Electrical and Computer Engineering, University of Pittsburgh, Pittsburgh, PA, USA

6.1 Power Grids Performance in Recent Natural Disasters

In the past years, as societies increasingly rely on electric power for sustaining healthy economic and other social systems, there have been an increasing interest for documenting and understanding the performance of electric power grids during extreme events. Power grids performance during extreme events has attracted significant attention from the public because of the extensive and significant power outages caused by those events and the disruptive effects that such outages have on communities. The last example of extensive and long power outages following a natural disaster was observed in September 2017 when Hurricane Maria left the entire island of Puerto Rico without electric power for several weeks with a significant portion of the island experiencing power outages lasting for months. However, such performance is not an isolated event. In addition to hurricanes, other natural disasters, such as earthquakes, floods or ice storms, can cause long power outages affecting large areas. Other natural disasters, such as tornadoes or wild fires, can also cause long power outages although the affected areas are relatively smaller. Other events, such as geomagnetic storms, have the potential for causing very long power outages over a very large area but they tend to be events with much lower probability of occurrence.

Among all of these natural disasters, tropical cyclones (i.e. hurricanes or typhoons depend on the area where they develop) tend to be the ones that receive the most attention because of the extensive power outages they cause. For example, in 2005, Hurricane Katrina caused almost 3 million power outages (i.e. about 3 million customers lost power), Hurricane Rita about 1.5 million power outages, and Hurricane Wilma about 3.2 million outages. In 2008, Hurricane Gustav caused about 1.3 million power outages and Hurricane Ike almost 3 million power outages in the Gulf Coast and about 1.5 million power outages more in the Ohio River Valley, more than 1000 mi from where it made landfall. In 2011, tropical storm Sandy affected the northeastern United States causing about 6 million outages and the following year Super-storm Sandy affected the same region causing 8.2 million outages following Hurricane Isaac, which two months earlier had left 1 million customers in the Gulf Coast without power. Recently, in 2017, Hurricane Irma

Resiliency of Power Distribution Systems, First Edition.
Edited by Anurag K. Srivastava, Chen-Ching Liu, and Sayonsom Chanda.

caused approximately 4.1 million outages in the southeast United States and, as indicated, Hurricane Maria left all of the about 1.5 million customers in Puerto Rico without power. Except in the case of Maria in which damage to transmission lines also played a critical role in the duration and extension of the power outages, in most natural disasters the power distribution portion of the grid is the part that takes longer to recover. Nevertheless, these effects are not limited to hurricanes affecting the United States. Similar extensive and long power outages are observed with other types of disasters, such as earthquakes, and in other places different from the U.S., such as in Chile in 2010, New Zealand in 2011, and Japan also in 2011, although in Japan severe damage also happened on the power generation portion of the grid.

Many of these outages caused by hurricanes could be attributed to the fact that, except in the case of densely populated areas, such as Manhattan, Houston or Miami, most of the power infrastructure was installed overhead, which is more vulnerable to tropical cyclone strong winds than underground infrastructure. As Figure 6.1 exemplifies, dense vegetation common in areas at risk of tropical cyclones, tend to make overhead infrastructure particularly vulnerable to storms and even when it is possible to mitigate this issue with tree trimming programs, they add a considerable cost to electric utilities, which, typically, have constrained financial resources. Still, during these events, it is possible to observe loss of service in areas with primarily underground power infrastructure. The reason for loss of service occurring in areas with underground power infrastructure is that strong winds are not the only damaging actions of hurricanes. Typically, the most intense hurricane damage is caused by storm surge (a large mass of water carried inland by the storm's strong winds and low pressure) but this damage is limited to a relatively small coastal area. As Figure 6.2 shows based on observations made during field damage assessments, in coastal

Figure 6.1 Power lines damaged by Hurricane Maria in Puerto Rico.

Figure 6.2 A coastal area near High Island, Texas, after Hurricane Ike, with power infrastructure destroyed by the storm surge.

Figure 6.3 High Island, Texas after Hurricane Ike, showing little damage to the power infrastructure.

areas affected by the storm surge it is possible to observe that a significant portion of the power grid infrastructure is damaged. This damage distribution is not uniform. As shown in Figure 6.3, although areas affected by storm surge may be completely devastated, it is possible to find that other coastal areas a few hundred meters away are not as significantly damaged. Nevertheless, as expected, the outage incidence – the peak percentage of power

Figure 6.4 A typical area inland in Monmouth County, New Jersey where more than 75% of the customers lost power after "Superstorm" Sandy.

outages – in areas affected by the storm surge is usually close to 100% [1]. Damage in areas located inland are primarily caused by strong winds and in some limited areas by flood waters due to intense rains. Although in most cases damage to power grid infrastructure in these areas is relatively light (see Figure 6.4), power outage incidence tend to still be considerably high [1]. This observation suggests that electric power grids tend to be relatively fragile systems in which relatively small damage may lead to significant power outages. Such conclusion should not be considered to be an issue originating in power grids general design because when power grids were conceptualized more than a century ago, the primary design concern was the development of an economically affordable power distribution system with high availability in normal conditions.

In general, a statistical analysis of the power outages caused by hurricanes from 2004 to 2013 [1] indicated that storm surge height, wind speeds, storm area, and time under the effects of tropical storm winds are all contributing factors characterizing the outage incidence of hurricanes at a given location. In these events, the most significant issues related to these power outages are usually observed at the power distribution level. Although many electric power distribution utilities operating in areas affected by tropical cyclones typically implement design and maintenance strategies, such as tree trimming programs, to reduce the damaging effects of tropical cyclones, comparison of the effects of hurricanes affecting the same area in different years, such as Louisiana affected by Katrina in 2005, Gustav in 2008 and Isaac in 2012, or Florida affected by Wilma in 2005 and Irma in 2017, show inconclusive improvements yielded by these hardening strategies as power outage incidence in these instances were higher in the later events. A commonly suggested hardening strategy is to install as much power infrastructure possible underground. However, underground infrastructure demands an investment several times higher than overhead infrastructure. Moreover, restoration times of underground infrastructure tends to be longer than for equivalent overhead systems. Thus, as it is explained in Section 6.3 of this chapter, underground infrastructure resilience may not be higher than that of overhead systems because of longer restoration times. Furthermore, such longer restoration times are observed also during normal operating conditions. Thus, often times, quantitative risk analysis planning process may not yield favorable results for underground infrastructure because its main advantage over overhead infrastructures may only be observed when a damaging extreme event happens, which probabilistically it is usually a relatively unlikely event. Additionally, underground cables may, actually, perform worse than overhead lines during earthquakes, particularly in areas with soil liquefaction. An example of this situation was observed in Christchurch, New Zealand, during the series of earthquakes that affected the city in 2010 and 2011 and that caused particularly long outages due to the complexity involved with locating multiple point of failures in each cable and, then, repairing those failures.

Section 6.3 of this chapter indicates that modern interpretations of grid resilience involve how well an electric power system is able to withstand an extreme event, recover quickly in case such event produces outages and also how well a power grid is prepared for an extreme event and adapts to potential disruptive events. While technical factors play a key role in improving all of these resilience aspects, other factors are also important. In particular, recovery speed is also influenced by resources availability, spares management, and logistical factors. For example, although technical solutions, such as the use of portable transformers, such as the one in Figure 6.6 contributes to reduce restoration times, mutual assistance contracts between electric utilities in the continental United States and extensive deployment of restoration crews, such as those in Figure 6.7, have a more significant effect on reducing power outages duration in many past natural disasters. One particularly interesting example is found with Hurricane Maria, when the more isolated location of

Figure 6.5 Portable transformer deployed in Galveston Island after Hurricane Ike.

Figure 6.6 A large number of trucks deployed to restore electric power after Hurricane Irene.

Figure 6.7 A substation with its communications microwave antenna.

Puerto Rico and economic and political factors significantly limited the available resources for service restoration activities and, as a result, caused much longer power outages when compared to those observed during other hurricanes that affected the continental U.S. Evidently, restoration speed depends on the extreme event characteristics. For example, during hurricanes, restoration activities do not usually begin until winds drop below tropical storm wind speed limits. The type of event also influences planning and preparation activities. While the general area that is affected by a hurricane can be anticipated many hours and, in several cases, even days before the effects of the hurricane affects such area, general areas to be affected by an earthquake cannot be anticipated. Hence, while it is possible to effectively pre-position additional resources in anticipation of a hurricane, earthquake responses are primarily reactive and, thus, potentially less effective than response activities after hurricanes.

6.2 Resilience Modeling Framework

The previous discussion about power grids performance during natural disasters suggest that a complete resilience modeling framework for power grids need to include the fundamental aspects that influence their operation during a disaster and in their aftermath, during the repair and restoration period. In most cases, power grids modeling frameworks focus on physical components, such as transformers and poles, and their interconnection based on transmission and distribution lines and substations. However, although electric power grids physical components and their interconnection are a fundamental aspect of any power grid resilience modeling framework, other aspects are also important. One of such aspects is human resources management, which plays a particularly important role during the service restoration phase after a natural disaster. Human resources management includes contracts, policies, procedures, and general electric grid companies personnel activities that could be formally written in official documents produced by electric power utilities or that could be the result of informally developed working culture at such companies. Important contracts and procedures that influence repairs, service restoration, and recovery period in the aftermath of a disaster includes logistical process, spare parts stocking procedures, personnel training, mutual assistance contracts with other electric power utilities and others. Human influence is, of course, not only affecting electric power grids performance after an extreme event happens but it also affect the performance during the event through, for example, design decisions of whether to have an underground or overhead infrastructure or maintenance activities to harden existing infrastructure through tree trimming programs or aging components replacement or maintenance activities. In addition to the indicated contracts, procedures, and policies, the human component of an electric power grid is, in essence, formed by humans (i.e. the electric power companies personnel) and their social interactions.

Another important aspect influencing electric power grids operations is the cyber component influencing power grids operations. This component includes information technology assets, such as computers, sensors, state variables measurements, supervisory control and data acquisition (SCADA) systems, and databases. The cyber components also include electric power system utilities-owned and operated communication networks and assets used for control and state sensing of the grid. A relatively common misconception is to believe that electric power grid operators use public communication network assets to transmit sensing and control signals to and from operation centers. Although this may be presently true in a very limited number of cases and use of public communication network assets may become more common in the future as increasing number of smart grid technologies, such as smart meters, are deployed by all utilities but, particularly, at the power distribution level, the present reality is that in almost all electric power grids communication assets are owned and operated by the companies operating such power grids. That is, as Figure 6.7 exemplifies, those communications assets are part of the electric power grid physical components and the data and information transmitted through these communication assets are part of the cyber component of electric power grids.

A broad resilience framework was introduced in [2]. This framework is applicable not only to infrastructure systems, such as electric power grids, but also to social systems [3]. Hence, from a broad perspective this modeling framework defines community systems as "structures that combine resources in an organized manner in order to provide services needed by communities." Community services are divided in social systems, which are "specific combinations of resources and processes developed to deliver services primarily through human interactions," and infrastructure systems which are defined as "specific combinations of resources and processes developed to deliver services primarily through a physical built environment or a cyber subsystem". In particular, electric power grids are infrastructure systems. Their service is the provision of electric energy. Also [3, 4] indicate that community systems can be considered to be composed of three domains:

(1) physical domain, made out of physical components necessary to deliver the services provided by the system.
(2) human/organizational domain, made out of processes, policies, contracts, procedures, as well as the human system operators and administrators, and their actions, necessary to manage, administrate, and operate the system.
(3) cyber domain, made out of databases, information, communications, and control and operations algorithms.

As Figure 6.8 exemplifies, this three-domain structure is in agreement with the introductory discussion in this section and, thus, applicable to electric power grids.

In [5] the general framework presented in [3] was expanded by detailing that each domain can be modeled mathematically as graphs made of a set of nodes and edges, in which each graph represents the provision of a service. Hence, a general model for a power grid is given by:

$$\mathcal{G}_{\mathrm{PG}}(t) = (N_{\mathrm{P}}(\boldsymbol{\varphi}(t), t) \cup N_{\mathrm{C}}(\boldsymbol{\varphi}(t), t) \cup N_{\mathrm{H}}(\boldsymbol{\varphi}(t), t), E_{\mathrm{P}}(\boldsymbol{\varphi}(t), t) \cup E_{\mathrm{C}}(\boldsymbol{\varphi}(t), t) \cup E_{\mathrm{H}}(\boldsymbol{\varphi}(t), t)) \tag{6.1}$$

where N_{P}, N_{C}, and N_{H} represent the set of nodes in the physical, cyber, and human domain, respectively, and E_{P}, E_{C}, and E_{H} are the set of edges in the physical, cyber, and human

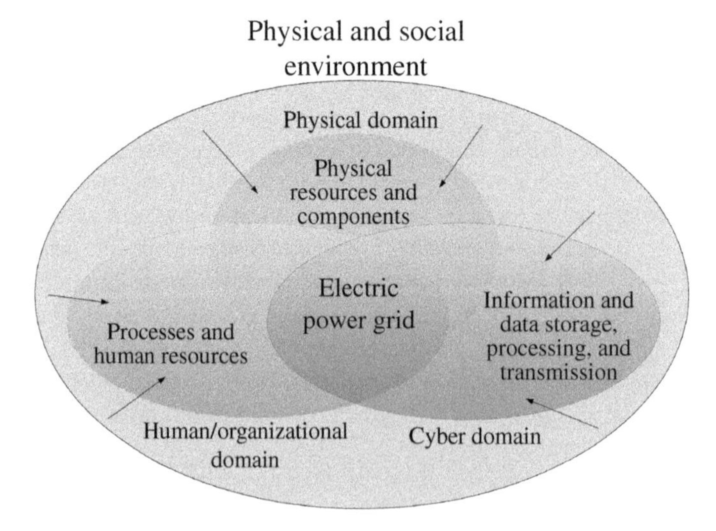

Figure 6.8 Model of an electric power grid based on its three domains.

domain, respectively. Each node and edge is characterized by a set of attributes indicated by $\boldsymbol{\varphi}(t)$ that may include a geographic location (in case a location could be identified) $\mathbf{x}(t)$, their service provision nominal rate, and others, some of them which will be discussed in Section 6.3. As it is indicated, the graph indicated in (6.1) is a function of time because not only the edge and nodes attributes may change over time, but also the graph configuration itself may change over time by adding or removing nodes and edges or by changing their attributes. For example, in the aftermath of a natural disaster, emergency portable transformers, like the one in Figure 6.5 could be deployed at a site of a substation with a previously existing transformer with a different capacity; or a new line, such as the one in Figure 6.9 could be installed to provide emergency service with an area left with no service. Changes often occur due to interactions with the physical and social environment. All domains of a community system interact with the physical and social environment. For example, by having the weather affecting operation of components in the physical domain or by having information affecting decisions people make in the human domain. Although graph changes usually occur during normal conditions as a result of normal power grid's evolution processes, in the case of resilience studies it is of particular interest the changes that occur due to interactions with the environment. Let us denote the environmental conditions at a point $\mathbf{x}$ as $W(\mathbf{x},t)$, thus, defining a vector space $\mathcal{W}$. Then, changes in the environmental conditions and the graph representing a power grid due to a hazard $\mathcal{H}$, such as a hurricane, ice storm, earthquake or another extreme event, is given by the mapping:

$$\mathcal{H} : \mathcal{W} \times \mathcal{G}_{\mathrm{PG}} \rightarrow \mathcal{W} \times \mathcal{G}_{\mathrm{Si}} \tag{6.2}$$

Figure 6.9 Two recently installed poles of a new 66 kV emergency line built to restore service to the area served by Bromley substation in Christchurch, New Zealand after the February 2011 earthquake.

Each node in a graph can be classified as a source, a sink or a transfer node. Edges, which are established between a pair of nodes, represent a service provision from one of those nodes to the other and they may or may not directly correspond to a physical connecting component, such as a transmission line or a distribution line. Source nodes represent the providing end of a service and, thus, they are part of the graph representing the provision of such service. In order to provide such service, source nodes transform services and resources into such output service. For example, a natural gas fueled turbine generator transforms natural gas provided by a natural gas distribution network into electric power. Hence, one attribute of source nodes is the function $\boldsymbol{f}_{\mathrm{SN},k}$ characterizing the transformation occurring from input services $s_{i,k}$ and resources r_k to a service output s_o, characterized by attributes $\boldsymbol{x}_k$ with vector components $\boldsymbol{x}_{j,k}$ corresponding to each of the J edges given by the subindex j leaving the source node k. That is,

$$\boldsymbol{x}_k = \begin{pmatrix} \boldsymbol{x}_{1,k} \\ \boldsymbol{x}_{2,k} \\ \vdots \\ \boldsymbol{x}_{J,k} \end{pmatrix} = \boldsymbol{f}_{\mathrm{SN},k}(s_{i,k}, r_k) \tag{6.3}$$

In addition to being inputs for source nodes, other resources contribute to support the provision of services by infrastructure systems. For example, in an electric power grid, resources also include poles, transformers and other materials, labor force used to operate the system and monetary assets used to procure fuel or pay employee salaries. These resources may be part of services provided by other systems. For example, work force education and qualification are services provided to power grid operators by an educational community system.

Source vertices, can also have service buffers on their input side. The function of these service buffers is to temporarily provide a given service to its corresponding node. For example, the natural gas fueled power plant may include a local natural gas storage tank that can supply natural gas to the turbine in case there is a loss of service in the natural gas distribution network. Sink nodes are the receiving end of a service represented by an edge. Sink nodes also belong to the graph providing such service. For example, electrical loads are sink nodes within the physical domain graph representing the service electric power provision. Transfer nodes transmit services without changing the service being provided. A transfer node without buffers could also be considered a back-to-back sink and source vertex. Although there is no transformation of services occurring in a transfer node because the input and output service of a transfer node are the same, transfer nodes may still require the provision of other services different from the one being modeled in the graph containing such node in order for the transfer node to transmit the service modeled in the graph.

In the same way that electric power grids require the provision of services, such as water for cooling power plants, for their operation, other infrastructure systems, such as water distribution systems, require electric power for delivering their services. Hence, infrastructure systems in general and power grids in particular are dependent on services for their operation. For example, as shown in Figure 6.10, electric power grids need services from an economic community social system in order to manage the capital needed for its operation. Such needed services could be provided by different community systems

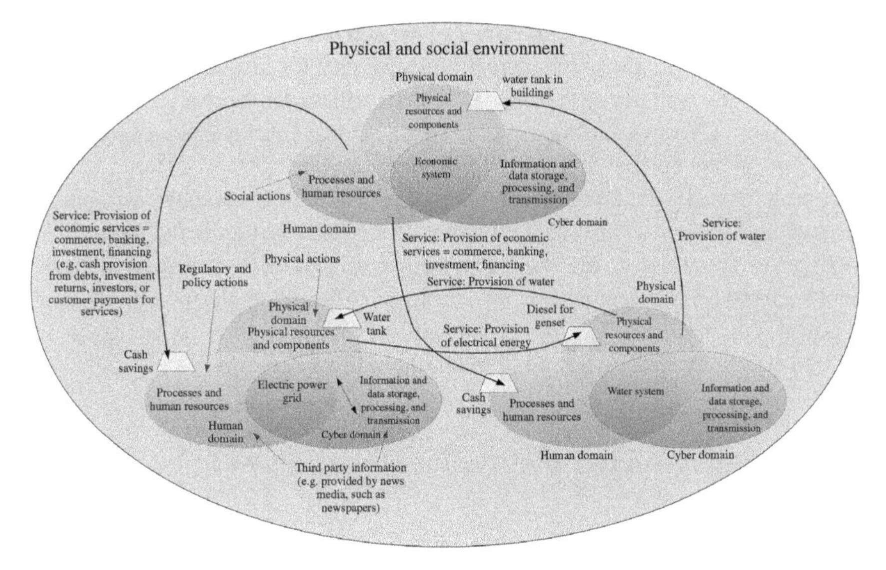

Figure 6.10 Interactions among electric power, water distribution, and economic community systems.

or in some cases it could be provided through harvesting of natural resources present in the environment as it is the case when generating electric power from renewable sources. Hence, it is possible to say that a dependency is established between community systems, such as power grids, and the services they need to operate. Moreover, it is possible to identify cases in which one community system depends on services provided by another community system and vice versa. These are the cases exemplified in Figure 6.10 in which an electric power grid depends on water provision by a water distribution company and the water distribution company depends on electric power provision from an electric power grid. Hence, in this case it is possible to characterize this mutual relationship as an interdependence condition. As Figure 6.10 also shows an analogous interdependent relationship exist between an electric power grid and an economic social system.

The existence of dependencies that, commonly, are the result of service needs that can only be provided on a continuous basis by a given infrastructure system, called a lifeline of the dependent infrastructure system, introduces an operational vulnerability for the dependent infrastructure system because a loss of service in a lifeline may propagate into its dependent infrastructure systems and cause them, in turn, to fail, too. Interdependencies may aggravate these cascading loss of services by causing further loss of service to loop around and worsen the failure in the outage-originating lifeline.

Needed services can also be temporarily provided by services buffers. Service buffers are defined as a local entity able to store the necessary resources and assets in order to temporarily provide a given service to the node where such buffer is located. In the particular case of electric power grids, since they provide electric energy, buffers are realized by energy storage devices, such as batteries. As explained in [5], the addition of service buffers, or energy storage in the particular case of power grids, allows for delaying the propagation of service losses both within a given infrastructure system and across infrastructure systems when a dependence relationship exists for service being buffered. Hence, dependencies of a system on externally provided services can be regulated with the use of internal service buffers within the system. That is, loads of electric power grids can regulate their dependency on the provision of electricity by having energy stored locally in batteries or other devices. Electric power systems can also regulate outage propagation within an electric power grid by also adding energy storage, for example, in substations. Electric power grids can also regulate their dependence on externally provided services, such as fuel for electric power generators, by storing fuel locally at the power plant or by building power plants, such as hydroelectric facilities, with the necessary energy to be transformed into electricity already locally stored in the water reservoir.

6.3 Quantitative Resilience Metrics for Electric Power Distribution Grids

One of the key needs in order to develop more resilient electric power grids is to identify quantitative resilience metrics. In particular, since long power outages following extreme events typically originate at the power distribution level of electric grids, it is of the interest of planners, system operators, utility regulators, and other stakeholders to identify quantitative metrics to characterize electric power distribution grids performance based on their resilience. It is important to identify such quantitative metrics, as opposed to just qualitative evaluations because quantitative approaches allow to perform objective evaluation and assessments, such as those needed to make objective decisions during planning processes or prioritize restoration efforts after a natural disaster. Nevertheless, resilience metrics not only need to be quantitative and be representative of the resilience definition under consideration, they should also allow to compare performance in different locations or different grids and be sufficiently simple to be applicable in practical contexts. Evidently, performance is also dependent on the type of event and its intensity. However, in order to allow for better comparison of different cases, it is desirable that resilience metrics are not heavily dependent on the event characteristics. Some desired features of resilience metrics include being able to reflect the effect of modern technologies, such as distributed generation and energy storage, or the effect of human actions.

Traditionally, the concept of resilience applied to electric power grids was limited to the idea of quick recovery after an extreme event. Such view is found in works like [6] that defines resilience as "the ability of a system to bounce back from a failure." A similar definition of resilience aspects discussed in [7] – it defines resilience as "the ability of a system to respond and recover from an event" – also considers distributed generation systems, but it does not consider the effect of interdependencies and it only considers service restoration as the only operational component associated to a resilience metric – i.e. it considers resilience based on the traditional definition of a system recovering from an event without considering how well the system withstands such event. However, an expanded view of resilience is gaining broader acceptance by both research and applied communities. This view of resilience accepted by Refs. [8–10] and others originates in the U.S. Presidential

Policy Directive 21 (PPD-21) [11] that indicates that "resilience is the ability to prepare for and adapt to changing conditions and withstand and recover rapidly from disruptions." Hence, this definition indicates that resilience has four components also recognized in definitions provided by Refs. [12, 13]:

- ability to prepare
- ability to adapt
- capability for withstanding disruptions
- speed of recovery.

Of these four components the last two are the most relevant ones in terms of realizing a resilience metric for electric power distribution grids because it is possible to say that the first two components are indirectly represented in the withstanding and rapid recovery capabilities. These two components are the focus of the resilience definition in [10], which indicates that resilience is the "robustness and recovery characteristics of utility infrastructure and operations, which avoid or minimize interruptions of service during an extraordinary and hazardous event."

Other works, such as [14] have also recognized the importance of these two last components in identifying resilience metrics by suggesting instead three relevant components to resilience: resistance, absorption, and recovery. Another similar definition of resilience to the one indicated in [11] originates in North American Electric Reliability Corporation (NERC), which in [15] indicates that "infrastructure resilience is the ability to reduce the magnitude and/or duration of disruptive events" and then adds that "the effectiveness of a resilient infrastructure or enterprise depends upon its ability to anticipate, absorb, adapt to, and/or rapidly recover from a potentially disruptive event." However, due to NERCs main interest on bulk power systems – i.e. the transmission and generation portions of a power grid – resilience is treated with a focus on adequacy and security. These two concepts are defined by NERC as "the ability of the electricity system to supply the aggregate electrical demand and energy requirements of the end-use customers at all times, taking into account scheduled and reasonably expected unscheduled outages of system elements" [16] and "ability of the bulk power system to withstand sudden disturbances such as electric short circuits or unanticipated loss of system elements from credible contingencies" [10], respectively. However, after September of 2001 NERC replaced the concept of security by that of "operating reliability" making the pre-September 2001 definition of security related to contingencies management equal to the post-September 2001 definition of operating reliability [16]. The reason for this change was to avoid confusions because after "September 2001 ... security became synonymous with homeland protection in general and critical infrastructure protection in particular" [17]. However, although such change may have avoided confusion with the term security, later on it created confusion by relating the concepts of reliability and resilience.

There exists substantial and fundamental differences between the concepts of reliability and resilience. In the context of electric power distribution grids, reliability "can be defined as the ability of the power system to deliver electricity in the quantity and with the quality demanded by users" [18]. That is, as also indicated in [18] a reliable power distribution grid is one in which users receive power "in a consistent manner," which implies "a binary view of system performance where systems are either functional or failed."

This notion is not necessary completely in disagreement with the aforementioned definition of resilience. However, differences between reliability and resilience become more evident once reliability metrics are examined. Typically, electric power distribution grids reliability is measured based on the IEEE Standard 1366 indexes. One of the objectives of this standard is to provide a common reference to compare reliability of different electric power distribution utilities. As a result, the so-called "major event days" – days when relatively uncommon events, such as natural disasters, happen causing power outages more significant and widespread than in usual "normal" operating conditions – are excluded when computing such indexes. Thus, from a practical perspective, reliability excludes the type of disruptions that are the focus in reliability definitions. This difference is in agreement with the math associated with reliability and availability theory in which reliability and availability measurements imply performing a very large number of tests (theoretically an infinite number of tests) under a standard set of conditions. However, resilience metrics involve dealing with few number of events or even single events under different set of conditions (e.g. different type of extreme event or different intensity). Moreover, the concept of reliability in its most strict and general sense is concerned with continuous operation and failures but not with repairs. However, the concept of resilience is closer to that of availability in which maintaining operation is one of the influencing factors. The other important influencing factor is to recover quickly from disruptions. That is, although it is not possible to say that an electric power distribution grid is reliable if it experiences frequent power outages, it is possible to say that it has high availability if those outages are restored very quickly or if that it is resilient even when experiencing significant outages after a natural disaster if service was restored quickly after such an event. An example of a power distribution grid that can be considered to be relatively resilient although it experienced a widespread power outage is the one that operated in the Tohoku region of Japan during the March 2011 earthquake. As it is explained in [19] although millions of its customers lost power, service to undamaged homes, businesses, and industrial facilities was restored comparatively faster than in other similar events.

In general, it is possible to recognize two main approaches to measure resilience. One is based on so-called fragility, quality or functionality curves. The other is based on analogous metrics to those found to measure availability considering the previously indicated binary approach for power grid operation. In this latter approach resilience measured at a single load equals [8]:

$$R_{\mathrm{SL}} = \frac{T_U}{T_e} \tag{6.4}$$

where T_U is the period of time that is part of the total event duration T_e when the load is receiving power. In this resilience metric, the total event duration, T_e, equals the sum of T_U and T_D (the period of time that is part of T_e when the load is not receiving power). Equation (6.4) can be extended to N loads into [8]:

$$R_N = \frac{\sum_{i=1}^{N} T_{U,i}}{NT_e} = \frac{\sum_{i=1}^{N} T_{U,i}}{\sum_{i=1}^{N} (T_{U,i} + T_{D,i})} \tag{6.5}$$

The same reference [8] also provides a metric for resilience of a single load that has some local energy storage, such as in batteries, that provides an autonomy of T_A in case its grid tie has no service. This resilience metric for a single load with energy storage is:

$$R_{ES} = 1 - (1 - R_{SL})e^{-\frac{T_A}{T_D}} \tag{6.6}$$

This equation has important implications because, as explained in [8], it provides a relationship between amount of energy storage and resilience and, thus, loads that depend on electric power for their operation can mitigate the negative impact of such dependence on resilience by having local energy storage devices.

This resilience metric analogous to that of availability has several advantages. A main advantage is that represents well the two main components of availability by corresponding the withstanding capability to T_U and the recovery speed to T_D. The other two components of resilience are also considered in this metric, but implicitly through their influence when a disruptive event happens on T_U and T_D. Moreover, since (6.5) is analogous to IEEE Standard 1366 System Average Interruption Frequency Index (SAIFI) index, this resilience metric provides a simple and direct transition from reliability metrics into resilience metrics. Another important advantage shown by Eqs. (6.4) and (6.5) is that it can be used for both individual loads and aggregated loads. Furthermore, as indicated by Eq. (6.6), this metric represents well the effect of energy storage and, in the same way, it can also represent well the effect of local distributed electric power generation sources, such as an engine-generator or grid islanded photovoltaic resources. This metric also allows for comparison among different grids, locations, and type and intensity of events – e.g. the same resilience value is obtained for a less intense event yielding a T_e of 2 and T_U of 1 than a more intense event with T_e of 10 and T_U of 5.

As indicated, another approach to measure resilience is based on fragility, quality or functionality curves, such as the one shown in Figure 6.11. Time is always indicated in the horizontal axis of this curve, whereas infrastructure performance is represented on the vertical axis. For example, in electric power grids, the vertical axis could represent the number of electric power customers served in a given area. As shown in Figure 6.11, when a disruption occurs, the functionality level drops. In simple curves, like that in Figure 6.11 and used in [20], after the initial drop, restoration activities immediately start and functionality starts to recover to the original level or a new normal level, forming what is called a resilience triangle. Hence, as indicated in [20] the area of the triangle serves as a measure of resilience.

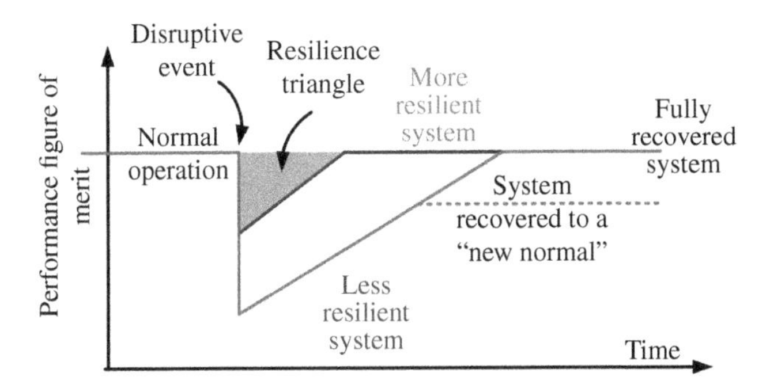

Figure 6.11 Conceptualization of resilience based on the resilience triangle.

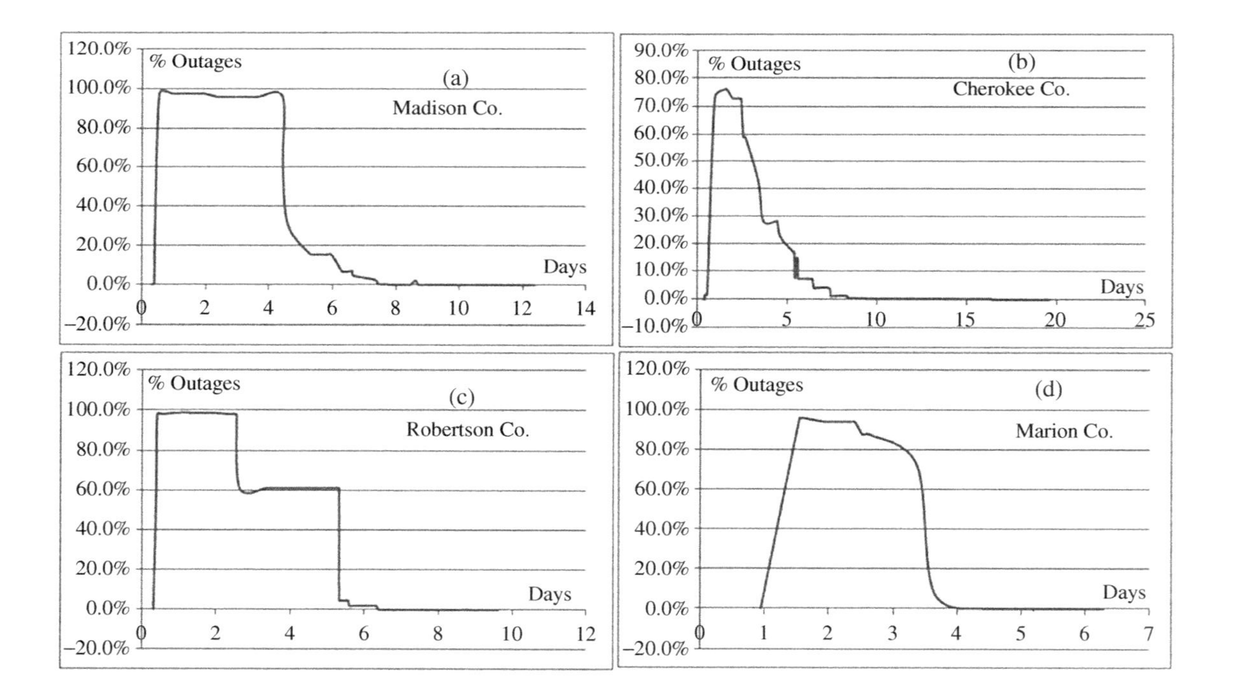

Figure 6.12 Percentage of customers without power in four Texas counties after Hurricane Ike. (a) Madison County, (b) Cherokee County, (c) Robertson County, and (d) Marion County.

However, as Figure 6.12 shows, although the resilience triangle concept is simple, it is not realistic of how actual functionality curves look unless power outages are aggregated to consider very extensive areas, such as an entire state. Additionally, since the size of the triangle depends on the disruptive event intensity and other factors, such resilience metric is difficult to apply when performing comparisons. One alternative metric was presented in [21] based on quality curves. In [21] quality $Q(t)$ is associated to the capacity of the system so for power grids, quality can be related to the complement to 100 of the percentage of customers experiencing loss of service over time – thus, the curves in Figure 6.12 are the complement to 100 of the quality of the system. Hence, from [21] resilience equals:

$$R_Q = \frac{\int_{t_1}^{t_2}(Q_\infty - (Q_\infty - Q_0)e^{-bt})\mathrm{d}t}{(t_2 - t_1)} \tag{6.7}$$

where Q_∞ is the capacity of the fully functional system, Q_0 is the post event capacity of the system, b is a parameter derived empirically representing the restoration speed assuming that the system quality increases exponentially after reaching a minimum value immediately after the event, t is the time in days post-event, and t_1 and t_2 are the endpoints of the time interval under consideration (typically, t_1 is the time when the event happens and t_2 is the time when service restoration is completed). Although (6.7) represents an improvement from the resilience triangle metric because it provides a better approach for comparison due to its normalization when dividing by $t_2 - t_1$, it still presents issues of how representative an exponential quality curve, such as the one in Figure 6.13 and similar to that in [21], is of the reality observed in past natural disasters an exemplified in Figure 6.12. Although the complement to 100 of the percentage of outages in Figure 6.12b can be approximated to an exponential quality curve, the other three curves differ considerably from exponential waveforms. Part of the reason for such difference is that, typically, during hurricanes restoration activities do not start until sustain wind speeds drop below tropical cyclone thresholds. As a result only aggregated data from very large areas, like an entire state, may approximate more commonly to exponential waveforms. However, neither aggregating many loads statistics nor the use of quality or functional curves are able to represent the effect of energy storage or local power generation on loads resilience. Hence, the effect of dependencies on resilience may not be included in such resilience metrics.

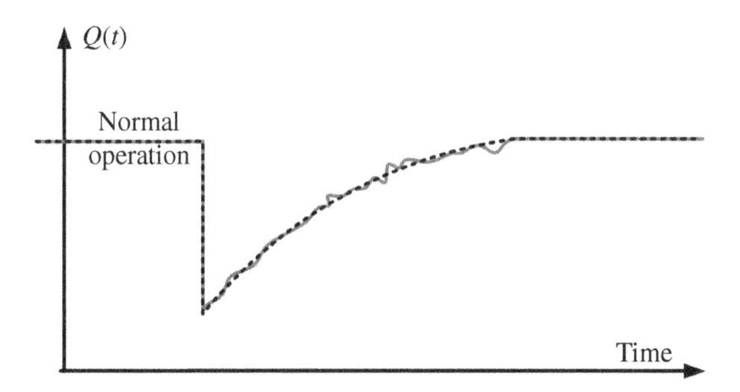

Figure 6.13 Quality curve similar to that representing power grid performance in Louisiana after Hurricane Katrina discussed in [21]. Source: Adapted from Reed et al. [21].

An alternative to the quality curves that allows having different restoration curves is the one presented in [6] based on functionality curves. In [6] functionality $F(t)$ is a figure of merit, such as percentage of electric power customers receiving power, that is used as an indicator whether the system or the part of the system under evaluation is in a functional state or not. Then, in [6] resilience is mathematically defined as the "ratio of recovered functionality to lost functionality" so it is calculated based on:

$$\Lambda(t) = \frac{F(t) - F(t_d)}{F(t_0) - F(t_d)} \tag{6.8}$$

where, based on Figure 6.14, $F(t)$ is the system functionality at a time t, $F(t_d)$ is the functionality value when catastrophic failure is assumed to have occurred, and $F(t_0)$ is the initial functionality when the system is assumed to be fully operational. Although the resilience metric in (6.8) is simple to calculate, can be used both at an aggregated level and for individual loads, allows for comparison in different conditions and is able to consider different shapes of functionality curves it presents an important flaw with respect to the aforementioned definition of resilience indicated above, which is that although it can be said that in some ways the metric in (6.8) is able to represent the system withstanding capability, this metric does not really represent restoration speed because the time that takes to reach the final functional state, shown in Figure 6.14, has no influence in the value of resilience given by Eq. (6.8).

Resilience metrics based on quality or functionality curves present other common issues. One is that such methods do not present limitations in how the figure of merit chosen to represent the quality or functionality of the system. As a result, such figure of merit could be chosen in order to show a metric that may show better results than others. One example of such choice was observed in Puerto Rico after Hurricane Maria when the figure of merit chosen was load level in MW instead of percentage of customers being powered. Since largest consumers were given priority over individual residential customers, functionality or quality curves with respect to the load level in MW yielded a much higher resilience value than if percentage of customers being powered would have been used. However, such measure of functionality or quality was not representative of the reality that many customers in very extensive areas of the island were not receiving electric power months after the hurricane made landfall on the island.

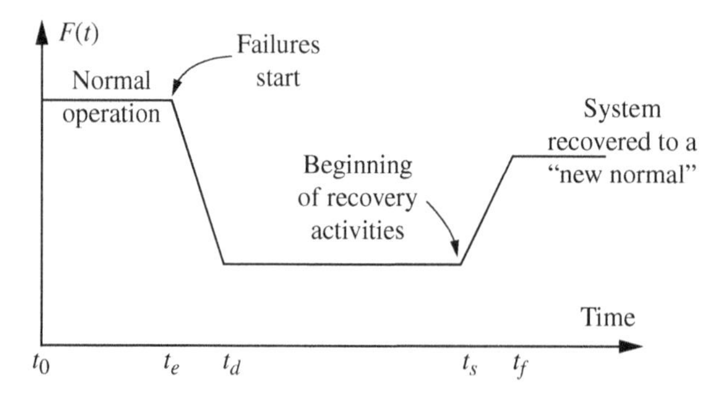

Figure 6.14 Fragility curve. Source: Adapted from Albasrawi et al. [6].

References

1 Krishnamurthy, V. and Kwasinski, A. (2013). Characterization of power system outages caused by hurricanes through localized intensity indices. In: *Proceedings of 2013 IEEE Power and Energy Society General Meeting*. IEEE, Vancouver, BC, Canada, pp. 1–5.

2 National Institute of Standards and Technology (2015). *Community Resilience Planning Guide for Buildings and Infrastructure Systems*. NIST Special Publication.

3 Kwasinski, A., Trainor, J., Wolshon, B., and Lavelle, F.M. (2016). A conceptual framework for assessing resilience at the community scale. In: *NIST GCR 16-001*.

4 Kwasinski, A., Francis, R., Joseph, T., Chen, C., and Lavelle, F. (2017). Further development of a conceptual framework for assessing resilience at the community scale. In: *NIST GCR 17-013*.

5 Kwasinski, A. and Krishnamurthy, V. (2017). Generalized integrated framework for modelling communications and electric power infrastructure resilience. In: *Proceedings of INTELEC 2017*. IEEE, Gold Coast, Australia, pp. 1–8.

6 Albasrawi, M.N., Jarus, N., Joshi, K.A., and Sarvestani, S.S. (2014). Analysis of reliability and resilience for smart grids. In: *Proceedings of 2014 IEEE 38th Annual Computer Software and Applications Conference (COMPSAC)*. IEEE, Vasteras, Sweden, pp. 1–6.

7 Momoh, J.A., Meliopoulos, S., and Saint, R. (2012). *Centralized and Distributed Generated Power Systems – A Comparison Approach*. PSERC Publication.

8 Kwasinski, A. (2016). Quantitative model and metrics of electrical grids' resilience evaluated at a power distribution level. *Energies* 9: 93.

9 Willis, H.H. and Loa, K. (2015). Measuring the resilience of energy distribution systems. RAND Corporation Document Number: RR-883-DOE.

10 Keogh, M. and Cody, C. (2013). *Resilience in Regulated Utilities*. The National Association of Regulatory Utility Commissioners (NARUC).

11 Presidential Policy Directive (2015). Critical infrastructure security and resilience. https://www.whitehouse.gov/the-press-office/2013/02/12/presidential-policy-directive-critical-infrastructure-security-and-resil (accessed 29 September 2015).

12 Haimes, Y.Y. (2009). On the definition of resilience in systems. *Risk Analysis* 29 (4): 498–501.

13 Committee on Increasing National Resilience to Hazards and Disasters, Committee on Science, Engineering and Public Policy (2012). *Disaster Resilience: A National Imperative*. The National Academies Press.

14 Ouyang, M. and Dueñas Osorio, L. (2012). Time-dependent resilience assessment and improvement of urban infrastructure systems. *Chaos* 22 (3): 1–11.

15 Severe Impact Resilience Task Force (2012). *Severe Impact Resilience: Considerations and Recommendations*. North American Electric Reliability Corporation (NERC).

16 North American Electric Reliability Corporation (NERC) (2013). Understanding the grid (August 2013).

17 North American Electric Reliability Corporation (NERC) (2007). Definition of "Adequate Level of Reliability" Version 2 – Draft 3b (31 October 2007).

18 Clark-Ginsberg, A. What's the difference between reliability and resilience?. *U.S. DHS, The Industrial Control Systems Cyber Emergency Response Team (ICS-CERT)*.

https://ics-cert.us-cert.gov/sites/default/files/ICSJWG-Archive/QNL_MAR_16/reliability%20and%20resilience%20pdf.pdf.

19 Kwasinski, A. (2015). Numerical evaluation of communication networks resilience with a focus on power supply performance during natural disasters. In: *Proceedings of INTELEC 2015*, Osaka, Japan (October 2015). IEEE, Osaka, Japan, pp. 1–7.

20 Oregon Seismic Safety Policy Advisory Commission (OSSPAC) (2013). The Oregon resilience plan: reducing risk and improving recovery for the next Cascadia earthquake and tsunami. https://www.oregon.gov/oem/documents/oregon_resilience_plan_final.pdf.

21 Reed, D.A., Kapur, K.C., and Christie, R.D. (2009). Methodology for assessing the resilience of networked infrastructure. *IEEE Systems Journal* 3 (2): 174–180.

7

Frameworks for Analyzing Resilience

Ted Brekken

School of Electrical Engineering and Computer Science, Oregon State University, Corvallis, OR, USA

There are many qualitative definitions of resilience. One of the most straightforward definitions is that a resilient system is one that is prepared, can absorb a shock, and can recover from a shock. This can be summarized as *prepare*, *absorb*, and *recover*, or PAR for short.

Qualitative definitions of resilience are useful, but how can we discuss resilience in more concrete, quantitative terms? Are there metrics with useful input and output information that can inform our choices on how to assess and improve the resilience of a system?

This chapter discusses some metrics useful for describing resilience in power systems, along with a description of a mathematical risk analysis modeling framework that can be used to incorporate several resilience concepts into an overarching determination of the risk to a power system asset to a shock or disaster.

7.1 Metrics

7.1.1 Traditional Power Systems Metrics

Standard electrical reliability metrics can be utilized for analyzing resilience. Indeed, there is significant overlap between the concepts of *resilience* and *reliability*. Some examples and definitions of applicable metrics are:

7.1.1.1 Loss of Load Probability (LOLP)

- LOLP $= p(C - L < 0)$, where C is the available capacity and L is the peak load; it is the probability that the load will exceed the available generation during a given period.
- It gives no indication as to how severe the condition would be when the load exceeds available generation.

7.1.1.2 Loss of Load Expectation (LOLE)

- Generally defined as the average number of days (or hours) on which the daily peak load is expected to exceed the available generating capacity.

Resiliency of Power Distribution Systems, First Edition.
Edited by Anurag K. Srivastava, Chen-Ching Liu, and Sayonsom Chanda.

7.1.1.3 Effective Load Carrying Capability (ELCC)

- Quantifies the incremental amount of load that can be served due to the addition of an individual generator (or group of generators) while maintaining the existing reliability level.
- Also known as capacity credit.

7.1.1.4 Capacity Margin

- Capacity margin = (total capacity – peak load)/(peak demand).
- Normalized capacity safety margin.

7.1.1.5 System Average Interruption Frequency Index (SAIFI)

- Total number of customers interrupted/total customers.

7.1.1.6 System Average Interruption Duration Index (SAIDI)

- Total customer interruption durations/total customers.

7.1.1.7 Customer Average Interruption Duration Index (CAIDI)

- Total customer interruption duration/total number of customers interrupted.

7.1.2 Performance Curve

Let the function P define a level of system performance. The units of the function depend on the system being evaluated. For example, the system performance could be economic output, or passengers per hour, or kWh of electricity delivered per day.

See Figure 7.1. The initial system performance level is P_0. At time t_{db} an event occurs and damage to the system begins to propagate. The damage accumulation continues until the damage ending time t_{de}, at which point the system is in a damaged state P_d. Recovery efforts begin at time t_{rb} and end at time t_{re} with the system having recovered to post event performance level P_e.

The key points on the performance curve can be used as metrics. For example, a system specification could be that the damaged level P_d not fall below a given threshold for a certain event, or that the assessment and recovery time $t_{re} - t_{de}$ must be below a specified interval.

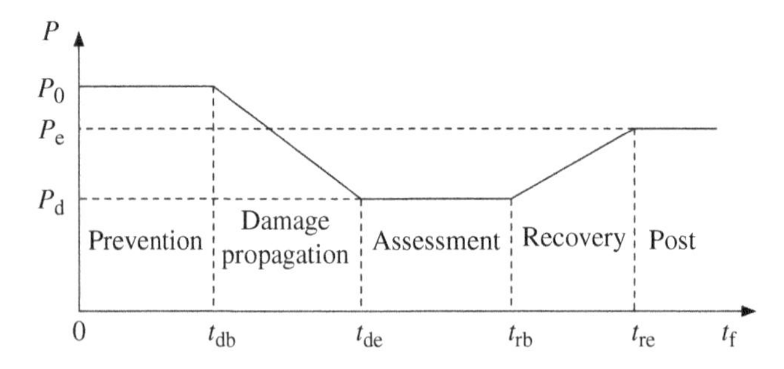

Figure 7.1 System performance curve as a function of time.

7.1.3 Resilience Metric

The area under the performance curve is an important metric. Generally, a smaller area under the curve represents a greater and more enduring level of damage. The area under the curve can be normalized by a target level of performance.

The resilience metric R is defined as the area under the curve of actual system performance, over the area under the curve of the target level of performance.

$$R(T) = \frac{\int_0^T P(t)\mathrm{d}t}{\int_0^T P_{\mathrm{T}}(t)\mathrm{d}t}$$

where P is the performance and P_{T} is the target, or desired, system performance (Figure 7.2).

7.1.4 Example: Performance Curve and Resilience Metric

Below is a performance curve adapted from data from the 2011 Tohoku Japan earthquake. The performance measure is homes with power within an affected district (Figure 7.3).

Referring to the key points in the performance curve, we can assign the following metrics:

- Initial performance $P_0 = 4.43 \cdot 10^6$ homes
- Damage performance $P_{\mathrm{d}} = 0$ homes
- Ending performance $P_{\mathrm{e}} = 4.31 \cdot 10^6$ homes
- Damage begin time $t_{\mathrm{db}} = 0$ days
- Damage end time $t_{\mathrm{de}} = 0$ days
- Recovery begin time $t_{\mathrm{rb}} = 0.54$ days
- Recovery end time $t_{\mathrm{re}} = 10$ days

Obviously, the actual performance curve does not perfectly match the idealized characteristics shown in Figure 7.1, so some judgment is required in assigning some of the key points (Figure 7.4).

To determine the resilience metric R, we need to specify a target level of performance. For the sake of illustration, let us arbitrarily define the target level of performance to be a 25% reduction in performance, a 1-day damaged period, and a 5-day recovery period to full function (Figure 7.5).

By inspection, we can see that the actual system performed worse than our target performance. But how can we quantify that? The resilience metric R is the area under the curve of actual performance P divided by the area under the target performance curve P_{T}.

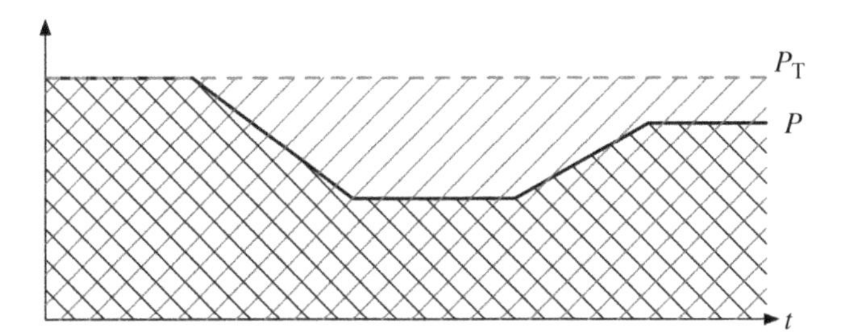

Figure 7.2 Example of actual system performance P and the target level of system performance P_{T}. The ratio of the area under the curves is the resilience metric R.

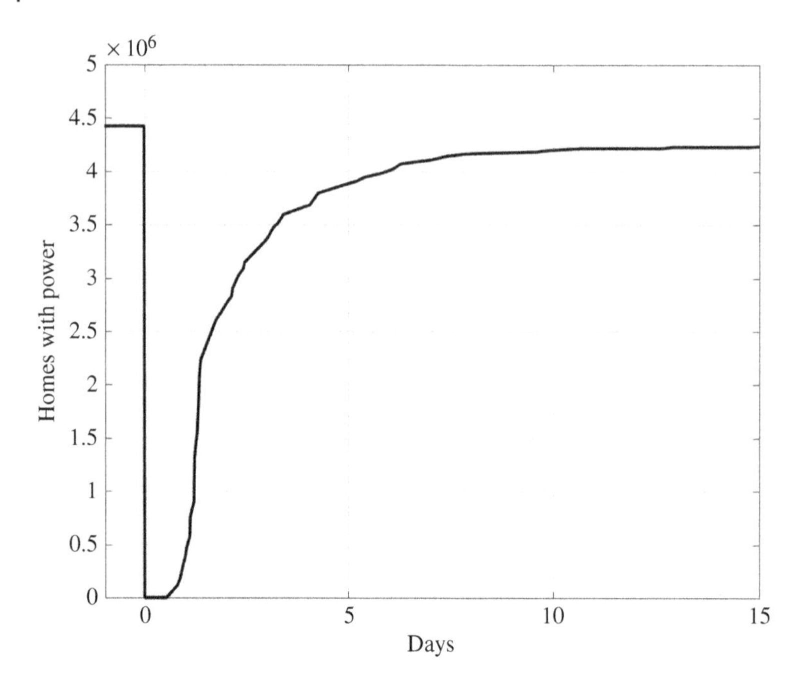

Figure 7.3 Performance of a section of the Japanese grid during and after the 2011 Tohoku earthquake in terms of homes with power (previous to the earthquake) that were temporarily lost during the earthquake. Source: Data adapted from Kazamaa and Noda [1].

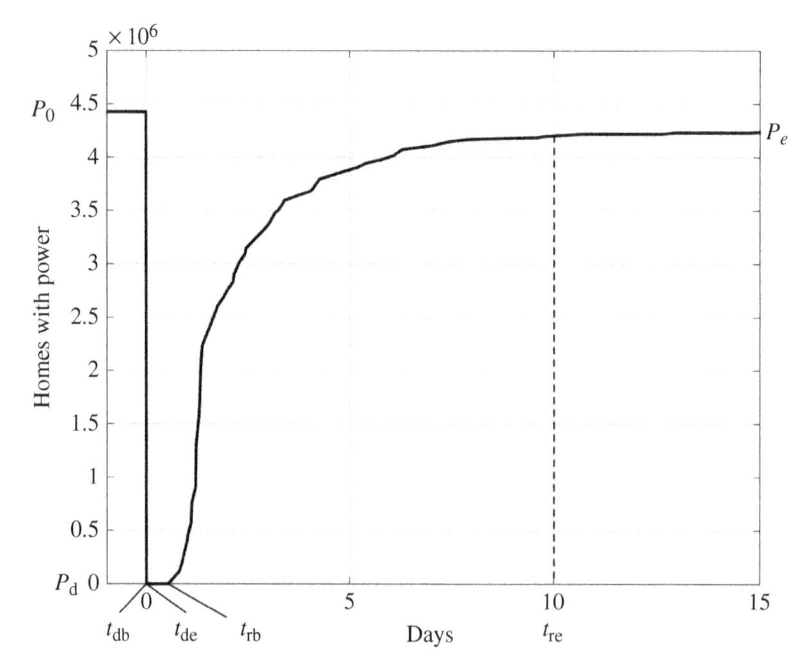

Figure 7.4 Tohoku earthquake performance including metrics.

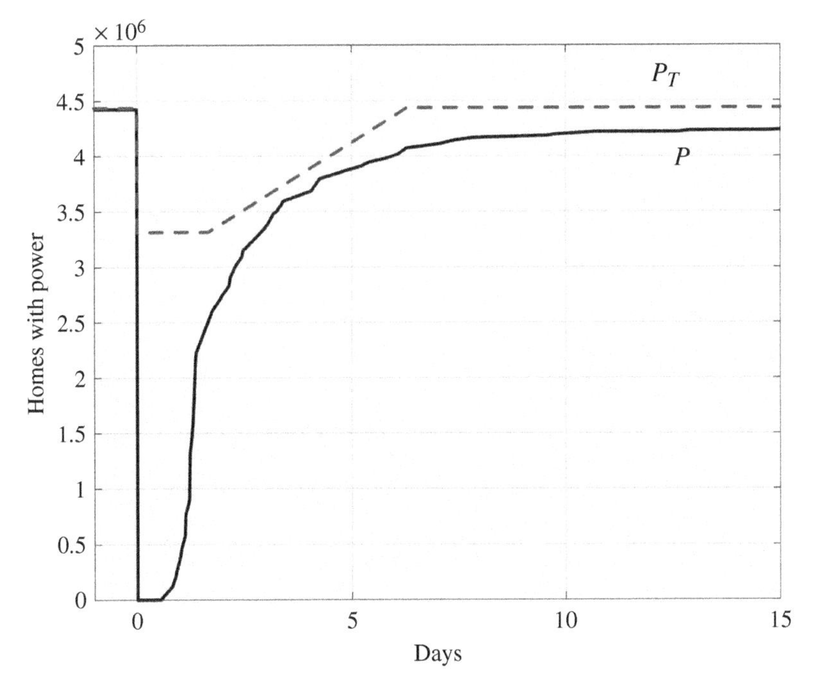

Figure 7.5 Tohoku earthquake electrical system performance P with example target performance P_T included.

The area under the target performance level P_T is

$$\int_0^{15} P_T(t)\,dt = 62.6 \cdot 10^6 \text{ home} - \text{days}$$

The area under the actual performance P is

$$\int_0^{15} P(t)\,dt = 54.8 \cdot 10^6 \text{ home} - \text{days}$$

The resilience metric is then

$$R(T) = \frac{54.8 \cdot 10^6 \text{ home–days}}{67.0 \cdot 10^6 \text{ home–days}} = 0.88$$

7.1.5 Vulnerability and Fragility Functions

Vulnerability and fragility functions have similar forms, but model two different things. Typically, both are functions of an excitation input. In the case of earthquakes and seismic hazards, a typical input is a measure of shaking magnitude, such as peak ground acceleration (PGA), peak ground velocity (PGV), peak ground displacement (PGD), or earthquake magnitude.

7.1.5.1 Vulnerability Function

A vulnerability function defines an estimated reduction in performance as a function of system excitation. For example, with reference to Figure 7.6, which is an observation of

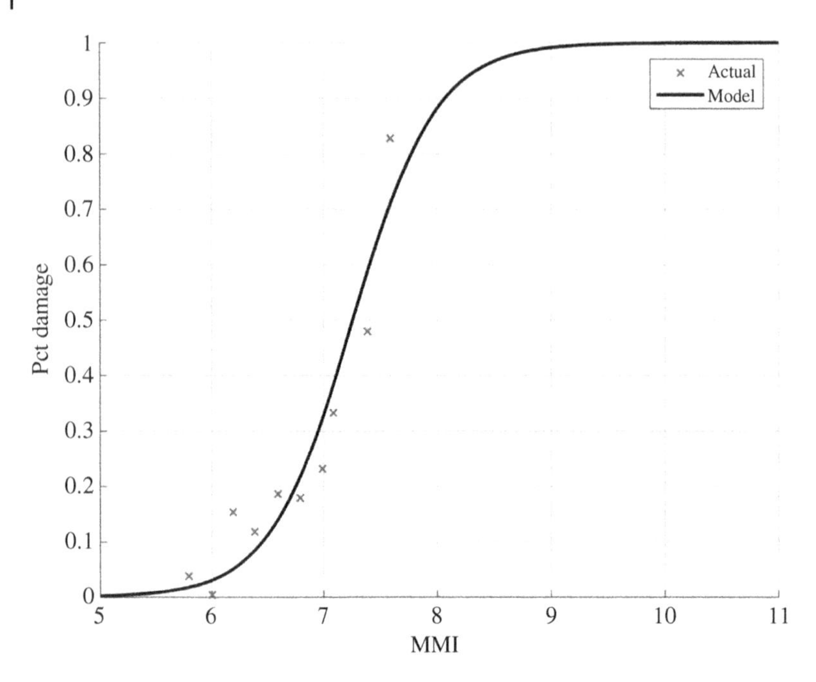

Figure 7.6 Example of a vulnerability function, showing percent damage to the Seattle area distribution system due to the February 2001 Nisqually earthquake. Source: Data adapted from Park et al. [2].

percentage damage to the Seattle distribution system during the Nisqually earthquake of February 2001.

The figure shows the actual observed damage to sections of the distribution grid as a function of shaking (as defined by the Modified Mercalli Intensity scale, a qualitative measure of shaking [3]).

A logistic model fit to the actual data is also plotted. The logistic model is of the form:

$$y = \frac{\exp(\beta_0 + \beta_1 x)}{1 + \exp(\beta_0 + \beta_1 x)}$$

A logistic model has a range of 0–1, and a domain of negative infinity to infinity. It works well for phenomena which naturally saturate at upper and lower limits. In this case, the amount of damage is clearly limited between 0 and 1 (i.e. 0% and 100%), and the amount of damage asymptotically approaches the upper limit (100%) for increasing shaking. For the data shown in Figure 7.6, the β_0 and β_1 parameters are −19.8 and 2.7, respectively.

A logistic model was used for this particular model, but many modeling functions are candidates. A key point for vulnerability and fragility functions are that they are not defined by the particular model used, but instead by what the models represent. In Section 7.1.5.2, we shall see a fragility function modeled with a log-normal function.

7.1.5.2 Fragility Function

A fragility function provides a probability of an asset transitioning to a certain state, given a certain input or a certain existing state. Fragility functions provide a likelihood of set membership.

Figure 7.7 Live-tank circuit breakers. Source: Jens Bender/Wikimedia Commons.

Porter [4] defines a fragility function as "... a mathematical function that expresses the probability that some undesirable event occurs (typically that an asset, facility or a component' reaches or exceeds some clearly defined limit state) as a function of some measure of environmental excitation (typically a measure of acceleration, deformation, or force in an earthquake, hurricane, or other extreme loading condition)."

For illustration, consider a live-tank circuit breaker as shown in Figure 7.7. Fragility functions for this circuit breaker were determined with finite element analysis of mechanical stresses [5]. The fragility functions are shown in Figure 7.8.

Notice the function input is again a measure of shaking intensity, although in this case it is PGA instead of MMI as it was for the vulnerability function example. However, whereas the vulnerability function gives a percentage damage, the fragility function is the probability of the asset being in a damage state. In the example, there are two damage states: moderate and severe.

It is important to note that damage states can be established to be either exclusive or non-exclusive. For example, consider if for the circuit breaker there are damage states "damaged bushing" and "damaged foundation." In this case, clearly the asset could be in either 0, 1, or 2 of those states. Alternatively, consider the damage states shown in Figure 7.8: "moderate" and "severe." These damage states are exclusive, in that if the asset is moderately damaged, it is not severely damaged, and if it is severely damaged, it is not moderately damaged. (A more in-depth discussion of exclusive vs. non-exclusive states can be found in [4].)

For the sake of simplicity, in this chapter, we shall work with exclusive damage states.

The fragility curves shown in Figure 7.8 present the probability of the asset being in *at least* that damage state. For example, if the PGA was 0.2 g, the probability of the asset being in the "moderate" *or greater* damage state is 85%, and the probability of the asset being

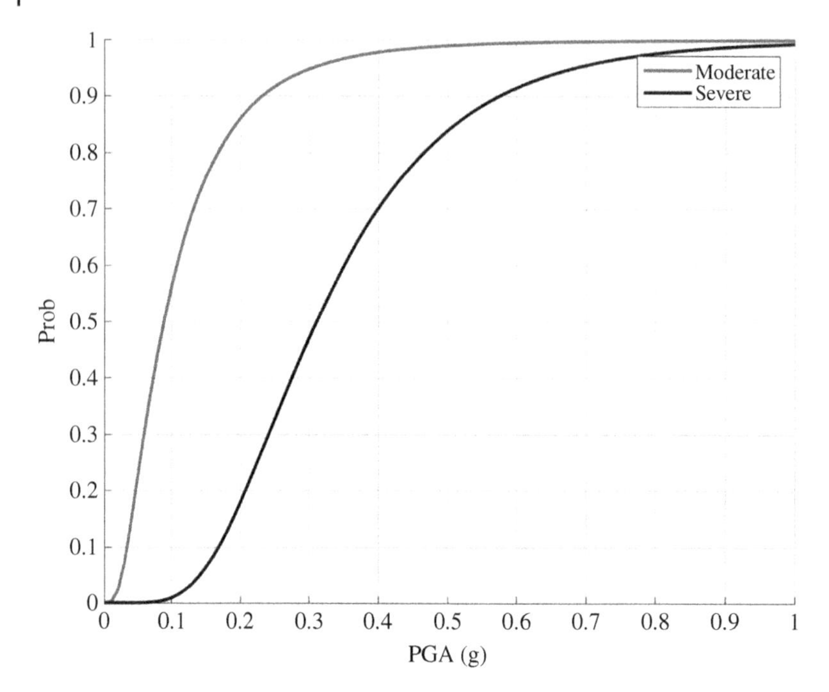

Figure 7.8 Example fragility functions showing probability of damage state membership as a function of excitation. Source: Data adapted from Zareeia et al. [5].

in the "severe" *or greater* damage state is approximately 17%. There is a 15% chance the asset is not at least in the "moderate" damage state, so therefore is a 15% chance of being undamaged.

So if the curves provide the probability of being in a damage state or greater, what is the probability of being in a specific damage state? It is the probability of being in that state or greater, minus the probability of being in the next higher state and greater. Consider damage states D_0, D_1, and D_2.

$$p[d = D_0] = 1 - p[d \geq D_1]$$

$$p[d = D_1] = p[d \geq D_1] - p[d \geq D_2]$$

$$p[d = D_2] = p[d \geq D_2]$$

Returning to our example at 0.2 g of PGA

$$p[d = D_0] = 1 - p[d \geq D_1] = 1 - 0.85 = 0.15$$

$$p[d = D_1] = p[d \geq D_1] - p[d \geq D_2] = 0.85 - 0.17 = 0.68$$

$$p[d = D_2] = p[d \geq D_2] = 0.17$$

This is illustrated in Figure 7.16.

The fragility functions in Figure 7.8 of a log-normal form. A log-normal distribution is the cumulative distribution function of a variable whose exponent is normally distributed.

Assume a random variable Y is normally distributed. Then the random variable $X = e^Y$ is log-normally distributed. The probability distribution function of X is:

$$f(x) = \frac{1}{x\sigma\sqrt{2\pi}} \exp\left(\frac{-(\ln x - \mu)^2}{2\sigma^2}\right)$$

where σ and μ are the standard deviation and mean of Y, respectively.

The cumulative distribution function, which is the form of the fragility functions shown in Figure 7.8, is:

$$F(x) = \int_0^x \frac{1}{\hat{x}\sigma\sqrt{2\pi}} \exp\left(\frac{-(\ln \hat{x} - \mu)^2}{2\sigma^2}\right) d\hat{x} = \frac{1}{2}\left(1 + \operatorname{erf}\left(\frac{\ln x - \mu}{\sigma\sqrt{2}}\right)\right)$$

A logistic function, as shown in the vulnerability function example (Figure 7.6), could also be used. The appropriate choice of function for either a vulnerability function or a fragility function depends on an understanding of the underlying process, the judgment of the analysist, and a consideration of ease of use.

Fragility functions are often not simple to determine. A clear and quantifiable damage state definition is required, and then detailed simulation, laboratory testing, or field observation is required to estimate the function itself. This is further complicated by the fact that the asset being tested or evaluated can be damaged or destroyed during the testing.

Simulation includes the use of FEA and stress analysis tools to determine the amount of expected stress given an amount of ground shaking. The example fragility functions shown above were determined with FEA analysis [5].

Laboratory testing is expensive, and may cause the failure of the component being tested, which can itself be a significant expense (e.g. in the case of high-voltage transformers). Laboratory testing typically involves the use of a high-capacity shake table which limits the testing locations within North America. Guidelines for testing electrical equipment for seismicity and determining qualification levels are defined in the IEEE Standard 693 [6].

Field observation is one of the most practical methods of estimating fragility curves. In this case, the failure rates of actual system components in an actual earthquake are observed, and a model is created and fit to the data.

7.2 Risk Analysis Modeling

This chapter presents a mathematical framework for quantifying the probabilities of hazard occurrence, system response, system damage, and system loss. Much of the framework and its components have been refined with efforts from the Pacific Earthquake Engineering Research Center under guidelines and methods for qualitative analysis called Performance Based Earthquake Engineering [4, 7–9].

7.2.1 Concept

A conceptual diagram of the framework is given (Figure 7.9).

The procedure is as follows: first define an asset. For an electrical power system, this could be a transmission line, a generator, a load, a substation, etc. Next, determine the hazards of

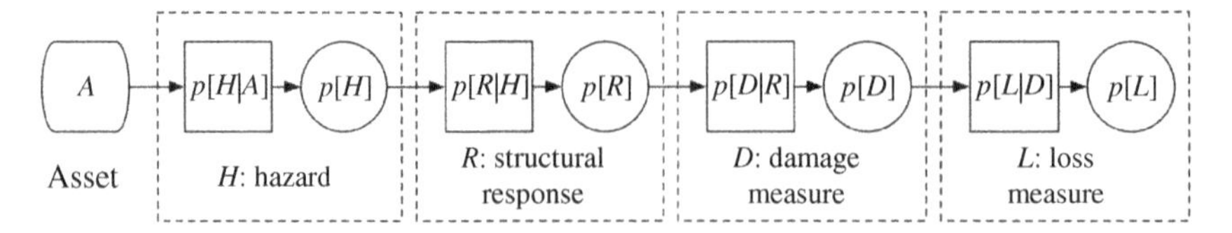

Figure 7.9 Risk analysis modeling.

interest for an asset at that asset's location. In this chapter we focus on earthquake hazards, but the hazard could be related to storms, cyberattacks, etc. In the case of earthquake threat analysis, PGA is a typical hazard measure. Next, determine the probability of those hazards occurring, given that asset type and location: $p[H \mid A]$. (This notation is read as the probability of a hazard, given that asset.) This will give the probability of a hazard occurring, $p[H]$. From there, determine the probability of the asset responding to the hazard, given that hazard ($p[R \mid H]$). This will give the probability of that asset's response $p[R]$. Note that the response should be defined as something meaningful for that asset, such as bending stresses, displacement, etc. A structural analysis will be necessary to link the hazard to the system response.

Once the probability of response is determined, the probability of damage given that response is determined, to give the probability of damage. Damage states are defined as certain undesirable states of the asset. For example, if the asset was a transformer, one damage state might be failed transformer bushings. A second damage state might be the transformer displaced from its pad. As discussed prior (Section 7.1.5.2), damage states can be exclusive or inclusive. In the examples presented here, we shall assume that the damage states are exclusive.

Lastly, the probability of damage is linked to the probability of loss. Loss measures could be defined many ways. For power systems studies, it may be convenient to define a loss state as a loss of equipment capacity. In another case, it may be useful to define a loss state as a monetary loss, or a loss of customers served, for example.

For some studies, damage state analysis may be enough and it may not be necessary to link damage to loss. This is a degree of freedom that is left to the designer of the analysis.

7.2.2 Alternative Risk Analysis Framework

Let us introduce an alternative definition of the risk analysis framework that will be useful for our power systems analysis example:

Note that the concept is the same, but instead of hazards we shall categorize events, and instead of response we shall categorize excitation. Events could be different levels and types of earthquakes, for example, Cascadia Subduction Zone 9.0, Cascadia Subduction Zone 8.0, Portland Hills Earthquake, etc. Excitation is the input to the electrical equipment, for example, PGA.

As before, a structural analysis is required for each asset (or asset type) to link excitation to probability of damage.

To determine the probability of excitation, damage, or loss at any point in the process, it is necessary to integrate over all probabilities along the process to that point.

$$p[X] = \int p[E|A]p[X|E]\,\mathrm{d}E$$

$$p[D] = \iint p[E|A]p[X|E]p[D|X]\,\mathrm{d}E\mathrm{d}X$$

$$p[L] = \iiint p[E|A]p[X|E]p[D|X]p[L|D]\,\mathrm{d}E\,\mathrm{d}X\,\mathrm{d}D$$

If the probabilities are discrete (as in the example in Section 7.2.3), then,

$$p[X_j] = \sum_i p[E_i|A]p[X_j|E_i]$$

$$p[D_k] = \sum_j \sum_i p[E_i|A]p[X_j|E_i]p[D_k|X_j]$$

$$p[L_m] = \sum_k \sum_j \sum_i p[E_i|A]p[X_j|E_i]p[D_k|X_j]p[L_m|D_k]$$

7.2.3 Probability Networks

One implementation of the general framework shown in Figure 7.10 is a discrete Bayesian probability network that defines an asset's probability of transitioning between event, excitation, damage, and loss states. An example of a discrete network is shown in Figure 7.11.

In this particular example, there are 2 event states, 3 excitation states, 2 damage states, and 2 loss states. The number of event, excitation, damage, and loss states is up to the analysist.

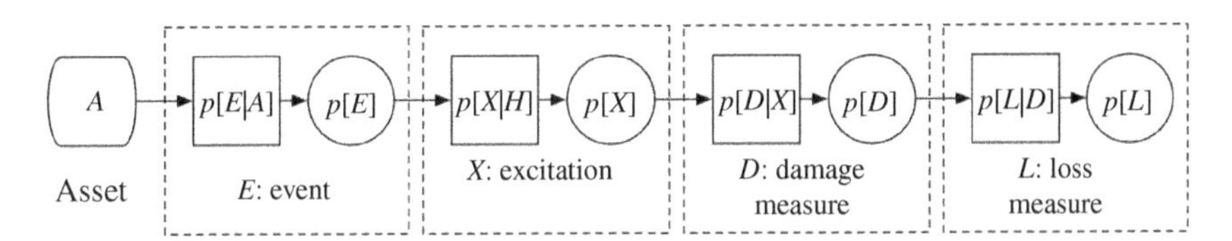

Figure 7.10 Alternative risk analysis modeling framework.

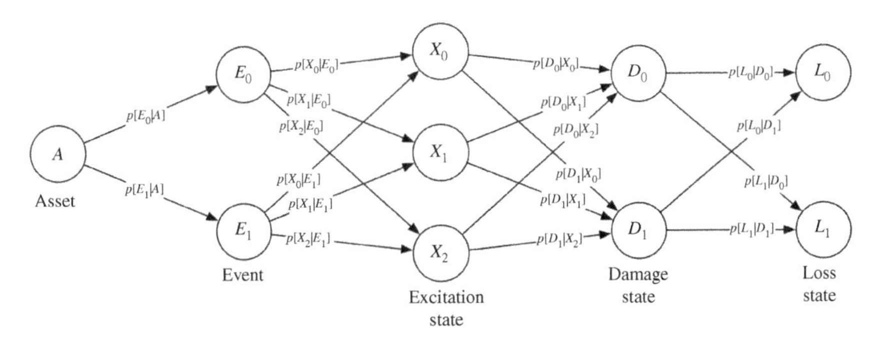

Figure 7.11 Example risk analysis framework as a discrete Bayesian network.

The probability of transitioning from one state to another is noted on the arrow connecting states. For example, the probability of this asset experiencing event state 0 (E_0) is $p[E_0 \mid A]$. If the asset is experiencing event state 0, then there is a $p[X_0 \mid E_0]$ probability that it experiences excitation state 0 (X_0), a $p[X_1 \mid E_0]$ probability that it experiences excitation state 1 (X_1), and a $p[X_2 \mid E_0]$ probability that it experiences excitation state 2 (X_2).

Consider the matrices for all the probabilities from one layer to the next:

$$P_{E|A} = \left[p[E_0|A] \;\; p\left[E_1|A\right]\right]$$

$$P_{X|E} = \begin{bmatrix} p[X_0|E_0] & p[X_1|E_0] & p[X_2|E_0] \\ p[X_0|E_1] & p[X_1|E_1] & p[X_2|E_1] \end{bmatrix}$$

$$P_{D|X} = \begin{bmatrix} p[D_0|X_0] & p[D_1|X_0] \\ p[D_0|X_1] & p[D_1|X_1] \\ p[D_0|X_2] & p[D_1|X_2] \end{bmatrix}$$

$$P_{L|D} = \begin{bmatrix} p[L_0|D_0] & p[L_1|D_0] \\ p[L_0|D_1] & p[L_1|D_1] \end{bmatrix}$$

Note that the sum of any matrix row must be equal to 1.

When the transition probabilities are given as discrete values, evaluation of the equations in Section 7.2.2 becomes simple matrix multiplication.

$$P_X = \sum_i p[E_i|A]p[X|E_i] = P_{E|A}P_{X|E}$$

$$P_D = \sum_j \sum_i p[E_i|A]p[X_j|E_i]p[D|X_j] = P_{E|A}P_{X|E}P_{D|X} = P_X P_{D|X}$$

$$P_L = \sum_k \sum_j \sum_i p[E_i|A]p[X_j|E_i]p[D_k|X_j]p[L|D_k] = P_{E|A}P_{X|E}P_{D|X}P_{L|D} = P_D P_{L|D}$$

7.2.4 Example

We wish to calculate the expected number of customers lost due to seismic damage to a circuit breaker in Portland, Oregon. Assume it has been determined that failure of the circuit breaker would cause an outage to 10,000 customers.

We shall use the probability network as shown here (Figure 7.12):

Asset

- Circuit breaker, like that shown in Figure 7.7.

Events

- E_0: null state. No event.
- E_1: magnitude 8.1 Cascadia Subduction Zone earthquake.
- E_2: magnitude 9.0 Cascadia Subduction Zone earthquake.

Excitation States

- X_0: null state. Small to no excitation, 0.00–0.05 g bedrock PGA.
- X_1: 0.05–0.10 g bedrock PGA.
- X_2: 0.10–0.20 g bedrock PGA.
- X_3: 0.20–0.30 g bedrock PGA.

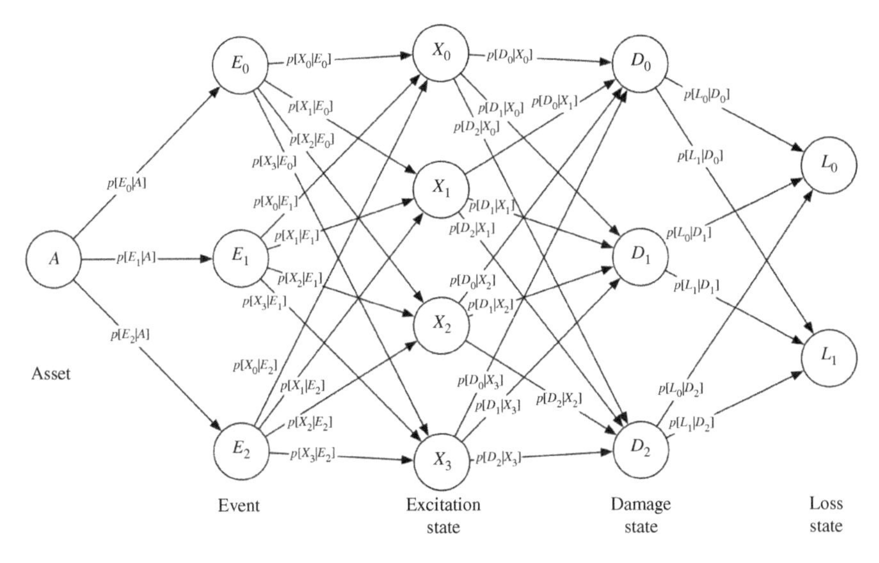

Figure 7.12 Circuit breaker example discrete probability network.

Damage States

- D_0: null state. No damage.
- D_1: moderate.
- D_2: severe.

Loss States

- L_0: null state. No loss.
- L_1: complete loss; 10,000 customers without power.

Assumptions:

- Probabilities are determined on a per-year basis. Daily or hourly would also be typical time frames considered for power system analysis.
- The relationship between the earthquake magnitude and the asset bedrock ground PGA is determined by geotechnical analysis, (such as [10]). Further site-specific analysis would be needed to determine the relationship between the bedrock PGA and the actual site PGA, which is highly dependent on soil type and the local geography. For the sake of illustration, we shall assume the ground to be stiff and not significantly liquefiable, in which case the bedrock PGA will be used as a direct estimate for the asset ground PGA.
- The relationship between the excitation state and the damage state is determined by structural analysis: in this case FEA analysis such as illustrated in [5], which determines the fragility curves necessary to determine the damage state probabilities. In other cases, engineering judgment, laboratory testing, or field observations and experience could be utilized to determine the fragility functions or transition probabilities directly.
- The relationships between loss state and damage state is determined by engineering expertise and judgment.

7.2.4.1 Determination of Probabilities

To determine the probability of event, a simplistic Poisson analysis of the estimated earthquake history of the Pacific Northwest was conducted. The image shows a history of estimated earthquakes in the Pacific Northwest

For the sake of simplicity and illustration, let us assume the earthquakes are modeled by a Poisson process, for which earthquakes are treated as independent events with a fixed yearly probability of occurrence. (In reality, because earthquakes are driven by the accumulation and release of energy in faults and plate motion, events are not independent.)

In the data shown in Figure 7.13, the yearly occurrence probability of a CSZ magnitude 8 earthquake is 0.23% and the yearly occurrence probability of a CSZ magnitude 9 earthquake is 0.20%.

Therefore,

$$P_{E|A} = \left[p[E_0|A]\ p[E_1|A]\ p[E_2|A]\right] = \left[0.9957\ 0.0023\ 0.0020\right]$$

Note that the probabilities sum to one across the row.

Next, we must determine the probability of excitation given an event. In the case of earthquakes, this can be determined by geotechnical analysis of the fault type, earthquake magnitude, and soil and geological structure between the fault and the asset. As noted above, for the purposes of simplified illustration, we shall assume the PGA response at the surface

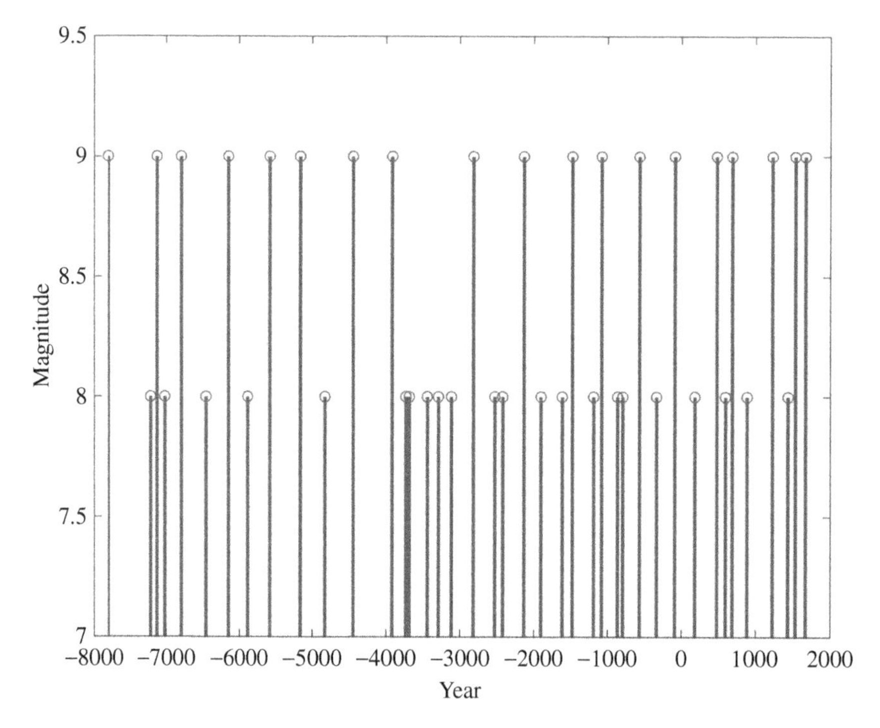

Figure 7.13 Estimated history of large (magnitude ~8) and very large (magnitude ~9) Cascadia Subduction Zone earthquakes in the past 10,000 years. Source: Data adapted from Oregon Seismic Safety Policy Advisory Commission (OSSPAC) [11].

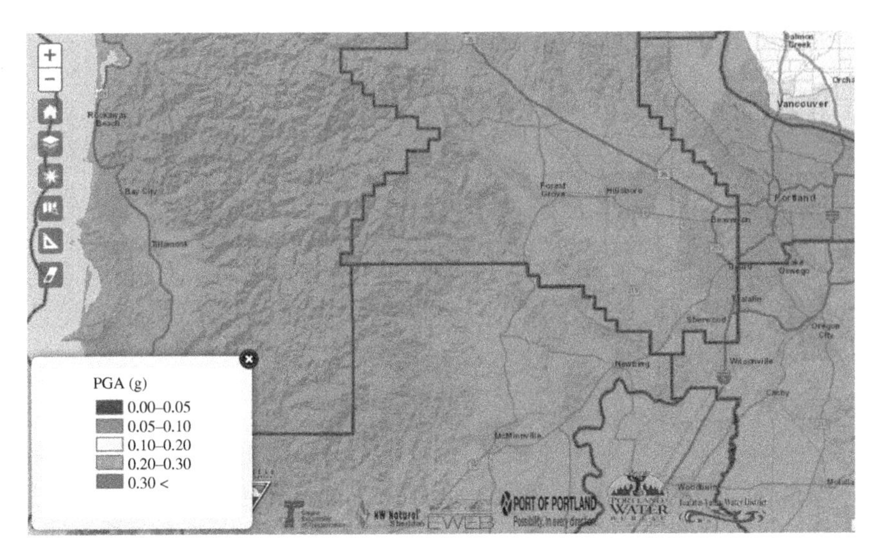

Figure 7.14 Estimated bedrock Peak Ground Acceleration (PGA) for the Oregon coast to Portland for a magnitude 8.1 Cascadia Subduction Zone earthquake. All of the Portland area to the coast is estimated to experience 0.00–0.05 g of PGA. Source: Image from the OHELP website http://ohelp.oregonstate.edu.

at the asset location to be the same as the bedrock PGA. (In reality, the PGA at the asset may be several times larger than the bedrock PGA, if the soil is soft.)

For this example, we shall use data from the Oregon Hazard Explorer for Lifelines Program website (http://ohelp.oregonstate.edu), shown in Figures 7.14 and 7.15.

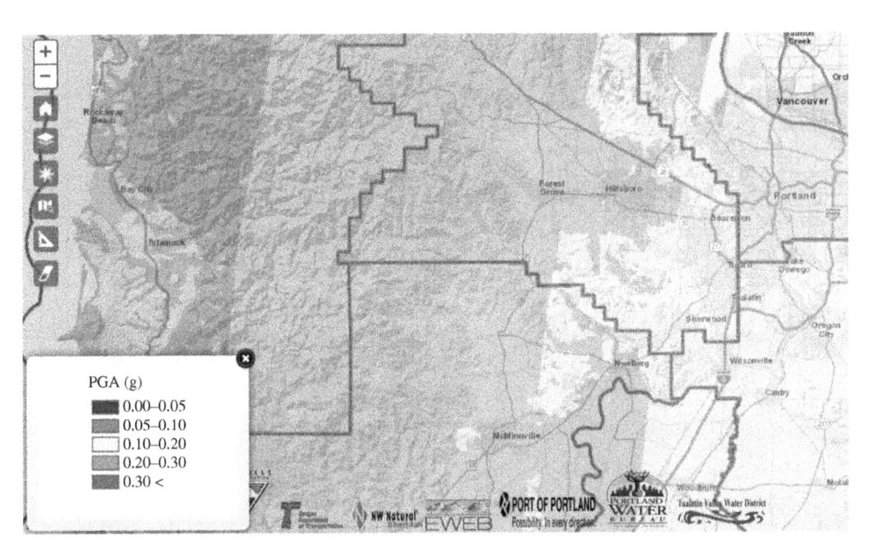

Figure 7.15 Estimated bedrock Peak Ground Acceleration (PGA) for the Oregon coast to Portland for a magnitude 9 Cascadia Subduction Zone earthquake. Most of Portland is estimated to experience 0.10–0.20 g of bedrock PGA, with the western suburbs experiencing perhaps 0.20–0.30 g of bedrock PGA. Source: Image from the OHELP website, http://ohelp.oregonstate.edu.

From Figure 7.14, all of Portland is expected to experience bedrock PGAs of 0.00–0.05 g for a magnitude 8.1 CSZ event (event state E_1), which corresponds to excitation state X_0. From Figure 7.15, the majority of Portland is expected to experience bedrock PGAs of 0.10–0.20 g (excitation state X_2) for a magnitude 9 CSZ event (event state E_2). The western suburbs may experience 0.20–0.30 g (excitation state X_3).

Therefore, let us assume that for the null no event case (E_0), there is a 100% chance of 0 g PGA (X_0), and 0% chance of any other PGA (X_1 through X_3). For a magnitude 8.1 CSZ event (E_1), there is a 100% chance of 0.05–0.10 g (X_1), and no chance of any other PGA. For a magnitude 9.0 CSZ event (E_2), let us assume there is a 50% chance of 0.10–0.20 g PGA (X_2) and a 50% chance of 0.20–0.30 g PGA (X_3).

$$P_{X|E} = \begin{bmatrix} p[X_0|E_0] & p[X_1|E_0] & p[X_2|E_0] & p[X_3|E_0] \\ p[X_0|E_1] & p[X_1|E_1] & p[X_2|E_1] & p[X_3|E_1] \\ p[X_0|E_2] & p[X_1|E_2] & p[X_2|E_2] & p[X_3|E_2] \end{bmatrix} = \begin{bmatrix} 1 & 0 & 0 & 0 \\ 0 & 1 & 0 & 0 \\ 0 & 0 & 0.5 & 0.5 \end{bmatrix}$$

To determine the probabilities for damage state given response, we need the fragility functions in Figure 7.8. For excitation state X_0, which is no shaking, there is a 100% chance of being in damage state D_0, and a 0% chance of being in the moderate (D_1) or severe (D_2) damage states. But what about X_1, which is 0.05–0.10 g? Let us use the upper end of that range, 0.10 g. Looking at the fragility function in Figure 7.16, we see that for a PGA of 0.10 g, there is a 1% chance of the damage being at least severe (D_2), a 55% of being at least moderate (D_1), and a 100% chance of being at least no damage (D_0). Assuming the damage states

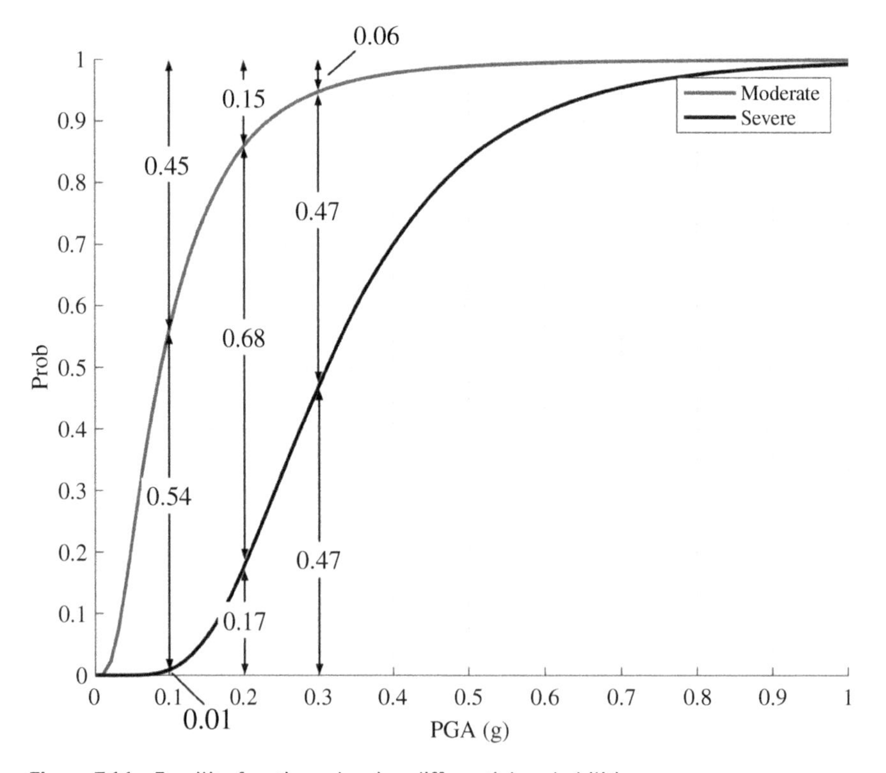

Figure 7.16 Fragility functions showing differential probabilities.

to be mutually exclusive, the probability of damage state membership is then the difference between these memberships. For example, if there is a 1% chance of being at least severe, and a 55% chance of being at least moderate, then the chances of being moderate is 55–1% = 54%

The damage state given response state matrix is then:

$$P_{D|X} = \begin{bmatrix} p[D_0|X_0] & p[D_1|X_0] & p[D_2|X_0] \\ p[D_0|X_1] & p[D_1|X_1] & p[D_2|X_1] \\ p[D_0|X_2] & p[D_1|X_2] & p[D_2|X_2] \\ p[D_0|X_3] & p[D_1|X_3] & p[D_2|X_3] \end{bmatrix} = \begin{bmatrix} 1 & 0 & 0 \\ 0.45 & 0.54 & 0.01 \\ 0.15 & 0.68 & 0.17 \\ 0.06 & 0.47 & 0.47 \end{bmatrix}$$

Lastly we need the state transition probabilities for loss state given damage state. In many cases we may be satisfied with ending our analysis at the damage state. As stated before, the loss state step provides a degree of freedom to link asset damage to a measure or unit that may be of more direct interest to our analysis. In this case, we shall have the loss states be the number of customers lost should the circuit breaker asset fail.

Often this step may require engineering judgment to inform how damage will be linked to a loss of functionality. For this example, let us assume that either moderate or severe damage (D_1 and D_2) results in a complete loss of the asset. As defined at the beginning of the section, there are two loss states: L_0, which is normal function, and L_1, which means 10,000 customers have lost power due to the loss of the circuit breaker.

$$P_{L|D} = \begin{bmatrix} p[L_0|D_0] & p[L_1|D_0] \\ p[L_0|D_1] & p[L_1|D_1] \\ p[L_0|D_2] & p[L_1|D_2] \end{bmatrix} = \begin{bmatrix} 1 & 0 \\ 0 & 1 \\ 0 & 1 \end{bmatrix}$$

7.2.4.2 Result

The probability of the asset experiencing excitation states X_0 through X_3 is:

$$P_X = P_{E|A}P_{X|E} = \begin{bmatrix} 0.9957 & 0.0023 & 0.0020 \end{bmatrix} \begin{bmatrix} 1 & 0 & 0 & 0 \\ 0 & 1 & 0 & 0 \\ 0 & 0 & 0.5 & 0.5 \end{bmatrix}$$

$$= \begin{bmatrix} 0.9957 & 0.0023 & 0.0010 & 0.0010 \end{bmatrix}$$

There is a 99.57% chance that the circuit breaker asset will see 0 g PGA in the year. There is a 0.23% chance the circuit breaker will see a 0.05–0.10 g PGA in the year, a 0.10% chance it will see a 0.10–0.20 g PGA in the year, and a 0.10% chance it will see a 0.20–0.30 g PGA in the year.

The probability of damage is calculated by integrating over the next set of probabilities: damage given excitation.

$$P_D = P_{E|A}P_{X|E}P_{D|X} = P_X P_{D|X} = \begin{bmatrix} 0.9957 & 0.0023 & 0.0010 & 0.0010 \end{bmatrix} \begin{bmatrix} 0 & 0 & 1 \\ 0.45 & 0.54 & 0.01 \\ 0.15 & 0.68 & 0.17 \\ 0.06 & 0.47 & 0.47 \end{bmatrix}$$

$$= \begin{bmatrix} 0.9969 & 0.0024 & 0.0007 \end{bmatrix}$$

The probability of the circuit breaker asset being undamaged in the year due to the two events considered is 99.69%. The probability of it being moderately and severely damaged is 0.24% and 0.07%, respectively.

Lastly we can determine the probability of the loss states, that is, loss of customers.

$$P_L = P_{E|A}P_{X|E}P_{D|X}P_{L|D} = P_D P_{L|D} = \begin{bmatrix}0.9969 & 0.0024 & 0.0007\end{bmatrix}\begin{bmatrix}1 & 0\\0 & 1\\0 & 1\end{bmatrix} = \begin{bmatrix}0.9969 & 0.0031\end{bmatrix}$$

Therefore there is a 99.69% chance there will be no loss of customers this year due to seismic damage to the circuit breaker. There is a 0.31% chance there will be a loss of 10,000 customers due to seismic damage to the circuit breaker.

7.3 Power System Monte Carlo Analysis

Using the framework in Section 7.2, we can determine state probabilities for different power system assets. But what about a probabilistic analysis of an entire power system? If the equations for power flow were simple, we may be able to analytically propagate the asset probability distributions through the power flow equations. But in almost all realistic cases this is too complex.

Instead we use a Monte Carlo analysis, in which we randomly sample our inputs (i.e. the damage or loss state of power system components) over many trials to develop a picture of the state of the overall system in a probabilistic sense.

To do this, we need to generate thousands or millions of randomly determined power system states, solve the power flow and log the result, then repeat with a new set of randomly determined asset states.

The risk analysis of Section 7.2 provides the tool we need to determine the states of the power system asset.

Consider a simple power system, as shown in Figure 7.17.

There are four classes of power system model assets: generator, load, bus, and branch (i.e. transmission or distribution line).

We can further classify the assets by region.

- Coast generator
- Coast load

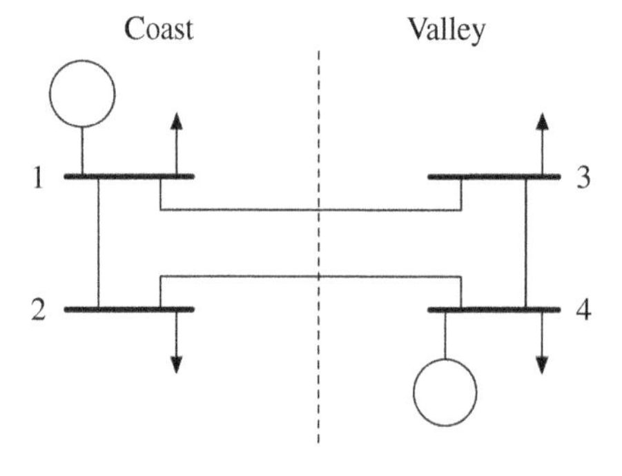

Figure 7.17 Sample 4 bus power system. There are two geographic regions: coast and valley.

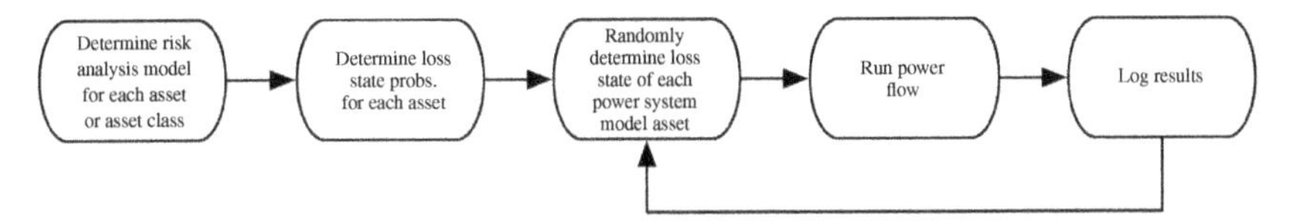

Figure 7.18 Flowchart for Monte Carlo analysis of power systems in conjunction with risk modeling.

- Coast bus
- Coast branch
- Valley generator
- Valley load
- Valley bus
- Valley branch

Following the methodology in Section 7.2, we can determine the loss (or damage) state probabilities for each of these eight asset classes. (What about the branches, which cross geographic regions? The most straightforward solution would be to assign it to whichever region it mostly occupies.)

For illustration, let us assume we have completed a risk model for the coast generator, and that its loss states are determined to be L_0 = no capacity loss, L_1 = half capacity loss, L_2 = full capacity loss, and that its loss probabilities are:

$$P_L = \begin{bmatrix} 0.70 & 0.20 & 0.10 \end{bmatrix}$$

At the beginning of one of the thousands or trials that will be run, we draw a random number between 0 and 1 and get 0.85. This is greater than 0.70, but not greater than 0.70 + 0.20. Therefore, our coast generator will be in loss state L_1 = half capacity loss. (We do this also for all other assets, each using their own unique random number draw to determine each of their unique loss states for this particular trial.)

Then in the power flow model we set the coast generator capacity to half of its original (and modify all other asset states), run the power flow, and log the result. Then we draw a new random number, determine the new loss state, and run a new power flow, repeating this process for thousands or millions of trials until the statistics of the logged results stabilize.

This process is illustrated in the flow chart of Figure 7.18.

7.4 Summary

In this chapter, we have:

- Reviewed common power systems reliability and resilience metrics, such as LOLP, LOLE, SAIFI, SAIDI, and CAIDI
- Introduced a system performance curve and defined the resilience metric R as the ratio of the area under the curve of the actual system performance to the target system performance.

- Introduced fragility and vulnerability functions. Fragility functions determine a probability of a system being in a certain state as a function of excitation or response. Vulnerability functions determine the loss in function (e.g. damage) has a function of excitation or response.
- Provided two examples of typical models for fragility and vulnerability functions: a logistic function, and a log-normal function.
- Introduced a typical risk analysis modeling framework, and worked through comprehensive example of using the framework for quantifying loss of a circuit breaker due to earthquake.
- Described the flow and structure of a Monte Carlo analysis for including probabilistic descriptions of asset loss (or damage) in power systems analysis.

References

1 Kazamaa, M. and Noda, T. (2012). Damage statistics (Summary of the 2011 off the Pacific Coast of Tohoku Earthquake damage). *Soils and Foundations* 52 (5): 780–792.
2 Park, J., Nojima, N., and Reed, D. (May 2006). Nisqually earthquake electric utility analysis. *Earthquake Spectra* 22 (2): 491–509.
3 United States Geological Survey (USGS). The modified Mercalli Intensity scale. https://www.usgs.gov/programs/earthquake-hazards/modified-mercalli-intensity-scale.
4 Porter, K. (2015). Beginner's guide to fragility, vulnerability, and risk. In: *Encyclopedia of Earthquake Engineering* (ed. M. Beer, I.A. Kougioumtzoglou, E. Patelli and S.K. Au). Berlin, Heidelberg: Springer https://doi.org/10.1007/978-3-642-35344-4_256.
5 Zareeia, S.A., Hosseinib, M., and Ghafory-Ashtianyb, M. (2017). Evaluation of power substation equipment seismic vulnerability by multivariate fragility analysis: case study on a 420 kV circuit breaker. *Soil Dynamics and Earthquake Engineering* 92: 79–94.
6 IEEE Recommended Practice for Seismic Design of Substations (IEEE Std 693), 2006.
7 Pacific Earthquake Engineering Research Center online library, https://nisee.berkeley.edu/elibrary/.
8 Porter, K. (2003). An overview of PEER's performance-based Earthquake engineering methodology. In: *Proc. Ninth International Conference on Applications of Statistics and Probability in Civil Engineering*.
9 Porter, K., Farokhnia, K., Vamvatsikos, D., and Cho, I. (2015). Guidelines for component-based analytical vulnerability assessment of buildings and nonstructural elements.
10 O-HELP oregon hazard explorer for lifelines program. http://ohelp.oregonstate.edu.
11 Oregon Seismic Safety Policy Advisory Commission (OSSPAC) (2013). The oregon resilience plan.

Part II

Enabling Resiliency

8

Resiliency-Driven Distribution Network Automation and Restoration

Yin Xu[1], *Chen-Ching Liu*[2], *and Ying Wang*[1]

[1] *School of Electrical Engineering, Beijing Jiaotong University, Beijing, China*
[2] *The Bradley Department of Electrical and Computer Engineering, Virginia Polytechnic Institute and State University, Blacksburg, VA, USA*

8.1 Optimal Placement of Remote-Controlled Switches for Restoration Capability Enhancement

The objective of distribution system restoration (DSR) is to restore the maximal amount of load as quickly as possible by a series of switching operations after the fault is isolated within a feeder section [1]. The shorter the time to conduct a DSR plan is, the more reliable the system will be [2]. The efficiency of a DSR plan is not only related to the number of switching operations but also associated with the time to complete each switching operation.

Remote-controlled switches (RCSs) enable the distribution system operator in the operating center to change the status of switches from remote. Compared with manual operations of switches by the field crew, remote control makes it much faster to implement switching operations [3]. Therefore, upgrading manual switches to RCSs improve the capability of distribution system restoration [4].

8.1.1 RCS Placement

Replacing manual switches by RCSs considerably decreases the outage time. However, the installation of RCSs is costly, so trade-offs should be performed between cost and benefit. Critical switches that are frequently needed in distribution system restoration should be identified and upgraded. In a distribution system, the problem of RCS placement is to determine the optimal locations and number of RCSs. The task of RCS placement is to maximize the benefit at minimum cost. In general, RCS placement should take into account the possible fault scenarios and associated DSR plans.

RCS placement is used in two scenarios. One is for distribution network planning and the other is for distribution network upgrading. In the former, the network topology is given, and the RCS can be installed anywhere it is needed. The status of switches depends on the operational requirements of the system [5]. In the latter, some of manual switches are selected and upgraded to RCSs to serve the intended purposes [6]. In this chapter, the RCS placement problem for the second scenario is formulated.

Resiliency of Power Distribution Systems, First Edition.
Edited by Anurag K. Srivastava, Chen-Ching Liu, and Sayonsom Chanda.

8.1.2 Problem Formulation

Let $\boldsymbol{S}$ represent the candidate switches to be upgraded to RCSs and $\boldsymbol{X} \subseteq \boldsymbol{S}$ denotes the selected set of switches to be upgraded. The symbol $|\cdot|$ denotes the number of elements in a set. Let $n_{cs} = |\boldsymbol{S}|$ and $n_s = |\boldsymbol{X}|$. A binary variable x_i is used to represent whether a switch $S_i \in \boldsymbol{S}$, $i = 1, 2, \ldots, n_{cs}$ is selected to be upgraded, that is, if $S_i \in \mathbf{X}$, $x_i = 1$; $x_i = 0$, otherwise.

Given a distribution system, a fault Z_f, and a set of switches $\boldsymbol{S}_R$, three non-linear functions, that is, $F_1(Z_f, \mathbf{S}_R)$, $F_2(Z_f, \boldsymbol{S}_R)$, and $F_3(Z_f, \boldsymbol{S}_R)$, are defined. The function $F_1(Z_f, \boldsymbol{S}_R)$ represents the DSR plan that operates switches in $\boldsymbol{S}_R$ to restore interrupted load after fault Z_f is isolated. If no feasible DSR plan is available, $F_1(Z_f, \boldsymbol{S}_R) = \text{NULL}$. The functions $F_2(Z_f, \boldsymbol{S}_R)$ and $F_3(Z_f, \boldsymbol{S}_R)$ return the amount of load restored by the DSR plan and the number of load-transfer actions in the DSR plan, respectively. The RCS placement is then formulated as follows:

$$\min\ n_s \tag{8.1}$$

subject to:

If

$$F_1(Z_f, \boldsymbol{S}) \neq \text{NULL, then, } F_1(Z_f, \boldsymbol{X}) \neq \text{NULL}, \forall Z_f \tag{8.2}$$

$$F_2(Z_f, \boldsymbol{S}) = F_2(Z_f, \boldsymbol{X}), \forall Z_f \tag{8.3}$$

$$F_3(Z_f, \boldsymbol{X}) \leq 1, \forall Z_f \tag{8.4}$$

$$x_i \in \{0, 1\}, i = 1, 2, \ldots, n_{cs} \tag{8.5}$$

$$\boldsymbol{X} = \{S_i | i : x_i = 1\} \tag{8.6}$$

The objective is to minimize the number of RCSs subject to constraints (8.2)–(8.6). Constraint (8.2) indicates that if there exists a feasible DSR plan that operates switches in $\boldsymbol{S}$, restoration can be accomplished by operating switches in $\boldsymbol{X}$. Constraint (8.3) means that the DSR plan using the selected switches can restore the same amount of load as the DSR plan using all available switches. Constraint (8.4) guarantees that load-transfer actions will not be implemented more than once in a DSR plan. Load transfer actions are not desirable because they will lead to a temporary violation of the radial configuration [7].

The RCS placement problem is a combination problem with non-linear constraints. The combination number will increase exponentially as the distribution system becomes larger. Finding an optimal plan for RCS placement is a challenging task.

8.1.3 A Systematic Method to Determine RCS Placement

In this section, a systematic method to determine RCS placement will be presented. The concepts of load group and switch group will be introduced. Using these concepts, the RCS placement problem is transformed into a weighted set cover problem, which is then solved by a greedy algorithm.

8.1.3.1 Load Groups and Switch Groups

A load group (LG) is a set of connected load zones on a feeder. For each LG, a corresponding switch group (SG) is defined as a set of switches consisting of a tie switch and all sectionalizing switches connected between load zones within and outside the LG. A LG–SG pair indicates a candidate restoration strategy, that is, load zones in a LG can be restored by operating switches in the corresponding SG. Note that operational constraints should be evaluated to ensure that the restoration strategy is feasible. A procedure is proposed in [4] to search for all feasible LG–SG pairs efficiently.

8.1.3.2 Problem Transformation

For a LG–SG pair, if load zones in a LG can be restored by operating the switches in the corresponding SG without violating any operational constraints, the LG is said to be *covered* by the corresponding SG. Define $\boldsymbol{S}$ as the set of candidate switches and $\boldsymbol{L}$ the set of load zones. Assume that $|\boldsymbol{L}| = n$ and the load zones are indexed by i. The number of LG–SG pairs is m and the LG–SG pairs are indexed by j. Use $\boldsymbol{G}_{\mathrm{L},j}$, $\boldsymbol{G}_{\mathrm{S},j}$, and $\boldsymbol{G}_{\mathrm{L},j} - \boldsymbol{G}_{\mathrm{S},j}$ to denote the jth LG, SG, and LG–SG pair, respectively. Suppose that each load zone is covered by at least one SG, then $\boldsymbol{L} = \cup_{i=1}^{m} \boldsymbol{G}_{\mathrm{L},j}$. The relationship of load zones and LGs is represented by a matrix $[a_{ij}]_{n\times m}$, where a_{ij} is a binary variable: if the ith load zone belongs to $\boldsymbol{G}_{\mathrm{L},j}$, $a_{ij} = 1$; otherwise, $a_{ij} = 0$. A binary variable z_j is introduced for $\boldsymbol{G}_{\mathrm{L},j} - \boldsymbol{G}_{\mathrm{S},j}$: if $\boldsymbol{G}_{\mathrm{L},j} - \boldsymbol{G}_{\mathrm{S},j}$ is selected, $z_j = 1$; otherwise, $z_j = 0$. Then, the RCS placement problem can be reformulated as a weighted set cover (WSC) problem as follows:

$$\min \ |\cup_{j:z_j=1} \boldsymbol{G}_{\mathrm{S},j}| \tag{8.7}$$

subject to:

$$z_j \in \{0,1\}, \quad j = 1,2,\ldots,m \tag{8.8}$$

$$\sum_{j=1}^{m} a_{ij} z_j \geq 1, \quad i = 1,2,\ldots,n \tag{8.9}$$

The objective is minimizing the number of switches in the selected SGs. Constraint (8.9) ensures that all the load zones are covered by at least one selected SG. Denote the set of selected SGs by $\boldsymbol{C}$. Once $\boldsymbol{C}$ is determined, the set of switches to be upgraded can be obtained by:

$$\boldsymbol{X} = \cup_{j\in \mathrm{C}} \boldsymbol{G}_{\mathrm{S},j} \tag{8.10}$$

8.1.3.3 Greedy Algorithm to Solve WSC

A greedy algorithm typically goes through a series of steps and makes the decision that achieves most progress toward the goal in each step. For the WSC problem defined by (8.7)–(8.9), a LG–SG pair is selected based on the following rules until all load zones are covered.

(1) Choose the LG–SG pair with the smallest $N_{\mathrm{s}}/N_{\mathrm{l}}$, where N_{s} and N_{l} denote the number of switches and load zones in the corresponding SG and LG, respectively.
(2) If two or more LG–SG pairs satisfy (1), choose the pair with the smallest $N_{\mathrm{s}}/S_{\mathrm{l}}$, where S_{l} denotes the total amount of load in the load zones belonging to the corresponding LG.
(3) Select the first LG–SG pair if the choice is not unique based on rule (1) and rule (2).

It can be proved that the above algorithm can produce a near-optimal solution of the WSC problem for a distribution system even when the number of feeders is large. More details can be found in [4].

8.2 Resiliency-Driven Distribution System Restoration Using Microgrids

A major disaster, such as a hurricane, may cause severe damages to transmission and distribution facilities. As a result, the utility power may not be able to access the interrupted loads. In this case, local resources, such as distribution generators (DGs) and microgrids (MGs), can be used to serve critical loads on distribution feeders. In this section, a resiliency-based restoration method using MGs to serve critical loads in a radial distribution network is introduced.

8.2.1 Challenges of Using Microgrids for Service Restoration

When using MGs as emergency sources for service restoration to critical loads, several issues should be considered [8].

A major issue is that the generation capacity of DGs within a MG is relatively small compared with generators in a bulk power system. Therefore, their ability to withstand disturbances is limited. Restorative actions, such as picking up loads and energizing transformers, can cause significant fluctuations in voltage and frequency, which may stall the prime mover of DGs or trigger protective relays. Such events are undesirable and should be avoided. Therefore, dynamic performance of DGs should be taken into consideration when making decisions on restoration strategies.

Another issue is the scarcity of generation resources. After an extreme event, fuels for DGs in a MG may not be sufficient to support the DGs to operate in full capacity during the entire outage period.

8.2.2 Critical Load Restoration Problem

Suppose that after an extreme event, the utility power becomes unavailable, faulted zones are isolated, and MGs operate in the islanded mode. A MG is considered available for service restoration if there is spare power capacity and generation resources in the MG after the critical loads within the MG are served. The interrupted loads on the distribution feeders can be connected to available MGs by closing tie switches and MG switches. The determination of the optimal strategy using available MGs to serve critical loads in the distribution system is referred to as the *critical load restoration problem*, which can be formulated as below:

$$\max \sum_{i \in \boldsymbol{L}} w_i \tau_i \tag{8.11}$$

subject to:

$$\underline{f} \leq f_k \leq \overline{f}, \quad k \in \boldsymbol{M} \tag{8.12}$$

$$\underline{f^{\text{tr}}} \leq f_k^{\text{tr}} \leq \overline{f^{\text{tr}}}, \quad k \in \boldsymbol{M} \tag{8.13}$$

$$\underline{V^{\text{tr}}} \leq V_g^{\text{tr}} \leq \overline{V^{\text{tr}}}, \quad g \in \boldsymbol{G} \tag{8.14}$$

$$I_g^{\text{tr}} \leq \overline{I^{\text{tr}}}, \quad g \in \boldsymbol{G} \tag{8.15}$$

$$\sum_{i \in \boldsymbol{R}_k} t_i P_i \leq E_k, \quad k \in \boldsymbol{M} \tag{8.16}$$

$$P_u^s - jQ_u^s = \left(\overline{V}_u^s\right)^* \sum_{v \in \Omega_u} \sum_t Y_{uv}^{st} \overline{V}_v^t, \quad u \in \boldsymbol{\Omega}, \quad s,t \in \{a,b,c\} \tag{8.17}$$

$$\underline{V} \leq V_u^s \leq \overline{V}, \quad u \in \boldsymbol{\Omega}, \quad s \in \{a,b,c\} \tag{8.18}$$

$$I_l^s \leq \overline{I}_l, \quad l \in \boldsymbol{\mathcal{E}}, \quad s \in \{a,b,c\} \tag{8.19}$$

$$\begin{cases} P_k \leq \overline{P}_k \\ Q_k \leq \overline{Q}_k \end{cases}, \quad k \in \boldsymbol{M} \tag{8.20}$$

The bold letters $\boldsymbol{M}$,$\boldsymbol{G}$, $\boldsymbol{L}$, $\boldsymbol{\mathcal{E}}$, $\boldsymbol{\Omega}$, $\boldsymbol{\Omega}_u$, and $\boldsymbol{R}_k$ denote sets of available MGs, DGs in service, load zones, lines in service, buses restored, buses adjacent to bus u, and load zones restored by MG k, respectively. MGs, DGs, load zones, lines, buses, and phases are indexed by k, g, i, l, u (and v), and s (and t), respectively. The variables τ, w, f, V, I, P, Q, and E denote the time period during which critical load is served, priority of critical load, frequency, voltage, current, real power, reactive power, and amount of generation resources (in kWh), respectively. The subscript tr indicates the transient value of a variable. The symbols $\overline{\blacksquare}$ and $\underline{\blacksquare}$ denote, respectively, the upper and lower limit of the corresponding variable.

The objective is maximizing the cumulative service time of critical loads considering their importance levels, as indicated by (8.11).

Constraints (8.12)–(8.15) are dynamic constraints. Constraint (8.12) indicates that MGs should maintain stable and the steady-state frequency must be within a preset range. Constraints (8.13)–(8.15) ensure that the fluctuations in frequency, voltage, and current are not too large when performing restorative actions.

Constraint (8.16) defines the limits on the amount of generation resources that each MG can provide to the external grid.

Constraints (8.17)–(8.20) are operational constraints. Constraint (8.17) represents the unbalanced three-phase power flow equations. Constraints (8.18)–(8.19) ensure steady-state bus voltages and line currents are within a preset range. Constraint (8.20) guarantees that the power outputs of MGs do not exceed the upper limits.

In addition, the topological constraint should also be satisfied during the restoration procedure. The radiality of the network topology should be maintained.

8.2.3 A Graph-Theoretic Algorithm

The dynamic constraints and the unbalanced three-phase power flow equations make the critical load restoration problem non-convex and difficult to solve. A graph-theory-based heuristic is proposed to determine critical load restoration strategy.

8.2.3.1 Critical Load Restoration Procedure

The procedure to find the critical load restoration strategy includes four steps:

Step 1: Construct Restoration Trees: For each MG, a restoration tree is construct. The MG is defined as the root. The "shortest" paths between the MG and critical load zones (CLZs) are identified. Here, the "length" of a path is evaluated by the total amount of load along the path. Feasibility of the path should be evaluated. All feasible paths together form the restoration tree.

Step 2: Form Load Groups: For a MG, load groups are sets of CLZs that can be restored at the same time by the MG. For each MG, there may exist several load groups. Feasible load groups will be identified based on the restoration trees.

Step 3: Problem transformation: With feasible load groups, the critical load restoration problem is transformed into a maximum coverage problem, which can be formulated as a mixed integer linear program and solved by off-the-shelf solvers.

Step 4: Determine Switching Actions: Step 3 determines the critical loads to be restored and the paths from MGs to CLZs. In this step, a sequence of switching operations is determined, which ensures that the dynamic simulations will not be violated during the restoration process.

8.2.3.2 Construct Restoration Trees

A distribution system can be modeled as an undirected graph, denoted by $G = (\boldsymbol{V}, \boldsymbol{E})$, where $\boldsymbol{V}$ and $\boldsymbol{E}$ are the set of nodes and arcs, respectively. Elements in $\boldsymbol{V}$ represent MGs and load zones, while elements in $\boldsymbol{E}$ represent switches. The amount of load in a load zone is defined as the weight of the corresponding node. The weight of a path is then defined as the sum of weights of nodes on the path. For a "MG-CLZ" pair, the *restoration path* between the MG and CLZ is the path with the smallest weight among all paths connecting them. Restoration paths are determined using the modified *Dijkstra*'s algorithm [9]. Power flow calculations and dynamic simulations are performed using GridLAB-D [10] to validate the feasibility of restoration paths. Unbalanced three-phase power flow is calculated to examine if operational constraints, that is, (8.17)–(8.20), are satisfied. If no operational constraint is violated, dynamic simulation will be conducted. Restorative actions are implemented to pick up load zones on the restoration path one by one. The time interval between two restorative actions is set to be one minute. Dynamic constraints (8.12)–(8.15) are evaluated. If the restoration path satisfies all constraints, it is considered a feasible path. All feasible restoration paths starting from a MG form the restoration tree of the MG.

8.2.3.3 Form Load Groups

For a MG, all possible combinations of CLZs in its restoration tree can be obtained, which is called the load groups with respect to the MG. For each load group, the feasibility is tested by performing power flow calculations and dynamic simulations using GridLAB-D. Load groups violating any operational or dynamic constraint is excluded from the candidate set of load groups. For each load group j remaining in the candidate set, a service time τ_j during which MG k can serve the critical loads in the load group can be calculated by $\tau_j = E_k / P_{\text{sum},j}$, where E_k is the amount of generation resources (kWh) available in MG k and $P_{\text{sum},j}$ is the total amount of load (kW) in load group j.

8.2.3.4 Maximum Coverage Problem Formulation

Assume that load group j is associated with MG k and contains CLZs $i_1, \ldots, i_\text{p} (p \geq 1)$. Then CLZs $i_1, \ldots, i_\text{p}$ can be restored by MG k through the restoration paths in the restoration tree. In other words, CLZs $i_1, \ldots, i_\text{p}$ is *covered* by load group j. The objective (8.11) can be achieved by choosing a set of load groups covering a maximum number of CLZs weighted by their priority. Therefore, the critical load restoration problem can be modeled as a *maximum coverage problem (MCP)*, that is:

$$\max \sum_{j:\mathbf{g}_j \in \boldsymbol{G}_{\text{uni}}} y_j \tau_j w_{\text{sum},j} \tag{8.21}$$

subject to:

$$y_j \in \{0,1\}, \quad \forall j \in \{j : \mathbf{g}_j \in \boldsymbol{G}_{\text{uni}}\} \tag{8.22}$$

$$\sum_{j:\mathbf{g}_j \in \boldsymbol{G}_k} y_j \leq 1, \quad \forall k \in \boldsymbol{M} \tag{8.23}$$

$$\sum_{j:i \in \mathbf{g}_j} y_j \leq 1, \quad \forall i \in \boldsymbol{L} \tag{8.24}$$

The set $\boldsymbol{G}_{\text{uni}}$ is the universal set of load groups, $\boldsymbol{G}_{\text{uni}} = \cup_{k\in \boldsymbol{M}} \boldsymbol{G}_k$, where $\boldsymbol{G}_k$ is the set of load groups corresponding to MG k. The set $\mathbf{g}_j$ denotes load group j, $\mathbf{g}_j \subseteq \boldsymbol{L}$ and $\mathbf{g}_j \in \boldsymbol{G}_{\text{uni}}$. The binary variable y_j indicates the status of load group j: if load group j is selected, $y_j = 1$; otherwise, $y_j = 0$. The parameter $w_{\text{sum},j}$ is the total weighting factor of load zones in load group j.

Constraint (8.23) ensures that at most one load group associated with the same MG is selected. Constraint (8.24) avoids a load zone being covered by two or more selected load groups.

The MCP defined by (8.21)–(8.24) is a mixed integer linear program and can be readily solved by commercial or open-source solvers. The solution of MCP meets the operational constraints, dynamic constraints, generation-resource constraints, and radiality constraint. Specifically, the feasibility of load groups has been evaluated by power flow calculations and dynamic simulations. The service time τ_j satisfies the limits on generation resources. The radiality constraint is satisfied because the restoration trees are radial in topology.

8.2.3.5 Determine Switching Actions

After solving the MCP, load zones to be restored and the restoration paths are determined. The final step is to determine the switching operations for the implementation of the restoration plan. It is necessary to balance the dynamic performance and efficiency (number of switching operations). Restoring loads zone by zone avoids large transients but takes more time. Conversely, restoring all zones at one time is more efficient, but the transient constraints may be violated. A greedy strategy is employed to decide the switching operations. In each step, the maximum number of load zones is picked up without violating dynamic constraints. The process continues until all zones on the restoration paths are restored.

8.2.4 Case Study

8.2.4.1 System Information

A modified 32-node network [11] with 4 MGs and 5 CLZs is used to validate the proposed method. Figure 8.1 shows the one-line diagram of the test system.

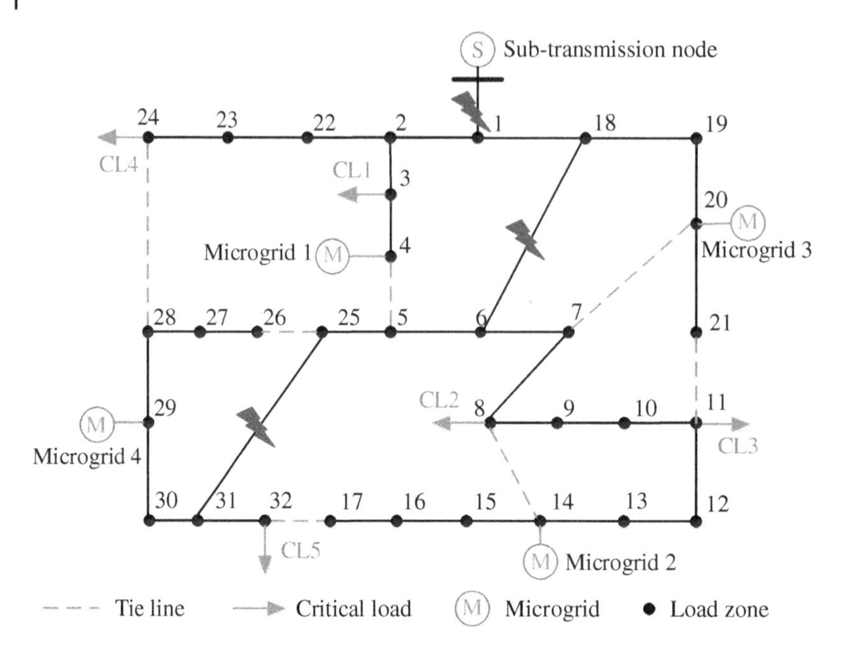

Figure 8.1 One-line diagram of the modified 32-node test system.

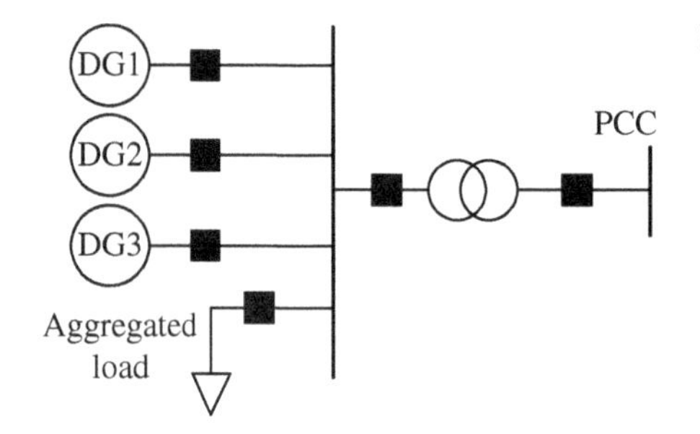

Figure 8.2 A simplified MG model.

A simplified MG model is used, which is presented in Figure 8.2. Each MG is modeled as a grid with a single bus, three DGs and an aggregated load representing critical load within the MG. The types of DGs considered in this study include synchronous generator (SG), microturbine generation system (MTGS), and battery energy storage system (BESS). Detailed generation and load information is shown in Table 8.1.

8.2.4.2 Results

There are 5 CLZs, that is, Z3, Z8, Z11, Z24, and Z32. The weighting factor of each load is 50, 50, 50, 30, and 30, respectively. The weighting factor of non-critical load zones is set to 1. It is assumed that three faults occur after an extreme event, as marked by the bolt symbol in Figure 8.1. The outage duration time is assumed to be five hours.

Step 1: Construct Restoration Trees Three restoration trees are constructed with two restoration paths for MG1, two for MG3, and five for MG4. The results are listed in Table 8.2. It is important to perform dynamic simulations to evaluate the feasibility of restoration paths. For instance, although MG2 is near CLZ3, the restoration path MG2-Z14-Z13-Z12-Z11 does

Table 8.1 Data of generation and load in MGs.

MG ID		1	2	3	4
DGs		2 SGs, 1 MTGS	2 SGs, 1 BESS	3 SGs	2 SGs, 1 BESS
Generation capacity	Real power (kW)	936	90	724	1324
Generation resource (kWh)		716	600	1700	4600
Critical load	Demand (kW)	182	18	145	164
	Power factor	0.9	0.9	0.9	0.9

Table 8.2 Restoration trees.

MG ID	Restoration paths
1	(1) M1 – Z4 – Z3 (2) M1 – Z4 – Z5 – Z6 – Z7 – Z8
2	None
3	(1) M3 – Z20 – Z7 – Z8 (2) M3 – Z20 – Z21 – Z11
4	(1) M4 – Z29 – Z28 – Z27 – Z26 – Z25 – Z5 – Z4 – Z3 (2) M4 – Z29 – Z28 – Z27 – Z26 – Z25 – Z5 – Z6 – Z7 – Z8 (3) M4 – Z29 – Z28 – Z27 – Z26 – Z25 – Z5 – Z6 – Z7 – Z20 – Z21 – Z11 (4) M4 – Z29 – Z28 – Z24 (5) M4 – Z29 – Z30 – Z31 – Z32

not meet dynamic constraints because the inrush current flowing through the BESS when energizing the transformer will trigger the overcurrent protection of the inverter. Note that the rated current of the inverter is 45.5 A and the threshold value of the overcurrent protection of the inverter is 68 A.

Step 2: Form Load Groups Three load groups are identified for MG1, three for MG3, and nine for MG4, as given in Table 8.3. For each load group, the sum of weighting factors $w_{\text{sum},j}$ and service time τ_j are calculated. In Table 8.3, only CLZs in the load groups are given.

Step 3: Formulate and Solve MCP The critical load restoration problem is reformulated as a MCP, which is solved by the MATLAB MILP Solver [12]. Load groups 1, 6, and 13 are selected. The corresponding service time is 3.98, 3.40, and 3.97 hours, respectively. Five CLZs are restored. The cumulative service time to loads weighted by their priority levels is 636.86 hours.

Step 4: Determine Switching Actions Load zones that will not be restored are first disconnected from the restoration paths. Critical load CL1 is picked up by MG1 through two actions. Critical loads CL2 and CL3 are picked up by MG 3 through five actions. Critical loads CL4 and CL5 are restored by MG4 through six actions. The restoration paths are indicated by green lines in Figure 8.3. As an example of dynamic performance, the frequency deviation of MG 3 during the restorative process is shown in Figure 8.4.

Table 8.3 Load groups.

Index j	Source	Critical load zones	$w_{\text{sum},j}$	τ_j (hour)
1	MG 1	Z3	51	3.98
2	MG 1	Z8	54	1.23
3	MG 1	Z3, Z8	104	1.02
4	MG 3	Z8	52	4.86
5	MG 3	Z11	52	5
6	MG 3	Z8, Z11	53	3.40
7	MG 4	Z3	57	5
8	MG 4	Z8	58	4.51
9	MG 4	Z8, Z11	110	3.88
10	MG 4	Z24	32	5
11	MG 4	Z32	33	5
12	MG 4	Z3, Z8	109	3.83
13	MG 4	Z24, Z32	64	3.97
14	MG 4	Z3, Z24	87	3.97
15	MG 4	Z3, Z32	89	3.97

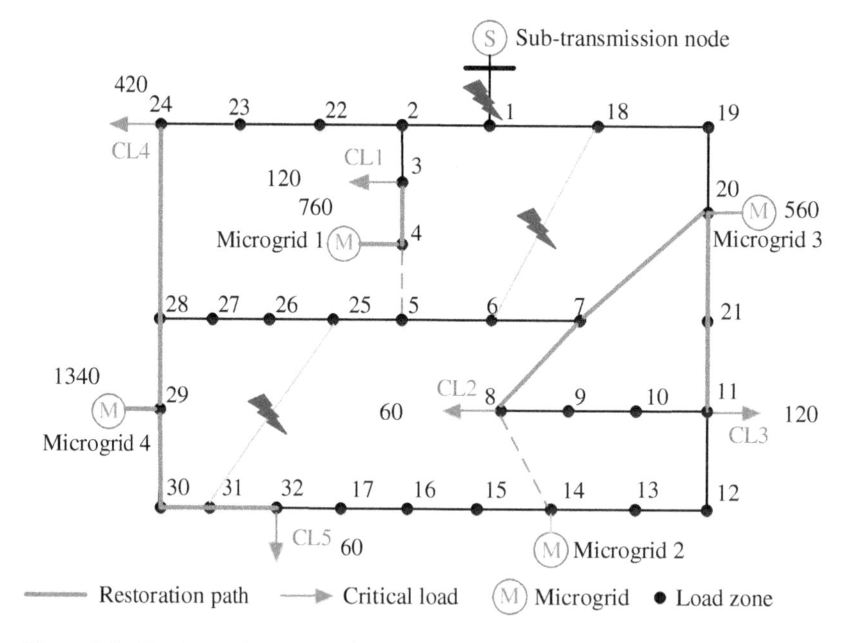

Figure 8.3 Topology after restoration.

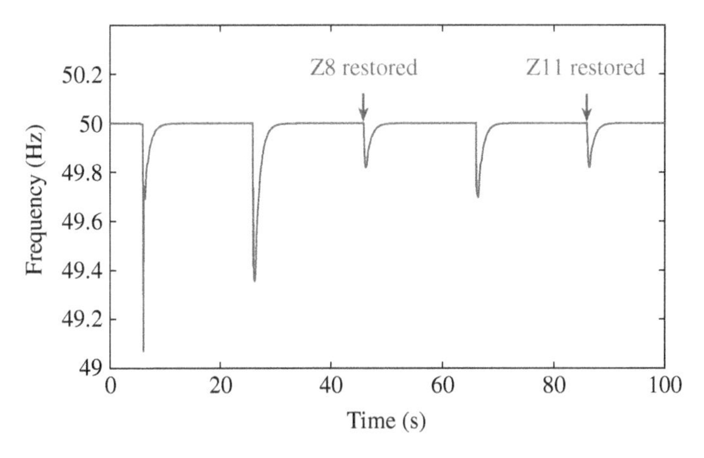

Figure 8.4 Frequency variations of MG 3.

8.2.4.3 Application

The proposed method is applied to the Avista distribution system that serves Pullman, including Washington State University (WSU). There are three critical loads on the WSU campus. The Hospital and Pullman City Hall are two critical loads on the distribution feeder. Three DGs are available on the WSU campus and can be used to restore critical loads. The presented method is used to obtain the restoration strategy for the Pullman-WSU system. Critical loads are restored through a restoration path with one transformer and nine zones as illustrated in Figure 8.5. The restoration strategy is implemented in seven switching actions. The time interval between two restorative actions is one minute. In Actions 1–3, critical loads within WSU are restored. The City Hall and Hospital are restored in Action 5 and Action 7, respectively. The dynamics of system frequency and generator voltages are shown in Figure 8.6, indicating that the dynamic constraints are satisfied.

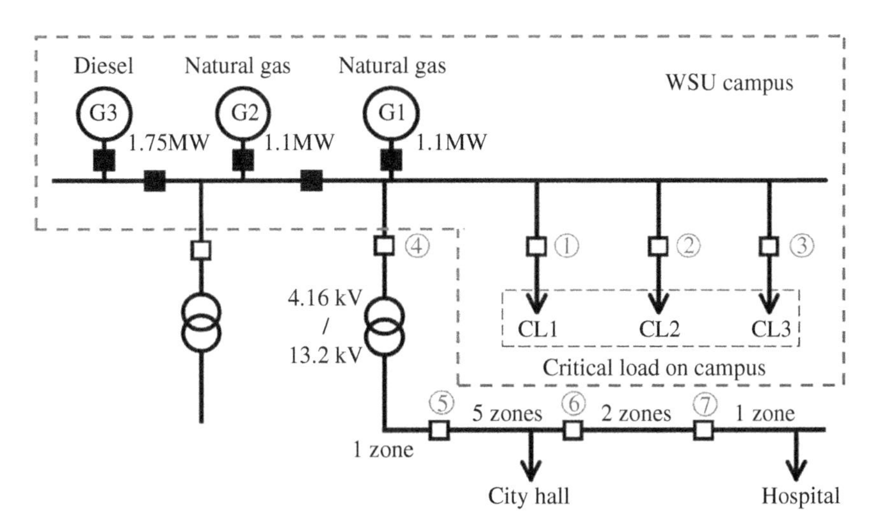

Figure 8.5 The restoration strategy for Pullman-WSU system.

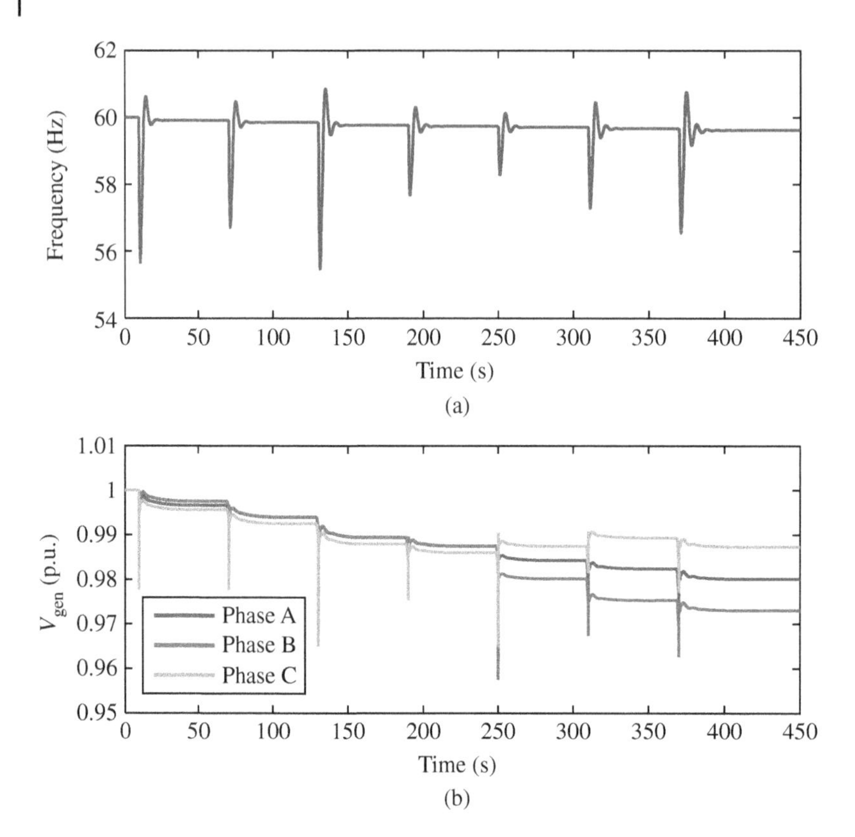

Figure 8.6 (a) System frequency and (b) generator voltages.

8.3 Service Restoration Using DGs in a Secondary Network

Secondary networks are widely used in the downtown areas of metropolitan and central business districts. In the previous section, a critical load restoration method for radial distribution networks is presented. This section introduces a service restoration method that uses DGs to serve critical loads in a secondary network (SN).

8.3.1 Features of Secondary Network

A SN consists of secondary mains arranged according to the geographic pattern of the actual area being served [13]. Figure 8.7 shows the typical topology of a SN distribution system. Feeders 1, 2, and 3 in the figure are the primary feeders (PFs), which are connected to the SN through network transformers and network protectors.

It can be seen that the SN is meshed, which enables it to provide uninterrupted power supply for customers in single-fault scenarios [14]. Therefore, the SN distribution system can achieve high reliability. Nevertheless, in multiple-fault scenarios, the power supply to the SN may be interrupted. In this case, DGs connected with PF or SN can be utilized to restore power service to critical loads in SN. However, SNs are not designed to integrate DGs. Several technical issues should be considered when using DGs for service restoration in a SN [15].

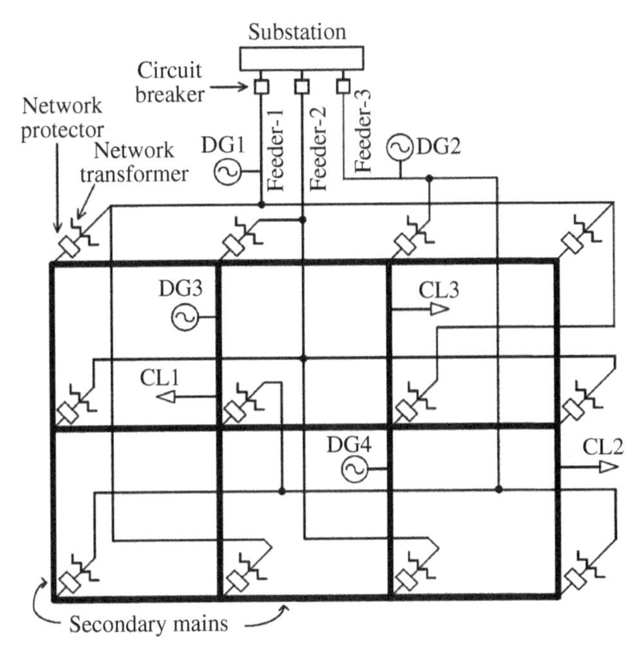

Figure 8.7 The typical topology of a SN distribution system.

8.3.2 Technical Issues

8.3.2.1 Operation of Network Protectors

Network protectors (NPs) is important in isolating the faults to protect network transformers and electric components at primary feeders. Undesirable tripping or closing operations of NPs [16] are more likely to occur when there are DGs on both the PF and SN, particularly in light-load conditions. In addition, NPs cannot be operated to separate or connect two dynamic systems and do not have synchronizing capability [17].

To avoid the aforementioned problems, a constraint is added to the optimization model of the critical load restoration problem. Reverse current is not allowed to flow through NPs to avoid improper tripping. Moreover, it is assumed that NPs between a PF and a SN already energized by DGs must not be closed. In order to use DGs on a PF to serve critical loads in a SN, one should first energize the PF with DGs on it. Then energize the SN by closing the corresponding NP. Finally, DGs in the SN are connected and loads are restored.

8.3.2.2 Energization Inrush of Network Transformers

The energization inrush occurs when energizing unloaded transformers [18]. The inrush may cause over-current and under-voltage at DG terminals, leading to the associated protection tripping and failure of restorative actions. When a DG is used to energize a PF for service restoration to a SN, all of the related network transformers connecting with the feeder will be energized, which may cause large inrush. Since DGs have a relatively small capacity, the energization inrush of network transformer may result in violation of DG transient thresholds and even instability.

In this study, transient simulations are conducted to obtain the maximum inrush current during the restoration process. If the maximum current exceeds the threshold value of the DG protection, the restorative actions are considered infeasible.

8.3.2.3 Synchronization of DGs

When using multiple DGs for restoration, synchronization conditions should be met, including phase sequence, frequency, and voltage differences. IEEE Std 1547-2003 [19] gives the detailed requirements on synchronization parameter limits. If the any condition is not satisfied, control actions will be needed to ensure a smooth connection of DGs. Transient simulations are performed in this study to make sure that no dynamic constraint is violated during the restoration process.

8.3.2.4 Circulating Currents Among DGs

The circulating currents among DGs will occur when DGs operate in parallel in the SN and the terminal voltages generated by different DGs do not match. Such undesirable currents may lead to certain DGs absorbing power in light-load conditions. In addition, circulating current may result in unbalanced power sharing among DGs, that is, certain DGs generate over-current while others still have spare capacity. Moreover, the equivalent impedance between two DGs in a SN is usually small. Excessive circulating currents are likely to occur even if the difference in DG terminal voltages is not too large. Proper control is required to avoid circulating currents. In this study, a centralized control system is assumed to be available to regulate the output power of DGs.

8.3.3 Problem Formulation

The assumptions below are made:

(1) There is a central controller to manage the system and it can monitor and control all the DGs and loads.
(2) Once a fault occurs, DGs and loads are disconnected from the network and NPs are open.
(3) It is not allowed to use NPs to connect or disconnect two dynamic systems.
(4) The real and reactive power of DGs are determined by the central controller. Due to limited generation resources, the real power output of DGs is considered as decision variables for the restoration problem. Reactive power is shared by DGs in proportion to their kVA capacity [20].

The outage duration is divided into T periods and the length of each period is T_{int}. The critical load restoration problem in a SN can be formulated as:

$$\max \sum_{t=1}^{T} \sum_{i \in \mathbf{L}} w_i x_{i,t} T_{\text{int}} \tag{8.25}$$

subject to

$$\underline{f^{\text{tr}}} \le f^{\text{tr}} \le \overline{f^{\text{tr}}} \tag{8.26}$$

$$\underline{V^{\text{tr}}} \le V_g^{\text{tr}} \le \overline{V^{\text{tr}}}, \quad g \in \mathbf{G} \tag{8.27}$$

$$I_g^{\text{tr}} \le \overline{I^{\text{tr}}}, \quad g \in \mathbf{G} \tag{8.28}$$

$$0 \le \sum_{t=1}^{T} P_{g,t} T_{\text{int}} \le E_g, \quad g \in \mathbf{G} \tag{8.29}$$

$$P_u^s - jQ_u^s = \left(\overline{V}_u^s\right)^* \sum_{v \in \mathbf{\Omega}_u} \sum_r Y_{uv}^{sr} \overline{V}_v^r, \quad u \in \mathbf{\Omega}, \quad s, r \in \{a, b, c\} \tag{8.30}$$

$$\underline{V} \leq V_u^s \leq \overline{V}, \quad u \in \boldsymbol{\Omega}, s \in \{a,b,c\} \tag{8.31}$$

$$I_l^s \leq \overline{I}_l, \quad l \in \mathcal{E}, \quad s \in \{a,b,c\} \tag{8.32}$$

$$\begin{cases} \underline{P}_g \leq P_{g,t} \leq \overline{P}_g \\ \underline{Q}_g \leq Q_{g,t} \leq \overline{Q}_g \end{cases}, \quad g \in \mathbf{G}, \quad t \in [1,2,\ldots,T] \tag{8.33}$$

$$\Delta P_g^- \leq P_{g,t} - P_{g,t-1} \leq \Delta P_g^+, \quad \forall g \in \boldsymbol{G}, \quad \forall t \geq 2 \tag{8.34}$$

$$P_{\text{NP}} \geq 0, \quad Q_{NP} \geq 0 \tag{8.35}$$

In the above model, the definition of variables and parameters in constraints (8.26)–(8.32) is the same as those in Section 8.2.2. $x_{i,t}$ is the status of load i in period t. $x_{i,t} = 1$ if load i restored in period t; $x_{i,t} = 0$, otherwise. $P_{g,t}$ is the real power generated by generator g in period t. E_g is the generation resource of DG g in kWh. ΔP_g^+ and ΔP_g^- are ramp-up and ramp-down limits of DG g, respectively. P_{NP} and Q_{NP} are the real and reactive power flowing through network protectors, respectively.

Constraints (8.26)–(8.28) are dynamic constraints, ensuring that the fluctuations in frequency, voltage, and current are not too large when performing restorative actions.

Constraint (8.29) represents the generation-resource constraint which indicates that the total amount of energy (kWh) provided by a DG is limited.

Constraints (8.30)–(8.35) are operational constraints. Constraint (8.30) represents the unbalanced three-phase power flow equations. Constraints (8.31)–(8.32) ensure steady-state bus voltages and line currents are within a preset range. Constraint (8.33) represents the limits on DG output power. Constraint (8.34) says that the ramp-up/ramp-down rates of DGs should be within a threshold. Constraint (8.35) indicates that reverse power flowing through NPs is not allowed.

8.3.4 A Heuristic Method for Service Restoration

Owing to the nonconvexity and complexity introduced by the unbalanced power flow equations and dynamic constraints, it is not practical to use mathematical-programming algorithms to solve the above problem. Instead, a heuristic method is designed to obtain a near-optimal solution [21].

The critical load restoration procedure includes three steps:

8.3.4.1 Step 1: Select a PF

To avoid improper operation of NPs, it is not allowed to close a NP between a PF and an energized SN. However, the DGs connected with PFs should be allowed to restore loads in the SN. Therefore, the following procedure will be applied: (i) use DGs to energize the PF; (ii) energize the SN by closing NPs; (iii) synchronize DGs and restore the critical loads in the SN. In this case, at most one PF can be used to serve loads in a SN.

PFs are sorted in a descending order according to the total kVA generation capacity. Select the PF from the top of list. Note that the inrush current may be too large to cause a failure in restoration because the transformers on the selected feeder will be restored simultaneously. To address the issue, a transient simulation is implemented to evaluate the feasibility of the selected PF. The feeder will be treated as an infeasible one if any violation is observed. Repeat the process until a feasible PF is found or there is no feasible PF.

8.3.4.2 Step 2: A MILP for Optimal Generation Schedule

To maximize the objective over a horizon with multiple time periods of the restoration process, power outputs of DGs and the loads to be restored in each period should be carefully decided. Without considering power flow equations and dynamic constraints, an optimal generation schedule can be formulated as follows:

MILP: (8.25)
over: $P_{g,t}, x_{i,t}$
subject to:
Eqs. (8.29), (8.33)–(8.34)

$$\sum_{t=2}^{T} |x_{i,t} - x_{i,t-1}| \leq 2 \tag{8.36}$$

Constraint (8.36) limits the number of changes in the load status. By introducing binary variables $y_{i,t}$, (8.36) can be reformulated as:

$$\begin{cases} -y_{i,t} \leq x_{i,t} - x_{i,t-1} \leq y_{i,t} \quad \text{for } t \geq 2 \\ \sum_{t=2}^{T} y_{i,t} \leq 2 \end{cases} \tag{8.37}$$

The optimization problem is a MILP because objective and constraints are linear, which can be readily solved by off-the-shelf optimization tools.

8.3.4.3 Step 3: Evaluate Operational and Dynamic Constraints

This step examines if the optimal generation schedule obtained in Step 2 satisfies the operational and dynamic constraints. For the evaluation of operational constraints, a power flow calculation is performed using GridLAB-D with all selected loads in the network. If any violation is observed, set the w_i value of the least important load to be -1 and go back to Step 2. The dynamic constraints are evaluated by transient simulations performed using PSCAD/EMTDC [22]. Loads are restored by their importance level from high to low. Set the w_i value of the load that causes the violation to -1 and go back to Step 2 if any violation is found. If all operational and dynamic constraints are satisfied, the restoration plan is found.

8.3.5 Case Study

Effectiveness of the proposed method is demonstrated on the modified IEEE 342-node low voltage networked test system.

8.3.5.1 Modified IEEE 342-Node Test System

Figure 8.8 shows the topology of the modified IEEE 342-node low voltage networked test system [23]. The system contains eight 13.2 kV PFs, which are in red color, and a 120/208 V grid network, which is in blue color. The PFs are connected to the SN through 48 network transformers with the capacity of 1 MVA. For convenience to validate the proposed method, 8 DGs labeled with DG1 to DG8 are added. Detailed information of DGs is listed in Table 8.4. For dynamic simulation, DGs are modeled as synchronous generators.

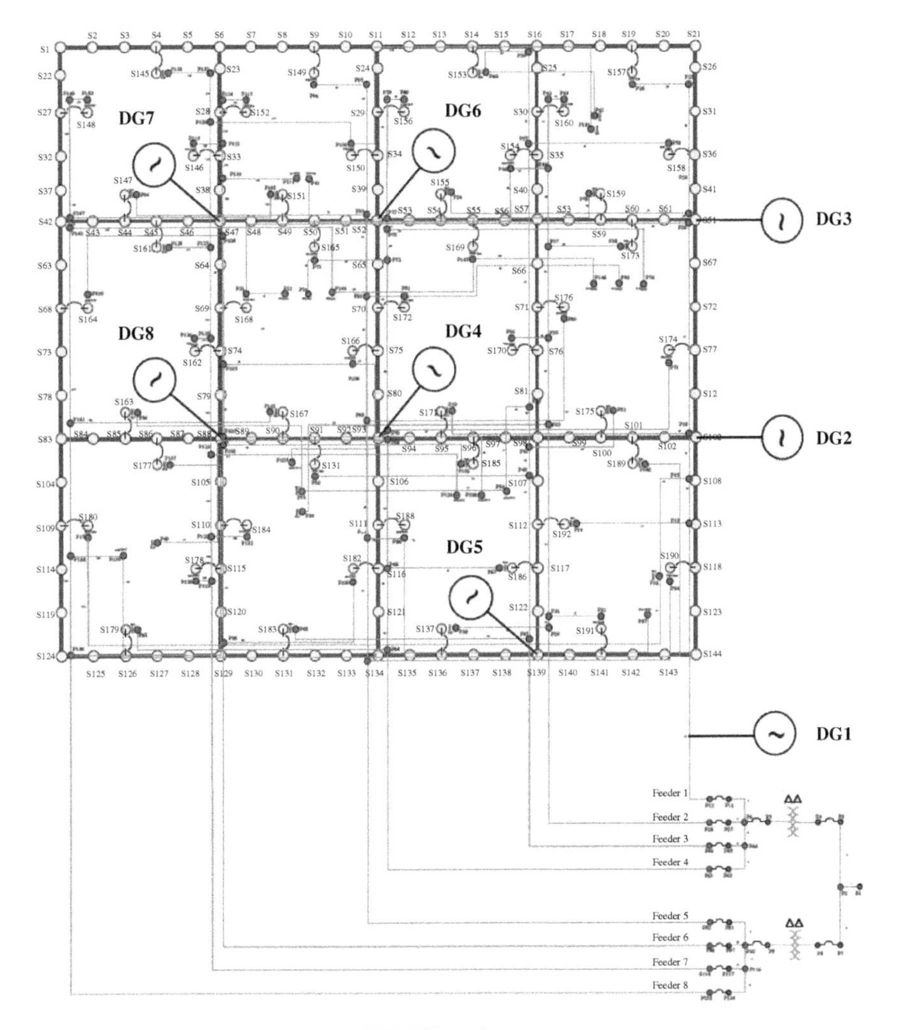

Figure 8.8 Topology of modified IEEE 342-node test system.

Table 8.4 The detailed information for DGs.

DG	Real power (MW)	Reactive power (MVar)	Generation resource (MWh)
1	5.10	3.16	40
2	1.70	1.05	14
3	2.55	1.58	15
4	3.40	2.11	20
5	2.55	1.58	10
6	2.13	1.32	7
7	2.13	1.32	8
8	2.55	1.58	20

A total of 96 unbalanced loads are served by the SN. The total demand is 26.03 MW + j16.08 MVar. The weighting factors w_i of loads are randomly selected in the range of [1, 3] ∪ [6, 10]. The weighting factors of critical/non-critical loads range in [6, 10] and [1, 3], respectively. The number of critical loads is 32 and the capacity is 9.78 MW + j6.03 MVar.

8.3.5.2 Results

Suppose that the duration of outage is 10 hours and set the time interval to be 1 hour, that is, $T_{int} = 1$ hour.

The restoration method is used to generate the restoration plan for the test system. The generation schedule is shown in Figure 8.9.

All the critical loads are determined to be restored and 22 non-critical loads are restored. This is because the generation capacity and resource are limited. About 49.7% of the load amount is restored while the generation resources are nearly used up, that is, 96.6% of the resources are used.

An unbalanced three-phase power flow calculation is conducted using GridLAB-D to evaluate the operational constraints. All operational constrains are satisfied. Take the node voltages as an example. Table 8.5 shows that the maximum and minimum voltages are within the preset limits. Figure 8.10 clearly demonstrates that the magnitude of node voltages satisfies the requirement.

Dynamic constraints are evaluated by transient simulations using PSCAD/EMTDC. PF1 is selected for service restoration and DG1 is chosen to energize PF1 as well as the six network transformers connected to PF1. Figure 8.11 gives the inrush current. It shows that the peak value of the inrush current is within the limit. Then NPs are closed to serve the SN. The transient current by energizing the network is small because the secondary mains are short.

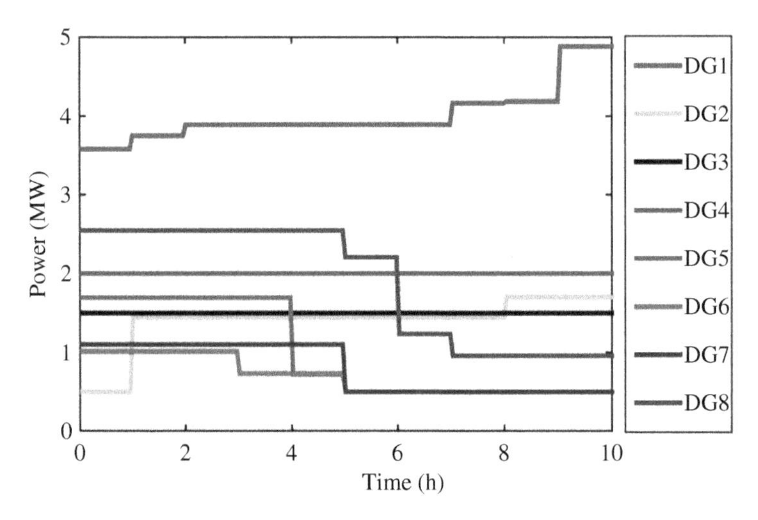

Figure 8.9 Generation schedule of DGs.

Table 8.5 Node voltages in the SN.

Phases	A	b	c
Minimum voltage (p.u.)	0.993	0.995	0.995
Maximum voltage (p.u.)	1.025	1.023	1.020
Average voltage (p.u.)	1.004	1.004	1.004

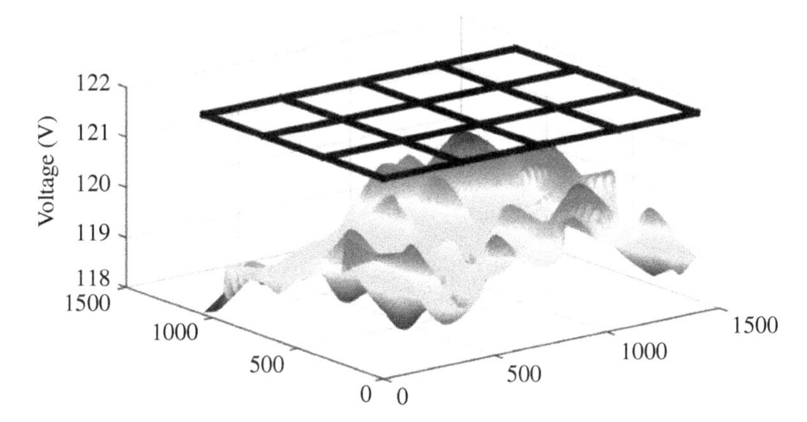

Figure 8.10 Voltage profile in the SN.

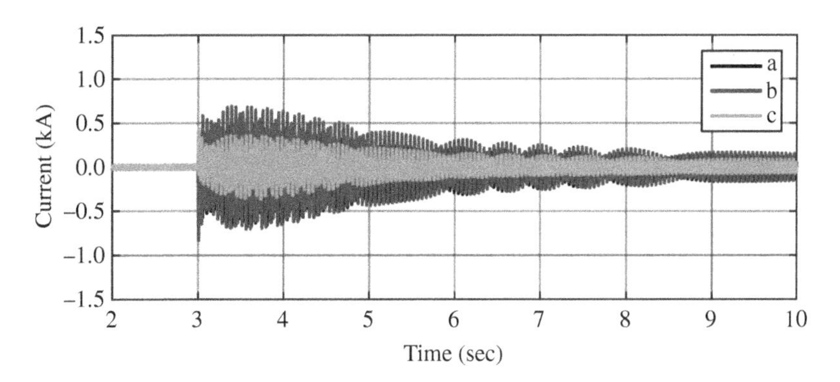

Figure 8.11 Inrush current by energizing transformers on feeder 1.

Other DGs are synchronized one by one, controlled by the central controller. Take DG4 as an example, the frequency curve and voltages at phase-A curves at the two terminals of DG4 breaker and the transient current at the terminal of DG4 are shown in Figure 8.12.

After synchronization of DGs, the selected loads are restored one by one based on their weighting factors. Picking up loads does not lead to excessive transients because the real power of each load is relatively small.

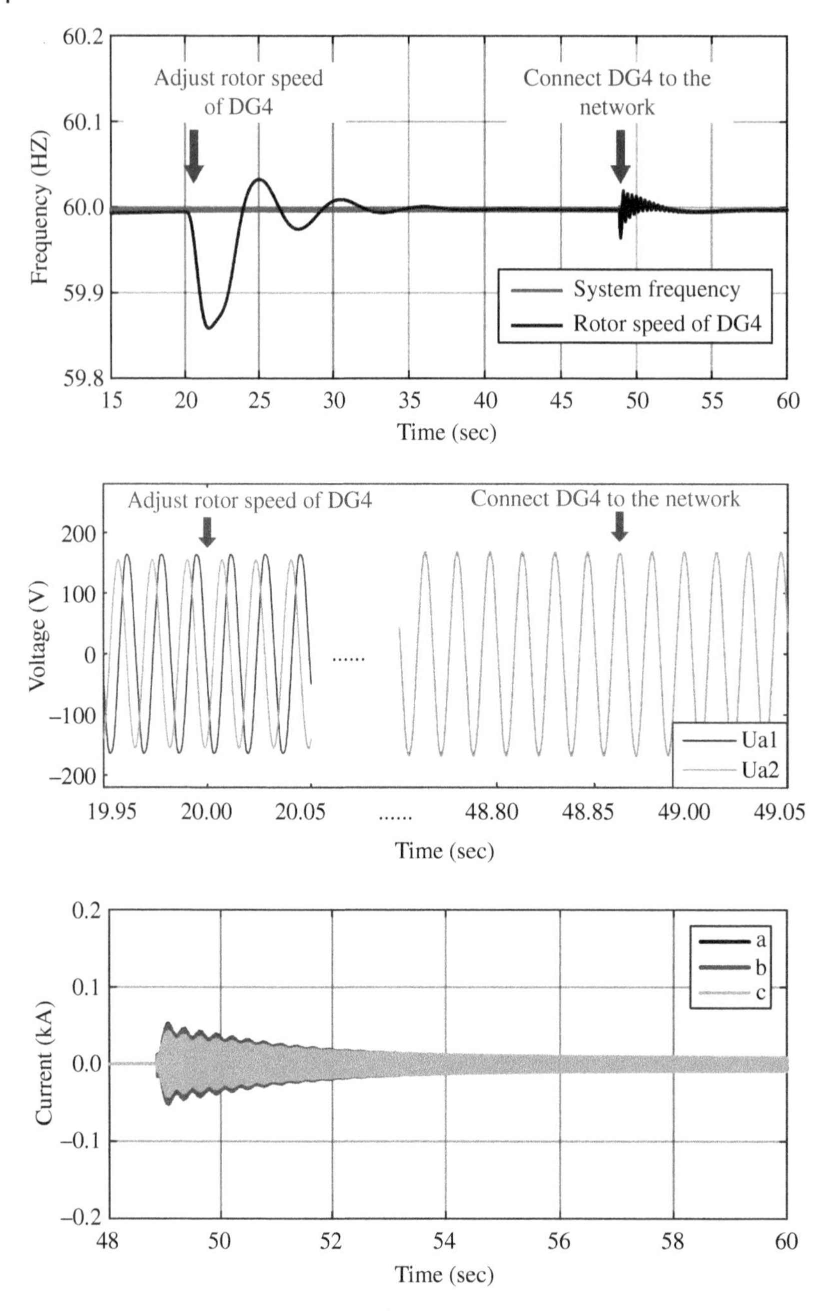

Figure 8.12 Synchronization of DG4 with the network: (a) frequency and (b) phase-A voltage at the two terminals of DG4 breakers and (c) transient current at the terminal of DG4.

8.4 Summary

This chapter introduces an optimal RCS placement method to enhance restoration capability of distribution systems, as well as critical load restoration methods using DGs and MGs after extreme events for radial and networked distribution systems. From the studies, it can be seen that dynamic constraints are essential when using DGs or MGs to serve critical loads. Dynamic and transient simulations play an important role in evaluating the feasibility of a restoration strategy. Other features affecting restoration strategies, such as intermittency of renewables, DG interface to the grid, and interdependency among multiple infrastructures, should also be studied in the future. Moreover, the role of electric energy storage, electric vehicles, and other local resources should be considered for resiliency enhancement.

Acknowledgment

This research was sponsored by U.S. Department of Energy, Office of Electricity, through Pacific Northwest National Laboratory. The support from Washington State University Facilities is greatly appreciated.

References

1 Li, J., Ma, X.-Y., Liu, C.-C., and Schneider, K.P. (2014). Distribution system restoration with microgrids using spanning tree search. *IEEE Transactions on Power Systems* 29 (6): 3021–3029.
2 Xu, Y., Liu, C.-C., and Gao, H. (2015). Reliability analysis of distribution systems considering service restoration. In: *Proceeding of the IEEE Power & Energy Society Innovative Smart Grid Technologies Conference (ISGT), Washington, DC, USA*, 1–5.
3 Smallwood, C.L. and Wennermark, J. (2010). Benefit of distribution automation. *IEEE Industry Applications Magazine* 16 (1): 65–73.
4 Xu, Y., Liu, C.-C., Schneider, K.P., and Ton, D.T. (2016). Placement of remote-controlled switches to enhance distribution system restoration capability. *IEEE Transactions on Power Systems* 31 (2): 1139–1150.
5 Lim, I., Sidhu, T.S., Choi, M.S. et al. (2013). An optimal composition and placement of automatic switches in DAS. *IEEE Transactions on Power Delivery* 28 (3): 1474–1482.
6 Carvalho, P.M.S., Ferreira, L.A.F.M., and Cerejo da Silva, A.J. (2005). A decomposition approach to optimal remote-controlled switch allocation in distribution systems. *IEEE Transactions on Power Delivery* 20 (2): 1031–1036.
7 Liu, C.-C., Lee, S.-J., and Venkata, S.S. (1988). An expert system operational aid for restoration and loss reduction of distribution systems. *IEEE Transactions on Power Systems* 3 (2): 619–626.
8 Xu, Y., Liu, C.C., Schneider, K. et al. (2018). Microgrids for service restoration to critical load in a resilient distribution system. *IEEE Transactions on Smart Grid* 9 (1): 426–437.

9 Xu, Y., Liu, C.-C., Schneider, K.P., and Ton, D.T. (2015). Toward a resilient distribution system. In: *Proceeding of the IEEE PES General Meeting, Denver, CO, USA*, 1–5.
10 U.S. Department of Energy at Pacific Northwest National Laboratory (2018). GridLAB-D, Power distribution simulation software. http://www.gridlabd.org/ (accessed 30 October 2018).
11 Baran, M.E. and Wu, F.F. (1989). Network reconfiguration in distribution systems for loss reduction and load balancing. *IEEE Transactions on Power Delivery* 4 (2): 1401–1407.
12 The MathWorks, Inc. (2018). MATLAB software. http://www.mathworks.com/products/matlab/ (accessed 13 September 2018).
13 ABB Power Systems Inc. Chapter 21: Primary and secondary network distribution systems. In: *Electrical Transmission and Distribution Reference Book*, 4e, 689–718.
14 Gonen, T. (2014). Chapter 6: Design considerations of secondary systems. In: *Electric Power Distribution Engineering*, 3e, 331–371. CRC Press.
15 Behnke, M., Erdman, W., Horgan, S. et al. (2005). Secondary network distribution systems background and issues related to the interconnection of distributed resources. In: *Technical Report, NREL/TP-560-38079*. http://www.nrel.gov/docs/fy05osti/38079.pdf.
16 Hardowar, R., Rodriguez, S., Uosef, R.E. et al. (2017). Prioritizing the restoration of network transformers using distribution system loading and reliability indices. *IEEE Transactions on Power Delivery* 32 (3): 1236–1243.
17 IEEE Recommended Practice for Interconnecting Distributed Resources with Electric Power Systems Distribution Secondary Networks. in IEEE Standard 1547.6-2011, pp. 1–38, 12 Sept. 2011.
18 Chiesa, N. (2010). Power transformer modeling for inrush current calculation, Ph.D. dissertation, Norwegian University of Science and Technology.
19 IEEE Standard for Interconnecting Distributed Resources with Electric Power Systems, IEEE Std 1547-2003.
20 Li, Y.W. and Kao, C.-N. (2009). An accurate power control strategy for power-electronics-interfaced distributed generation units operating in a low-voltage multibus microgrid. *IEEE Transactions on Power Electronics* 24 (12): 2977–2988.
21 Xu, Y., Liu, C.C., Wang, Z. et al. (2017). DGs for service restoration to critical loads in a secondary network. *IEEE Transactions on Smart Grid* 10 (1): 435–447. https://doi.org/10.1109/TSG.2017.2743158.
22 Manitoba HVDC Research Center (2018). PSCAD/EMTDC simulation tool. https://hvdc.ca/pscad/ (accessed 21 September 2022).
23 Schneider, K.P., Phanivong, P., and Lacroix, J.-S. (2014). IEEE 342-node low voltage networked test system. In: *Proceeding of the 2014 IEEE PES General Meeting, Harbor, MD*, 1–5.

9

Improving the Electricity Network Resilience by Optimizing the Power Grid

EngTseng Lau[1], Sandford Bessler[2], KokKeong Chai[1], Yue Chen[1], and Oliver Jung[2]

[1] *School of Electronic Engineering and Computer Sciences, Queen Mary University of London, Mile End Road, London, UK*
[2] *Digital Safety & Security Department, AIT Austrian Institute of Technology GMBH, Donau-City-Straße 1, Vienna, Austria*

9.1 Introduction

The invention of smart grid technology and the increase in power supplied by distributed energy resources (DERs) enables local administration in cooperation with distribution system operators to introduce new approaches for mitigating the impact of power outages. The smart grid is an enabler for the establishment of a microgrid, an integrated low-voltage (LV) power delivery system consisting of customer loads, localized DERs, and associated controllers that has the ability to isolate its LV grid portion from the bulk transmission grid or the partial isolation within the LV network with "islanding" concept, in the case of outage/contingency events [1, 2]. Contributions of several microgrid communities address the resilience of the power grid due to the impacts of climate change incidents, natural disasters, and social disruptions [3].

The sophisticated features of DERs in a microgrid such as photovoltaic (PV) arrays, backup generations, microturbines, fuel cells, as well as combined heat and powers (CHPs) can support grid clients both in time of normal operation and in crisis scenarios. According to [4], DERs provide the opportunity to improve resiliency through the impact mitigation of an emergency while ensuring critical infrastructures (CIs) operate without any interruptions. Ensuring the operability of CIs during or after a natural disaster will reduce response times of emergency workers in responding to community needs, and allow the appropriate amount of resources for other post-recovery efforts, resulting in a quicker recovery of the social welfare.

The microgrid's energy management system is based on the localized controllable DERs, consumer loads, and centralized microgrid controller modeled using the unit commitment (UC) and optimal power flow [5]. The controller optimally allocates DERs that will minimize the monetary cost and increase the performance of grid operations. In addition, the grid also needs to be capable to form islands during emergencies at pre-selected locations. In case of controlled islanding [6], a balance between the load and generation before the isolation from the main grid is created (including changing the topology) and the island is

Resiliency of Power Distribution Systems, First Edition.
Edited by Anurag K. Srivastava, Chen-Ching Liu, and Sayonsom Chanda.

isolated from the system. Both the IT and physical components of microgrids are involved in this process. While operating in a prolonged outage, failures may occur within the microgrid. The high-reliability distribution system (HRDS) introduced by the Illinois Institute of Technology [7] compensates for such a shortcoming where each building in a microgrid is connected to a loop cable using switches. In case of a fault cable, the direct isolation of the cable is accomplished that further avoiding disconnecting all other connected loads. This is one of the approaches to preventing cascading failures. The automatic switches in HRDS are able to sense the cable faults and isolate the fault with no impact on the remaining microgrids. The master controller, in turn, monitors the status of each HRDS switch using the supervisory control and data acquisition (SCADA) system and was responsible for the economic operation of the microgrid.

As planning for outages requires substantial knowledge of the power grid, with the tools presented in this chapter, city planners or utility engineers will be able to evaluate the resilience of different grid settings to consider these results in particular in the spatial planning of CIs. This chapter presents two simulation tools that evaluate a complete grid distribution network that composes an MV and LV level of an urban grid with a supply-and-demand balancing mechanism. Such tools further mitigate the impact of supply failure during disasters (i.e. supply outage due to an earthquake or flooding) through the identification of the respective node that has enough local generation capacity to be self-sufficient and the demand level that is required for the remaining nodes during an outage. This utilizes the decentralized nature of future energy generation to make urban power grids more robust and resilient against threats from cyberattacks and natural disasters, and on minimizing the impacts of power outages on associated CIs.

The organizational structure of this chapter is as follows. Section 9.2 presents the Microgrid Evaluation tool for the LV microgrid where the topology is radial and consists of several feeders. The modeling level is detailed going down to buildings, individual loads, and generators. Outages of arbitrary duration are caused by single line failure simulation in the grid topology. The resulting outage events trigger demand management and interruptible load mechanisms that lead to the reduction of consumption in the microgrid. Section 9.3 presents an overall grid modeling (OGM) tool for an MV grid of the region in consideration in which the failing components are power plants, lines, stations (SCADA), current breakers, and DERs. The energy balancing optimization based on a cost minimization function is presented for normal and outage contingency grid operations. Two types of outages – complete and partial grid outages are simulated and the described schemes are applied (i) to isolate grid portions or (ii) to connect alternatively a grid region that has been disconnected from their main supply. Finally, Section 9.4 concludes the chapter.

9.2 Microgrid Evaluation Tool

9.2.1 Outage Types to Be Simulated

[8, 9] describe the anatomy of power outages in the grid, emphasizing the cascading effect of outages that resulted in blackouts. The usual method to explore the effect of outages is to

simulate single-line contingencies, to re-calculate the resulting power flows, and identify the overloaded lines. Although power plants, SCADA systems, and transformers may also be damaged or attacked, the majority of outages caused either by natural disasters (floods, storms, and earthquake) or by cyber-attacks results in the disconnection of a power line. This leads to a drop in the supply, not necessarily to a total loss of power.

At the edges of the grid, so-called microgrids [10] have the capability to disconnect from the main grid (islanding). Planning a microgrid against total loss of power includes adding dispatchable local generation. Once the outage is detected, the microgrid goes in islanding mode and the generators ramp up their power. A detailed grid resilience study in the New York state [11] estimates, however, that the costs of migration to a microgrid with islanding capability, including backup generators (during an outage), often exceed the benefits. The authors arrive although to the conclusion that, if the microgrid participates in a demand response program, such that the peak demand is reduced, then less generation capacity is needed and a microgrid resilient solution becomes economically feasible.

In order to explore the effects of single power line failure in a simulation environment, that specific line is removed from the topology, the computed power flows provide information about the node voltages and line currents. Lines that exceed the nominal current (are overloaded) would trigger the circuit protection after some time and trip (disconnect), causing other lines to be overloaded, and so on. This cascading effect observed during the emergence of a blackout could be avoided, if the demand in each of the grid nodes could be reduced, once the outage is identified, and an event is sent to all relevant microgrids.

Mechanisms for load reduction (curtailing) exist and are considered as part of incentive-based demand response programs [12]. Typical customer sizes to be eligible for an interruptible/curtailable (I/C) load program range between 200 kW and 3 MW. Customers that participate in the I/C program agree to reduce their load to a predefined level, for which they receive discounts. Failing to reduce the load leads to penalties.

In this work, we propose to integrate an I/C mechanism with a direct load control (DLC) scheme, which uses the flexibility of thermal loads, electric storage, or EV charging loads. As we will see in the system architecture section, the microgrid controller uses the received flexibility information to avoid overloading the transformer by power peak shifting.

9.2.2 Proposed Microgrid-Based Architecture

We assume that the studied smart grid consists of microgrids at its edge. The microgrid architecture has been proposed in previous works due to its advanced control mechanisms for local generators (DER), to its flexibility, reliability, and islanding capability [10, 13, 14]. Referring to the classification of control mechanisms described by Olivares et al. [14], we adopt a secondary control centralized architecture, in which the time horizon is in the range of a few hours. The overall microgrid planning and simulation system in Figure 9.1. is based on a demand management control architecture that includes one microgrid (MG) controller and several customer energy management controllers (CEMS, EMS) for each building in the microgrid. The controllers use flexibility information, demand management, and to cope with the changes in the power supply [15]. The flexibility concept applies to those loads which are tolerant (in certain limits) to an increase, decrease or shift in time, such

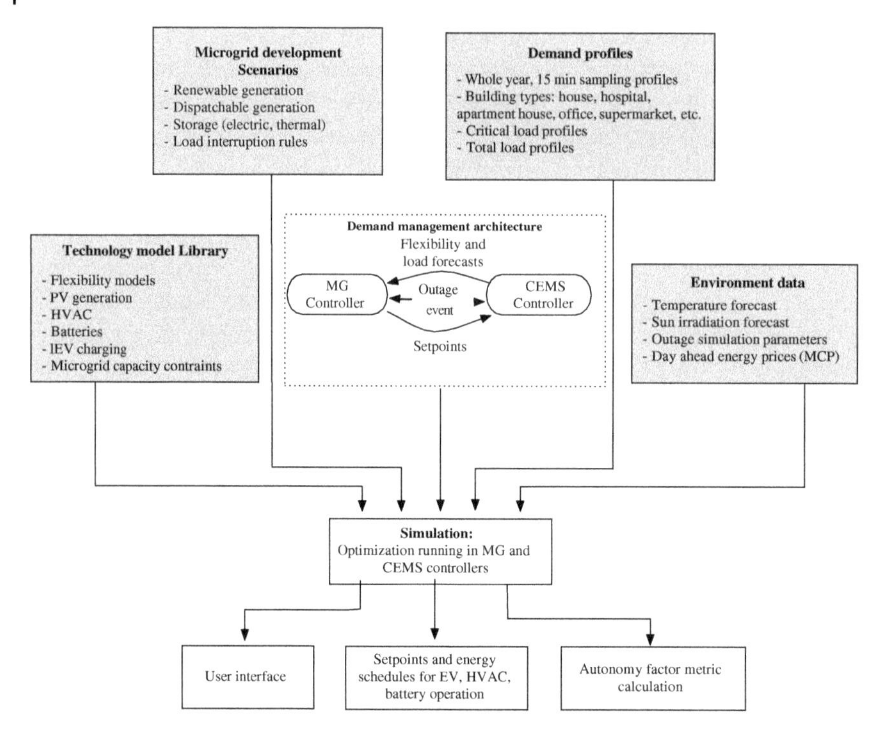

Figure 9.1 Microgrid planning system overview.

as thermostat-controlled cooling, battery charging, etc. The novelty is to extend this control loop to the outage period, where loads and set points are recalculated according to I/C rules. It is clear that the controllers and the communication infrastructure have to operate also during the partial outage.

For the realization of the control loop, the model predictive control (MPC) technique [16, 17] is used, meaning that power consumption (and generation) is predicted for a certain time horizon (e.g. six hours), however, the actuation is performed only for the next period. The MPC mechanism requires the periodical exchange of flexibility information between controllers. In Figure 9.2, we illustrate several energy flexibility models such as for HVAC (heating, ventilation, air conditioning), electric vehicle charging, and battery storage, which provide their consumption prediction to the EMS. The latter optimizes multiple local objectives such as set point following, minimum costs for charging EV and battery, maximizing the self-consumption of PV power, etc. The MG controller reads the latest flexibility and consumption plans from the EMS and computes updated set points (for the whole time horizon). In case the EMS proposed consumption is too high, the set point following the objective in the EMS optimization has the effect that some flexible loads are reduced (within their flexibility limits). For details on the optimization models, see [18].

In Figure 9.2, the EMS controller receives information from the models and static load profiles. As a result, various control variables for the local flexible loads are updated, and new values for PV generation, consumption, and flexibility are estimated.

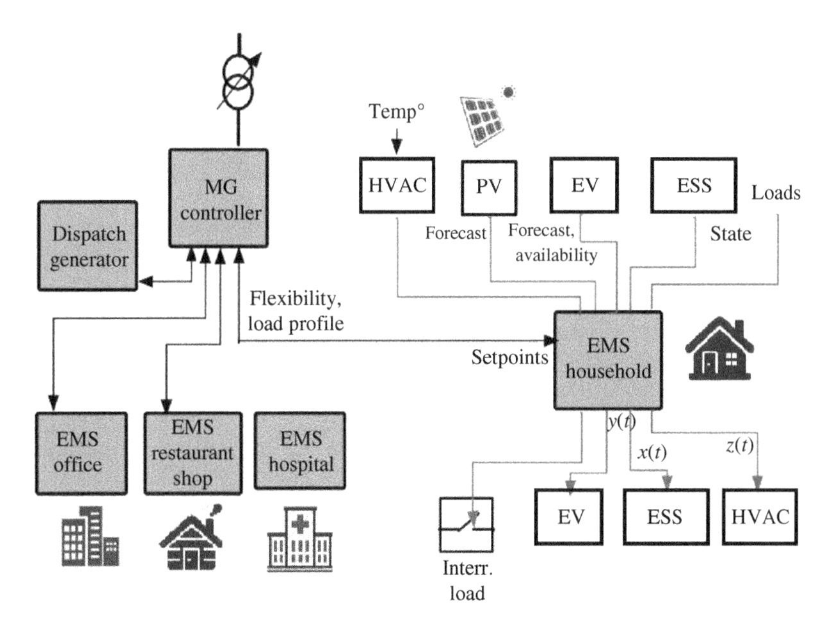

Figure 9.2 Main components of the MGE tool: the building controllers (EMS).

9.2.3 Demand Characterization

For the purpose of planning the consumption of a microgrid in both normal operation and outage mode, annual demand profiles (see Figure 9.1) are essential input data. We decided to use publicly available data from the Energy Information Administration (EIA), which provides high qualitative annual consumption data on an hourly basis, for various climatic regions in the United States. The Chicago area has been selected, as the climate resembles to that of northern Europe.

Fortunately, the consumption data of residential and commercial buildings in the portal [19] has been de-aggregated in the categories ventilation, cooling, heating, lights, and equipment. To these categories, we added flexible, model-based loads such as EV charging, home battery storage, as well as PV generation. The cooling/heating consumption has been modeled separately in order to exploit the flexibility due to thermal storage. We used a simplified thermodynamic model of the building and thermostat-based control. The HVAC models have been then calibrated to match the yearly consumption in the profiles.

As mentioned before, an additional characterization into critical, interruptible demand needs to be made. The critical demand has to be defined for each building type in advance and consists of loads and appliances that have to operate during an outage.

In the following, we give a few examples of critical loads: partial lighting in houses, commercial or industrial places, local energy controllers (CEMS), the microgrid controller, ICT wireline and wireless infrastructure for internet access, cellular nodes and antennas, public cash dispensers, refrigerators in food stores, pharmacies, hospitals and storage houses, water pumps of the district/town, gasoline pumps in gas stations, lifts and automatic doors in residential and commercial buildings, full supply of hospitals, pharmacies, police and fire stations, gas-based space and water heating (which needs electronics to operate).

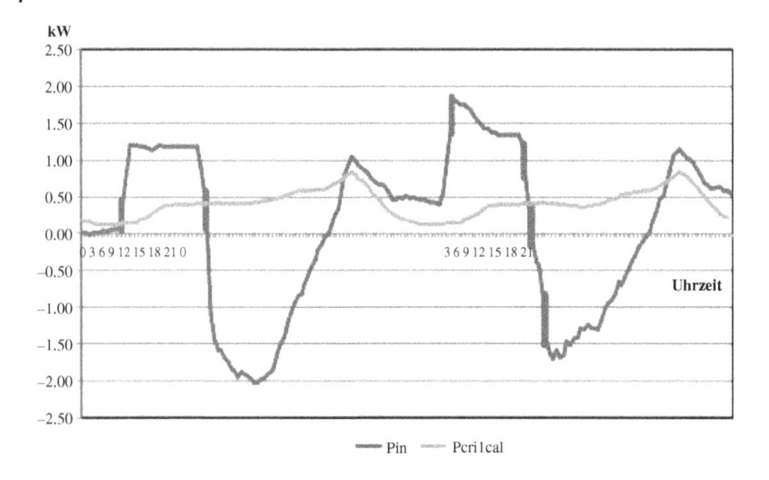

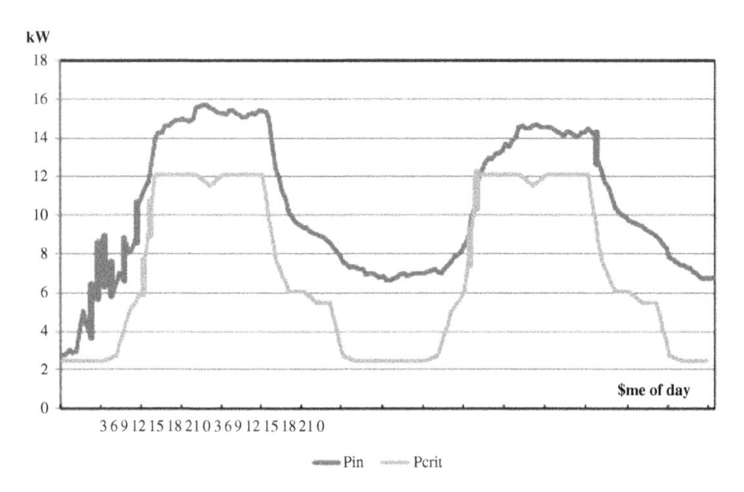

Figure 9.3 Consumption profile examples: residential house (left) and small office (right).

The remaining load is interruptible and will be discontinued during the outage. Examples of interruptible loads in the household are loads in the kitchen, entertainment equipment, washing machine, vacuum cleaner, air conditioning, and EV charging. In the EIA dataset and also throughout Europe (depending on building type and climate region), space and water heating are often done with natural gas. Therefore, the visible electricity consumption during the summer due to air conditioning is higher than in winter.

In Figure 9.3, the obtained critical consumption and the total consumption are depicted for a house and a small office. In case of the house, P_{in} is the total consumption, obtained by adding to the critical load (fan, light, and equipment) the HVAC, EV charging load, and PV generation such that $P_{in} = P_{critical} + PHVAC + PEV + PPV$.

9.2.4 Microgrid Evaluation Tool Operating Modes

The MGE tool is an event-based simulation in which CEMS controllers and the MG controller exchange periodically data and compute the updated consumption plan. Thus, each

Table 9.1 Summary of building characteristics.

Building type	Size [sqm]	Critical: fan(f), ICT(i), light(l)	Interruptible	PV (kWp)
Residential flats	3100	f,i,l	HVAC, EV	
Small office	511	f,i,l	HVAC	8
Residential house	250	f,i,l, battery	HVAC, EV	5
Supermarket	4180	refrig.,f,i,j		
Clinic	3804	f,i,l,HVAC		
Battery 100 kWh		x		

HVAC, EV, and battery load model in a building produces the updated load prediction and flexibility information which is combined with the "static" load profile. The CEMS optimization updates the local control actions. On the MG controller side, the optimization of the load "distribution" consists of the set point update for each building controller. Once an outage event is received by the controllers in the microgrid, each CEMS controller activates the rules defining the I/C load characterization. The rules can be restrictive or more relaxed, depending on the energy balance, that is the amount of dispatchable generation available and the societal needs in the different building types, see columns 3 and 4 in Table 9.1. In addition, the CEMS optimization model [18] includes rules activated by the outage event. Thus, all economic, price-dependent optimization criteria are disabled in the CEMS optimization, the load continues to follow the set points and keeps the strict balance between supply and demand, as mentioned by [10]. Furthermore, shedding the PV generation is not allowed, the PV output is maximized, interruptible loads are disconnected and the air conditioning/heating may be switched off in certain buildings to save energy. In any case, the thermostat range is relaxed to increase flexibility.

Through the outage simulation, the city planner tries to estimate the required power that has to be injected into the microgrid during the outage period in order to satisfy the critical load. Furthermore, the simulation shows the effect of existing renewable (PV) generation and storage resources on microgrid resilience.

In Section 9.2.5, we will present numerical results to illustrate the answers to the planning questions above.

9.2.5 Simulation Setup and Numerical Results (MGE)

9.2.5.1 Grid Autonomy Factor

The behavior of the grid in extreme situations, natural calamities, or cyberattacks is usually characterized by a resilience metric [20, 21]. This measure covers all the phases of the outage, including the restoration of the grid state and the repair of damaged equipment. If parts of the grid continue to operate as islands, the resilience of the whole grid increases. In order to assess the quality of an outage response with the mechanisms we have described in the previous sections, we are searching for a more specific metric that can be computed following the simulations.

We consider therefore the energy E^0_{in} that still has to be supplied to the microgrid during an outage. The demand of the microgrid during normal operation D^n is the sum of critical and interruptible demand, whereas the energy difference in the storage at the end of the considered interval $\text{ESS}^n_{\text{dif}}$

$$D^n = E_{\text{crit}} + E_{\text{interr}} + \text{ESS}^n_{\text{dif}} \tag{9.1}$$

The demand D^n has to be satisfied by the renewable generation E_{RES} and the power injected into the microgrid E^n_{in}.

$$D^n = E^n_{in} + E_{RES} \tag{9.2}$$

Similarly, in outage mode $D^0 = E_{\text{crit}} + ESS^0_{dif}$ and $D^0 = E^0_{\text{in}} + E_{\text{RES}}$ Let us define the autonomy factor α:

$$\alpha = 1 - E^0_{\text{in}}/D^n = 1 - E^0_{\text{in}}/\left(E^n_{\text{in}} + E_{\text{RES}}\right) \tag{9.3}$$

Using the definitions above, we obtain

$$\alpha = \frac{E_{\text{interr}} + \text{ESS}^n_{\text{dif}} - \text{ESS}^n_{\text{dif}} + E_{\text{RES}}}{E_{\text{crit}} + E_{\text{interr}} + \text{ESS}^n_{\text{dif}}} \tag{9.4}$$

For a non-smart grid where all the loads are considered critical, $E_{\text{interr}} = 0$, no renewables, that is $E_{\text{RES}} = 0$, thus $\alpha \approx 0$. Increasing E_{RES} can theoretically achieve a self-sufficient microgrid. In reality, renewable power fluctuates strongly, therefore dispatchable generation must be used. In our experiments, the terms in expression Eq. (9.3) can be measured, therefore the autonomy factor will be calculated.

A random microgrid configuration of buildings consisting of 8 offices, 4 midrise apartment blocks, and 26 residential houses has been simulated for a duration of two days. Outages of different durations such as 6, 12, and 24 hours have been simulated, starting on the second day at 9 am. The power limitations at the transformer of 300 and 200 kW for the outage, were set high enough not to constrain the power allocation for these tests. After 36 hours of normal operation, a simulated outage of 24 hours leads to a sharp drop in consumption. A series of experiments have been conducted to evaluate the impact of different system parameters on the proposed grid autonomy factor. Thus, we varied the outage duration, the amount of renewable generation, and battery storage in the microgrid. In the baseline scenario (see Table 9.2), the outage duration is 24 hours, each small office

Table 9.2 Impact of the system parameters on the autonomy factor.

Scenario	E^n_{in}	E^0_{in}	E^n_{RES}	α
Baseline (24 hours)	4.67	2.46	1.15	0.58
No-PV(offices)	5.11	2.90	0.71	0.50
Less storage	4.77	2.40	1.04	0.58
6 hours-outage	1.23	0.55	0.47	0.67
24 hours-outage-120 kW	4.67	2.39	1.15	0.54
6 hours-outage-120 kW	1.23	0.59	0.47	0.65

has 50 m^2 PV panels, and each residential house has a 10 kWh battery. The other scenarios are compared to this baseline.

More PV generation clearly improves the autonomy factor, as the difference between the scenario No-PV and the baseline indicates.

One would expect that the increase of battery storage in the grid would increase the autonomy factor. Surprisingly, the latter does not change. The reason is that both for smaller and larger batteries, the charging and discharging due to PV generation make no difference in the net consumption. An eventual PV power surplus from a certain household is immediately consumed by other buildings that have a net demand. Therefore, the net energy needed by the microgrid during the outage does not change with the storage capacity. A completely different situation is of course when an outage is expected with fully charged batteries.

The outage duration and its timing do influence the autonomy factor: thus, a shorter outage of six hours (9 am to 3 pm) improves α from 0.58 to 0.67.

Finally, we tested the effect of limiting, even more, the microgrid consumption during the outage. This reduction can be accommodated up to a certain degree by the demand management mechanisms in the microgrid. Thus, if the maximum consumed power is limited to 120 kW instead of the unconstrained consumption of 200 kW in the baseline, α decreases slightly.

9.2.5.2 Single Line Failure Impact on Microgrids Using MGE

For the simulation experiments, we use as the benchmark grid, the IEEE 14-node test topology in Figure 9.4. The original capacity of 280 MW has been scaled down, such that each of the nodes can be configured as a microgrid with the total load values P_{load} given in Figure 9.4.

A node in the IEEE 14 node grid is generally modeled as a whole microgrid. For instance, the total load of Node 3 is 940 kW and the microgrid has been populated with the following

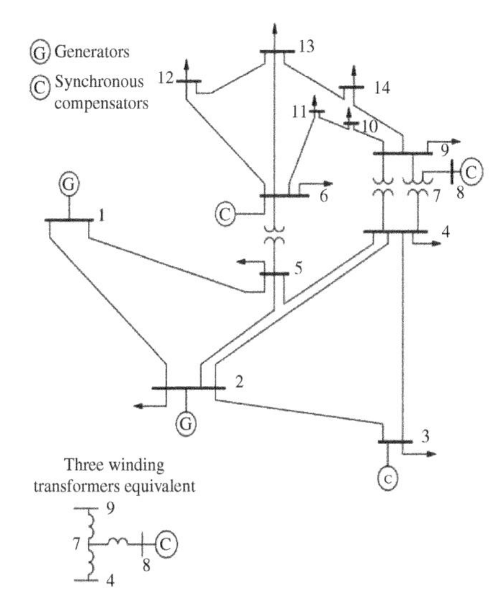

Node	Type	Pgen	Qgen	P_{LOAD}	$P_{REDUCED}$
1	3	2.32	0.0	0.00	0.000
2	2	0.40	0.0	0.217	**0.160**
3	2	0.00	0.0	0.942	**0.550**
4	0	0.00	0.0	0.478	**0.250**
5	0	0.00	0.0	0.076	0.076
6	2	0.00	0.0	0.112	0.112
7	0	0.00	0.0	0.00	0.000
8	2	0.00	0.0	0.00	0.000
9	0	0.00	0.00	0.295	**0.200**
10	0	0.00	0.00	0.090	0.090
11	0	0.00	0.00	0.035	0.035
12	0	0.00	0.00	0.061	0.061
13	0	0.00	0.00	0.135	0.135
14	0	0.00	0.00	0.149	0.149

Figure 9.4 Left: Selected overall grid topology. Right: Loads in the adapted 14-node test grid.

Table 9.3 Results of single failure tests.

Failed line	Overloaded line: current kA (regular loads)	Overloaded line: current kA (reduced loads)
No failure	1–2:3.5	
1–2	1–5:5.6, 4–5:2.8, 5–6:2.1	1–5:3.5
1–5	1–2:5.4	
2–3	1–5:2.2, 3–4:2–4, 5–6:2.1	
2–4	1–5:2.2, 2–3:2.1	
2–5	1–5:2.2	
3–4	2–3:2.3	
4–5	5–6:2.2	
4–9	5–6:2.4	5–6:2.26
5–6	4–5:2.4	

components: an EV charging point, 17 single houses, 8 small offices, 10 apartment blocks, 1 supermarket, 1 ESS, etc.

We perform contingency tests [22] by simulating single-line failures. If the node loads do not react to the outage event, then each line failure may cause overloads elsewhere in the grid, see the middle column of Table 9.3.

In Figure 9.4, the total nominal load of the nodes is shown. Following the outage, only the largest nodes 2, 3 4, and 9 reduce their demand (for simplicity reasons). In order to calculate the reduced load, we run the MGE tool. Figure 9.5 shows a snapshot of the simulation (total load curve) of Node 3 during normal, whereas Figure 9.6 is a snapshot of an outage between 9 am and 3 pm. On the right side of the actual simulation time (green vertical line) the total predicted load is displayed.

Following the outage simulation for Node 3, the required reduced demand can be determined for instance to 550 kW. Once the reduced demand values are obtained for all participating nodes, the single line failure simulation is run again. In the third column of Table 9.3, most overload situations have been removed.

9.3 Overall Grid Modeling Tool

The main aim is to optimize the distribution of DERs in the grid in a way that: (i) grid operation cost is reduced and at the same time and (ii) grid performance and stability are increased. However, less attention has been paid to maintaining grid resilience in case of adverse events. The investment in smart technologies may induce cost-efficient grid solutions but the degree of grid resilience is uncertain [20]. Various resilience-based grid operations were proposed, such as time-dependent resilience assessment of urban infrastructures [23], optimal-resilience-based scheduling model at a point of supply interruption [3], economic and emergency operation of ESS in electricity grid [24] or multiple smart microgrids to ensure low-carbon generations and grid survivability [25]. Yet, the

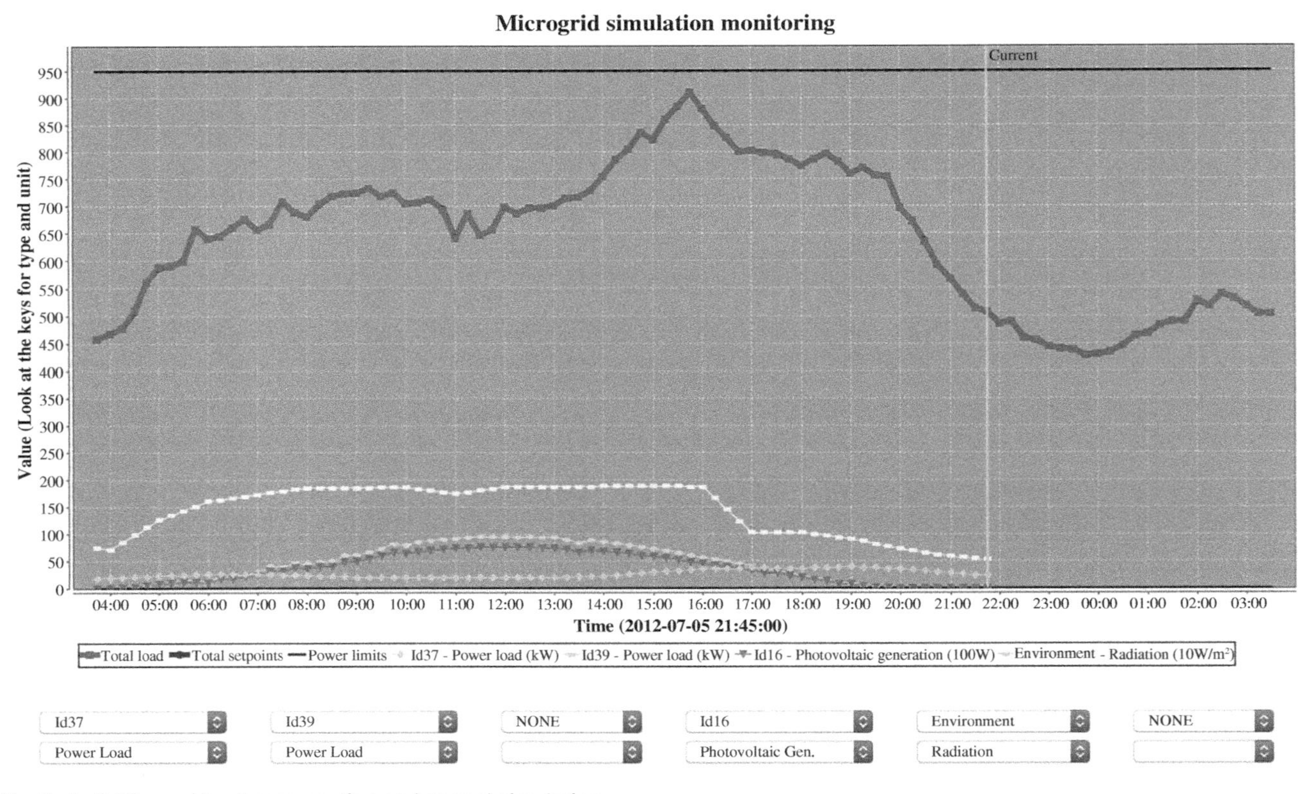

Figure 9.5 Node 3, Microgrid net consumption and normal simulation.

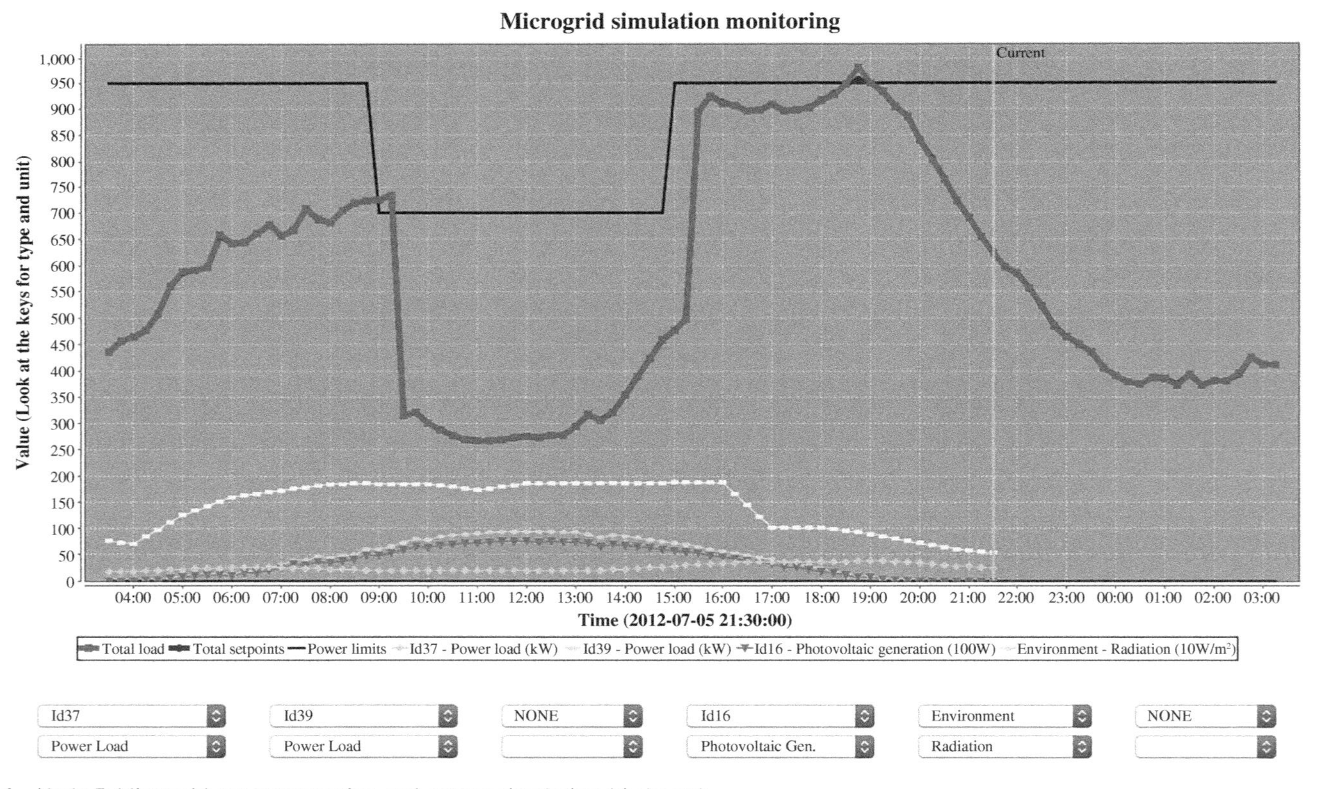

Figure 9.6 Node 3, Microgrid net consumption and outage simulation (six hours).

resilience-based operation was only introduced without quantifying the grid resilience further. Also, the exploitation of low-carbon energy resources may not guarantee grid resilience.

Henceforth, there is a need for resilience analysis to observe the grid evolvement, the process improvements, and the possibility of grid underperformance issues [23]. In this regard, the OGM tool is introduced. The underlying model supports supply and demand, wholesale electricity market prices, DER integration, and the simulation of outage events. The outage event is included to evaluate the capability of the grid to sustain the outage through operating in islanded mode by grid proportional isolation from the main grid, or by isolating grid portions and dropping the load (normal grid-connected operation for unaffected grid nodes). The ability to sustain the islanded operation allows the evaluation of grid resilience. The most economical approach to satisfy the demand shortage is implemented subsequently while ensuring grid resilience.

9.3.1 Overall Grid Modeling Architecture

An overall grid configuration is presented in Figure 9.7. Each node (Node 1, Node 2, ..., Node $M + N$) consists of mid-scale power plants (in the MV level), electrical loads for the end consumers, and DERs (distributed generations (DGs) such as thermal units, ESS, and RES) that are connected via electrical interconnectors for the LV level. Both M and N denote the number of nodes for the MV and LV levels correspondingly. Consumer profiles in each node are denoted as Profile 1, Profile 2, ..., Profile P. DERs in the microgrid are typically represented as distribution substations that are connected to feeders [26]. The feeders transfer the power from DERs to distribution transformers for the end customers [26]. The microgrid is connected to the MV grid through a transmission line. A circuit breaker (CB), *CB*1 acts as the protection mechanism to connect or disconnect the MV grid from the high-level grid. Similarly, *CB*2 is used to protect the LV microgrid from the MV grid.

A grid can have two operation modes: grid-connected mode (normal) and isolated grid mode (islanding). In the normal state, the microgrid is interacting with the MV grid. On the contrary, during an outage event, the microgrid can operate in an islanding mode by opening *CB*2 as in Figure 9.7. This is to enable the continuous supply of power within the microgrid without interruptions, especially to the critical loads. The DERs are activated by the microgrid controller to provide power to the LV microgrid. Similarly, the MV grid can still operate by transferring power to the LV grid by opening *CB*2. In addition, an islanded operation can also occur within the LV microgrid. For example, all the components in Node 3 of the microgrid can be operated in islanded mode if *CB*5 is opened. The remaining nodes interact with the MV grid in the normal operation mode.

9.3.2 Optimization Model

The model concentrates on the centralized perspective of the grid controller in optimizing the localized energy productions. The central controller in this case has the capability to perform switching by isolating the LV network from the grid due to the sophisticated capability of CBs. Additionally, during an outage event, the robust design feature of electricity line connections within the islanded node is assumed where there are no significant energy losses in the delivery of power, regardless of the location.

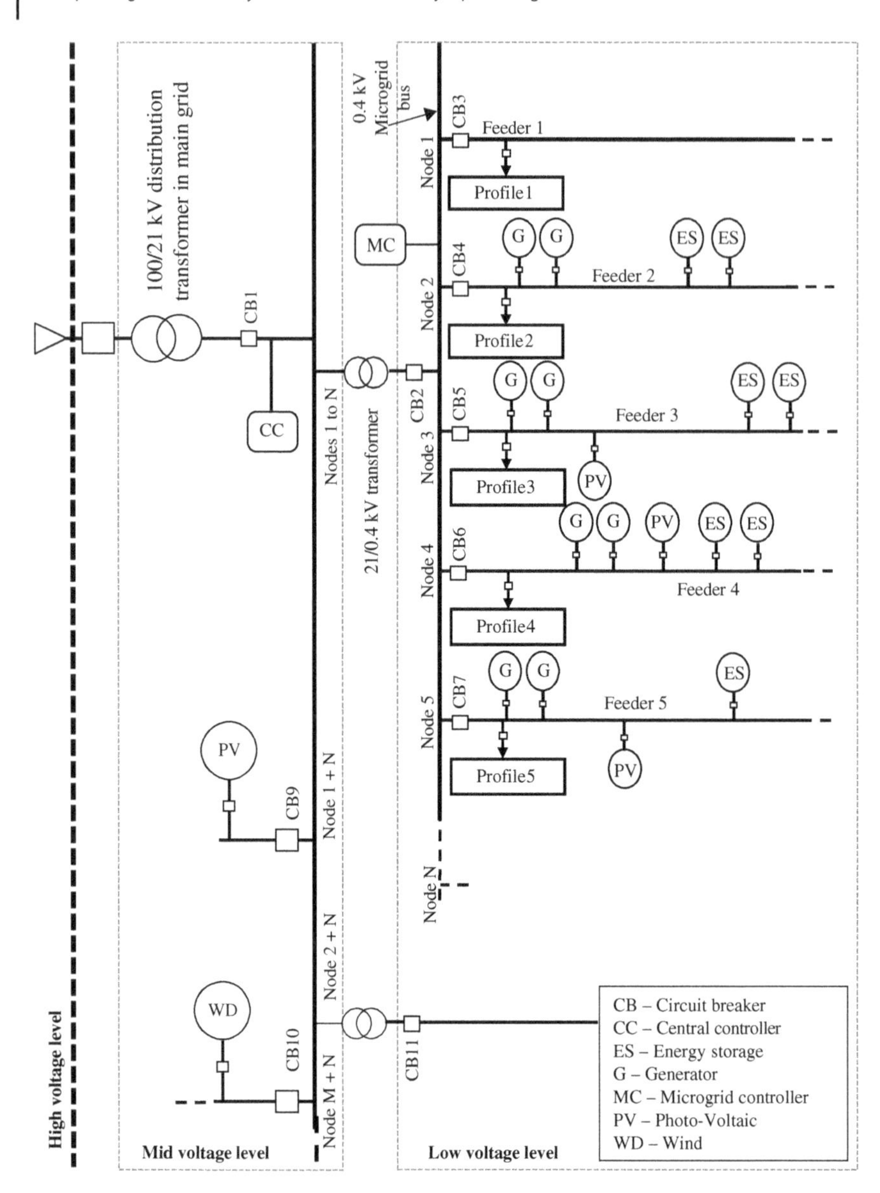

Figure 9.7 Grid configuration.

The optimization formulation is based on supply/demand economic dispatch and unit commitment problem. The problem formulations from [3, 24], and [27] are followed. A single objective function optimization based on linear programming problem is applied to enable the adaptive and better quality of the decision that is critical peculiarly in an emergency situation [28].

9.3.3 The Objective Function for Normal Operation

The objective function for normal operation as formulated in Eq. (9.5) is based on three important terms: the operating cost of the dispatched units, the ESS cost, and the cost of the electricity purchased from the main grid (high level):

$$\text{Min}\sum_{t=1}^{N_t}\sum_{g=1}^{N_g}[F_g(P_{G,gt})I_{gt} + \text{SU}_g + \text{SD}_g] + \sum_{s=1}^{N_s}[C_t P_{S,st}] + C_t P_{M,t} \tag{9.5}$$

The g, s, and t index are the corresponding generating units (thermal units), ESS, and the time period. $P_{M,t}$ is the transmission line power between the LV and MV grid. $P_{G,gt}$ is the power dispatch of corresponding unit g. SU_g is the start-up cost. $P_{S,st}$ is the power generated in s at time t. SD_g is the shutdown cost. C_t is the electricity market price at the time when there exists an electricity connection from LV to MV level. I_{gt} is the commitment state binary value of unit g.

For the first term, the generation cost is denoted by $F_g(P_{G,gt})$. For thermal units, the generation cost is presented in a quadratic function and is simply approximated by a piecewise linear approximation. In [29], the authors provide details of the fuel consumption and cost function. For the second term, ESS cost is imposed when the energy is drawn from the main grid (charging mode) at times of low electricity market prices C_t, whereas during the generation (discharging mode), zero generation cost is imposed by ESS. In the final term, depending on the operation state, there will be electricity costs when there is an interaction between the present and the main grid. In contrast, there will be zero electricity costs when no power is drawn from the main grid.

In this case, the grid is subject to the following operational constraints, where constraints involve the power supply and demand balance (Eq. (9.6)), capacity limits (Eq. (9.7)), ramped-up and down limits (Eqs. (9.8) and (9.9)), transmission (Eq. (9.10)) and distribution (Eq. (9.11)) limits, spinning reserve (Eq. (9.12)), and the number of hours to run and off (Eqs. (9.13) and (9.14)):

$$\sum_{g=1}^{N_g} P_{G,gt} + P_{M,t} + \sum_{r=1}^{N_r} P_{R,rt} + \sum_{s=1}^{N_s} P_{S,st} = P_{D,t} \tag{9.6}$$

$$P_{G,g}^{\min} I_{gt} \leq P_{G,gt} \leq P_{G,gt}^{\max} I_{gt} \tag{9.7}$$

$$P_{G,gt} - P_{G,g(t-1)} \leq UR_g \tag{9.8}$$

$$P_{G,g(t-1)} - P_{G,gt} \leq DR_g \tag{9.9}$$

$$|P_{M,t}| \leq P_M^{\max} \tag{9.10}$$

$$|P_{G,gt}| \leq P_{G,g}^{\max} \tag{9.11}$$

$$\sum_{g=1}^{N_g} P_{G,gt}^{\max} I_{gt} + P_M^{\max} \leq P_{D,t} + R_t \tag{9.12}$$

$$I_{gt} = 1, (t_1 \le t \le t_2) \tag{9.13}$$

$$I_{gt} = 0 \cdot (t_1 \le t \le t_2) \tag{9.14}$$

The $P_{D,t}$ is total demand, r indexes RES, and $P_{R,rt}$ is the power generated from RES. UR_g and DR_g is the ramp up and down rate limit. R_t is the reserve requirement at time t. The spinning reserve ensures a sufficient amount of generation due to the imbalance in demand during an emergency situation.

9.3.4 The Objective Function for Outage Operation

A $N-1$ contingency criterion is considered during an outage event. The $N-1$ system compliance ensures the grid can survive any single outage in any transmission and distribution links.

The objective function for the outage operation is similar to Eq. (9.5) but is intended only for outage events. This enables the supply of power to critical loads from DERs in an economically efficient manner:

$$\text{Min}\sum_{t=1}^{N_t}\sum_{i=1}^{N_i}\left[F_i(P_{G,it})I_{it}N1^c_{G,it} + SU_{it} + SD_{it}\right] + \sum_{i=1}^{N_i}\left[C_tP_{S,it}N1^c_{S,it}\right] + C_tP_{M,t}N1^c_{M,t} \tag{9.15}$$

The i refers to the generating units in the $N-1$ state. $N1^c_{G,it}$ refers to the $N-1$ state of unit i at time t. $N1^c_{S,it}$ is the $N-1$ state of ESS.

The power balance equation in the $N-1$ state is:

$$\sum_{i=1}^{N_i} P^c_{G,it} + P^c_{R,it} + P^c_{S,it} + P^c_{M,t} = P^c_{D,t} \tag{9.16}$$

The generation limit (Eq. (9.7)) and transmission capacity (Eq. (9.10)) during $N-1$ state are constrained by:

$$P^{\min}_{G,i} I_{it}N1^c_{G,it} \le P^c_{G,it} \le P^{\max}_{G,i} I_{it}N1^c_{G,it} \tag{9.17}$$

$$\left|P^c_{M,t}\right| \le P^{\max}_{M} N1^c_{M,t} \tag{9.18}$$

The superscript c denotes the $N-1$ state. $N1^c_{G,it}$ denotes the binary parameter for $N-1$ state of i at time t. Similarly, $N1^c_{M,t}$ denotes the binary parameter for $N-1$ state of the transmission line between the MV and LV level. The remaining operational constraints (for unaffected nodes) are applied using Eqs. (9.7)–(9.14).

9.3.5 ESS Operation

ESS stores (charges) energy at times of low energy market prices (off-peak electricity demand) and generates (discharges) the stored energy during high electricity market prices, low grid generation, and $N-1$ state. For the economic operations of ESS, we use the models in [24, 27].

There are two important operation modes in ESS, which are the charging and discharging modes:

$$P_{\text{SOC},st} = P_{\text{SOC},s(t-1)} - P_{S,st}\,\Delta t \tag{9.19}$$

$$P^{\min}_{\text{SOC},s}[= 20\%] \leq P_{\text{SOC},st} \leq P^{\max}_{\text{SOC},s}[= 80\%] \tag{9.20}$$

The energy stored in ESS is the state of charge (SOC). $P_{\text{SOC},st}$ is the SOC of ESS, Δt is the time interval in the SOC, s indexes ESS unit. $P_{S,st}$ is with positive and negative magnitude to indicate the charge and discharge mode. $P_{S,st}$ is negative during the charging mode where the ESS stores the energy, and thus the value of $P_{\text{SOC},st}$ increases. In contrast, during the discharging mode, $P_{S,st}$ is positive and hence the value of $P_{\text{SOC},st}$ decreases.

The current state of ESS is with an operating margin capacity of 80% upper limit and 20% lower limit [27]. In this case, the ESS is therefore allowed to discharge the power until 20% and charge at 80% of the maximum capacity. Additionally, the instant charging/discharging capability is assumed as soon as a commitment signal is sent by the grid controller. The charging and discharging modes of ESS have remained the same during a contingency event. It is also assumed that there is no associated energy lost when the ESS is in idle mode. For $N-1$ state, the ESS will discharge according to the available capacity.

9.3.6 Resilience Metric

According to [3, 30], resilience is the ability of a power system to withstand/remain in a state during a failure in an efficient manner and to quickly restore to the normal operating state. The grid resilience in this case is the extent to of the energy demand within the consumer is met when there is a disturbance in the grid. The performance metric to calculate the resiliency is based on the fraction of demand served, or the amount of energy produced by the microgrid during an outage [30]. The resilience metric (RI) in this case is therefore the fraction of demand served at dth consumer $\left(P^{c}_{D,t}\right)$ divided by the overall demand $\left(P^{c}_{D,t}\right)$ in the $N-1$ state:

$$\text{RI} = \sum_{t=1}^{P^{c}} \frac{P^{c}_{D,dt}}{P^{c}_{D,t} + P^{c}_{D,dt}} \tag{9.21}$$

9.3.7 Simulation Setup and Numerical Results (OGM)

The OGM module simulates the capability of the islanding operation by optimizing the available DGs and ESSs to provide electricity within the outage periods. RESs are however not optimized by the OGM, due to their uncontrollable intermittent behavior. Hence the imbalances in renewable outputs are compensated by DGs and ESSs. The required dispatching of DG and ESS units are known upon the simulation and the associated cost of operations (with or without savings) and the grid resilience during the outage are determined.

The grid-consuming components are aggregated and loads are descaled in order to obtain the total demand for the microgrid for individual nodes that are similar to the component

Table 9.4 Distributions of DG, ESS, and RES installed.

Node	Number of generators		ESS	Load (MW)
	Non-renewable	Renewable		
2	2	0	2	0.24
3	2	1	2	1.00
4	2	1	2	0.50
9	2	1	2	0.30

specifications as illustrated in Section 9.2.5 (i.e., Node 3 ≈ 940 kW, Node 2 ≈ 217 kW, Node 4 ≈ 478 kW, Node 9 ≈ 295 kW). Similarly, the demand profiles are from the portal [19]. A single RES, ESS, and three different types of DGs are used. Their distributions in IEEE- 14 node grid topology from Figure 9.4 are presented in Table 9.4. Two case studies are presented to illustrate the grid operation in sustaining different outages.

9.3.7.1 Case 1 – Complete Grid Outage

At first instance, a complete outage of the main grid is assumed. Consequently, the microgrid level was isolated from the main grid and the islanding capability of the microgrid level is triggered. Similar to Section 9.2.5, the outage is considered between 9 am and 3 pm.

The top left panel of Figure 9.8 shows the resultant optimized/balanced energy generation. The legends *Main grid, Nominal,* and *Balanced* denote, respectively: (i) the normal generation contributed by mid- or high-level grid; (ii) the expected generation in responding to the total demand without islanding capability; and (iii) optimized generation dispatches with the islanding capability from DGs, ESS, and RES. In the case of a complete grid outage, the main grid load drops to zero at hours 9 am to 3 pm. Therefore, the microgrid is operated in islanded mode during the outage period, where the required critical loads' supply and demand are successfully balanced based on localized generation. When the outage is completely solved, the islanded operation terminates and instantaneous main grid re-connection is carried out allowing to return to normal grid operation. In this case, the specifications and installations of DG, ESS, and RES in the IEEE-14 node grid are adequate in responding to the complete outage.

The bottom left panel in Figure 9.8 illustrates the operation costs for the conventional and optimized grid. Marginal cost savings are achieved (£66.54) through the optimized generation dispatches, even though in some instance the dispatching of generation units are costly in order to balance the demand during the outage. The overall resilience metric RI is based on the demand served during an outage event and is computed as 1.0. The highest RI metric is expected due to the complete outage mitigation in this case.

9.3.7.2 Case 2 – Partial Grid Outage

It is assumed that the main electricity connecting to Node 2 is broken down and therefore Node 2 is isolated from generation. The outage is considered between 9 am and 3 pm. Therefore, the needed demand in Node 2 is obtained from the localized generating units. The remaining grid will operate as usual.

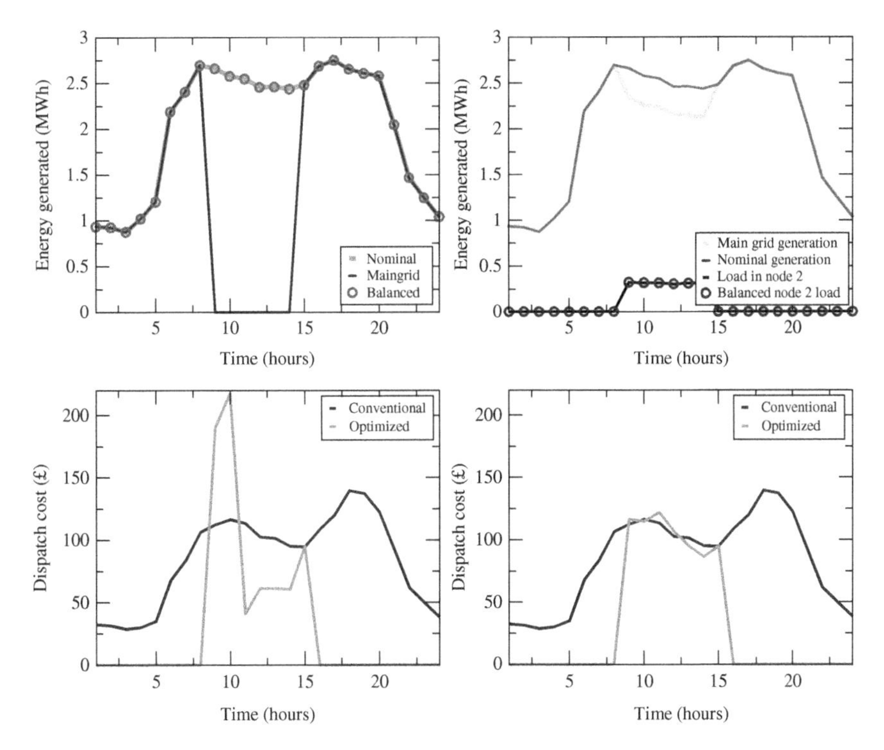

Figure 9.8 Plots for: [Case 1: Top left: The energy generation and its balanced/optimized generation during a complete grid outage; Bottom left: Cost of operations between the conventional and optimized solution during the complete grid outage]; [Case 2: Top right: The energy generation and its balanced/optimized generation in the case of an electrical connection breakout in Node 2. Bottom right: Cost of operations between the conventional and optimized solution in the case of an electrical connection breakout in Node 2].

The top right panel of Figure 9.8 shows the resultant energy generation and its optimized/balanced generation. The legends *Load in Node 2*, *Balanced Node 2 load*, *Main grid generation*, and *Nominal generation* are the current demand in Node 2, the balanced/optimized supply amount for Node 2, the main grid generation with the partial drop of loads during the outage period from Node 2, and the expected nominal generation where no outage occurred. Overall, the amount of generation required to compensate the load in Node 2 is secured during the outage.

The bottom right panel of Figure 9.8 presents the operation cost of the conventional and optimized grid solution. Additional costs (£15.83) are incurred for operating RES, DG, and ESS without monetary savings in this case. Overall, the calculated costs saving should not be related to the economical operation of dispatchable units during off-peak periods, as the environmental impacts from DGs such as emissions from air particles must take into consideration. The deployment of DGs that affects the severity level of environmental impacts is not implemented in the OGM module.

Figure 9.9 demonstrated the ESS operation state (top panel) and the *RI* metric distribution (bottom panel). The ESS is assumed charged with full capacity prior to the simulation. The ESS is in "idle" mode, awaiting the generation instruction. At 9 am, the outage occurred

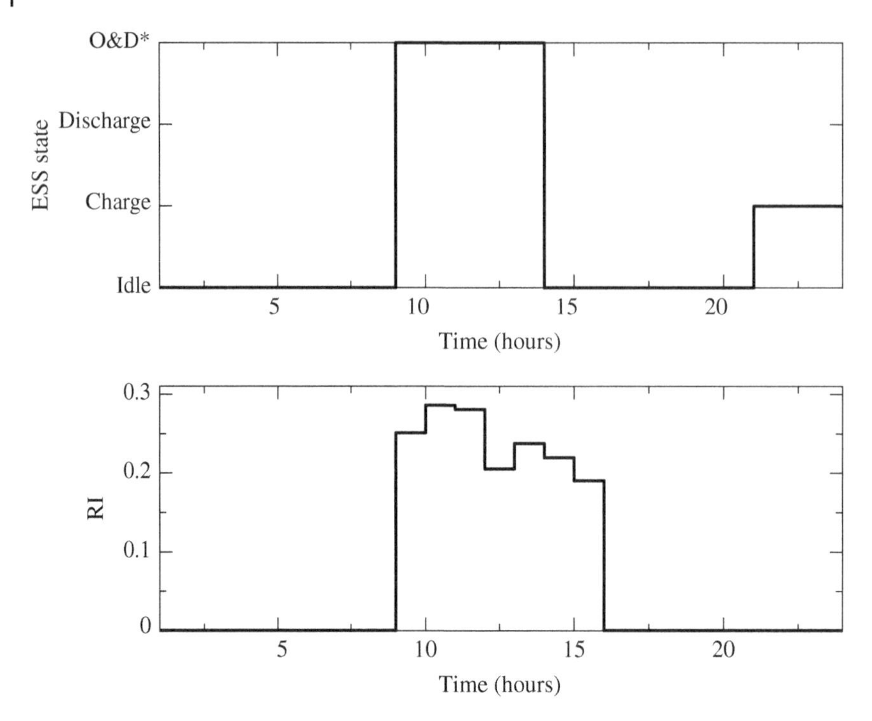

Figure 9.9 Top panel: ESS operation state. Bottom panel: Resilience metric distribution. Note: O&D* – Outage and discharge.

and instantaneous generation of ESS is required. Hence, ESS discharges (supplies) energy to Node 2.

The ESS stops discharging at 3 pm when the electricity connection of Node 2 to the main grid is fixed and ESS begins to charge at times of low electricity price (from 9 pm onwards), in order to prepare for the next stage of the scheduling horizon.

The computed distributions of *RI* metric in the bottom panel of Figure 9.9 during the outage are average-high. The zero percentage of resilience indicates the normal state of grid operation. Using Eq. (9.21), the *RI* is calculated as 0.245. Such calculation demonstrates the capability and resiliency of the Node 2 grid to sustain in an efficient manner. In contrast, *RI* is low when huge amount of critical loads are not met during the outage.

9.4 Conclusions

In this chapter, a complete smart grid and microgrid are modeled in normal and outage situations. The developed tools provide the architectural basis for the evaluation of the effects of outages in two different levels of the grid.

In both tools simulation experiments, outage operations are performed and the results are evaluated based on similar criteria: resilience, respectively, and autonomy factor in the grid setting. The MGE tool in Section 9.2 proposes a demand management scheme realized in central and building (CEMS) controllers at the microgrid level. The OGM tool in

Section 9.3 provides the integration of the main grid supply, electricity connections, DERs, energy demand profiles (large loads, substations, and hospitals), and power outage simulations. The contingency simulation is also considered in the model, where scenarios of power outage due to disasters or any other valid circumstances can be simulated. In this model, the $N-1$ contingency criterion is applied.

The MGE tool determines the reduced load capacity of the affected nodes, in order to prevent the overloaded nominal grid capacities. The simulation of the OGM tool, on the other hand, demonstrates the capability of the supply/demand balancing mechanism based on the commitment of DERs in compensating the demand shortage during outages.

As the chapter concentrates on single-line failures, a meshed grid topology for the model is essential to expand the analysis to examine cascading failures across the grid. Future work should extend the capabilities of the proposed method.

References

1 Chaouachi, A., Kamel, R.M., Andoulsi, R., and Nagasaka, K. (2013). Multiobjective intelligent energy management for a microgrid. *IEEE Transactions on Industrial Electronics* 60 (4): 1688–1699.

2 J. M. Guerrero, M. Chandorkar, T-L. Lee, and P. C. Loh. Advanced control architectures for intelligent microgrids part i: decentralized and hierarchical control. *IEEE Transactions on Industrial Electronics*, 60(4):1254–1262, April 2013.

3 A. Khodaei. Resiliency-oriented microgrid optimal scheduling. *IEEE Transactions on Smart Grid*, 5 (4):1584–1591, July 2014.

4 Hampson, A., Bourgeois, T., Dillingham, G., and Panzarella, I. (2013). Combined heat and power: enabling resilient energy infrastructure for critical facilities. Technical report, ICF International, March 2013. https://www.energy.gov/sites/prod/files/2013/11/f4/chp_critical_facilities.pdf (accessed 28 July 2017).

5 Elsied, M., Oukaour, A., Gualous, H., and Brutto, O.A.L. (2016). Optimal economic and environment operation of micro-grid power systems. *Energy Conversion and Management* 122: 182–194.

6 Diao, R., Vittal, V., Sun, K. et al. (2009). Decision tree assisted controlled islanding for preventing cascading events. In: *Proceedings of 2009 Power Systems Conference and Exposition (PSCE), 15–18 March*, 1–8. IEEE https://doi.org/10.1109/PSCE.2009.4839985.

7 Bollinger, L.A. (2012). Microgrid at illinois institute of technology. http://www.iitmicrogrid.net/microgrid/index_all.htm (accessed 28 July 2017).

8 Albasrawi, M.N., Jarus, N., Joshi, K.A., and Sarvestani, S.S. (2014). Analysis of reliability and resilience for smart grids. In: *IEEE 38th Annual Computer Software and Applications Conference (COMPSAC)*, 529–534.

9 Estebsari, A., Pons, E., Huang, T., and Bompard, E. (2016). Techno-economic impacts of automatic undervolt- age load shedding under emergency. *Electric Power Systems Research* 131 (9): 168–177.

10 Lopes, J.A.P., Vasiljevska, J., Ferreira, R., Moreira, C., and Madureira, A. (2009). Advanced architectures and control concepts for more microgrids.

11 NYSERDA, DPS, and DHSES (2014). Microgrids for critical facility resiliency in New York state. Technical report, NYSERDA. Technical Report Number: 14–36.

12 Aalami, H.A., Moghaddam, M.P., and Yousefi, G.R. (2010). Demand response modeling considering interruptible/curtailable loads and capacity market programs. *Applied Energy* 87: 243–250.

13 Katiraei, F., Iravani, M.R., and Lehn, P.W. (2005). Microgrid autonomous operation during and subsequent to islanding process. *IEEE Transactions on Power Delivery* 20: 248–257.

14 Olivares, D.E., Mehrizi-Sani, A., Etemadi, A.H. et al. (2014). Trends in microgrid control. *IEEE Transactions on Smart Grid* 5 (4): 1905–1919.

15 Lopes, J.P., Hatziargyriou, N., Mutale, J. et al. (2007). Integrating distributed generation into electric power systems: a review of drivers, challenges and opportunities. *Electric Power Systems Research* 77 (9): 1189–1203.

16 Chen, C., Wang, J., Heo, Y., and Kishore, S. (2013). Mpc-based appliance scheduling for residential building energy management controller. *IEEE Transactions on Smart Grid* 4: 1401–1410.

17 Parisio, A., Rikos, E., and Glielmo, L. (2014). A model predictive control approach to microgrid operation optimization. *IEEE Transactions on Control Systems Technology* 22: 1813–1827.

18 Bessler, S. and Jung, O. (2016). Energy management in microgrids with flexible and interruptible loads. In: *IEEE Power & Energy Society Innovative Smart Grid Technologies Conference (ISGT)*, 1–6.

19 US Energy Information Administration (2012). Cbecs. http://www.eia.gov/consumption/commercial/data/2012/ (accessed 28 July 2017).

20 Panteli, M. and Pierluigi, M. (2015). The grid: stronger, bigger, smarter? *IEEE Power and Energy Magazine* 13 (3): 58–66.

21 Sandia National Laboratories (2015). Conceptual framework for developing resilience metrics for the electricity, oil, and gas sectors in the United States. Technical report, Sandia National Laboratories. Sandia Report, SAND2014-18019.

22 Ejebe, G.C. and Wollenberg, B.F. (1979). Automatic contingency selection. *IEEE Transactions on Power Apparatus and Systems* PAS-98: 97–109.

23 Ouyang, M. and Dueñas-Osorio, L. (2012). Time-dependent resilience assessment and improvement of urban infrastructure systems. *Chaos: An Interdisciplinary: Journal of Nonlinear Science* 22 (033122): 320–330.

24 Bahramirad, S. (2012). Economic and emergency operations of the storage system in a microgrid. Degree project, School of Electrical Engineering, KTH Royal Institute of Technology, Stockholm.

25 Erol-Kantarci, M., Kantarci, B., and Mouftah, H.T. (2011). Reliable overlay topology design for the smart microgrid network. *IEEE Network* 25 (5): 38–43.

26 Liang, H. and Zhuang, W. (2014). Stochastic modelling and optimization in a microgrid: a survey. *Energies* 7 (4): 2027–2050.

27 Howlader, H.O.R., Matayoshi, H., and Senjyu, T. (2016). Distributed generation integrated with thermal unit commitment considering demand response for energy storage optimization of smart grid. *Renewable Energy* 99: 107–117.

28 Edurite (2015). Advantages and disadvantages of linear programming. http://www.edurite.com/kbase/advantages-and-disadvantages-of-linear-programming (accessed 30 January 2017).

29 Lau, E.T., Yang, Q., Taylor, G.A. et al. (2016). Optimisation of costs and carbon savings in relation to the economic dispatch problem as associated with power system operation. *Electrical Power Systems Research* 140: 173–183.

30 Bollinger, L.A. (2015). Fostering climate resilient electricity infrastructure. http://repository.tudelft.nl/islandora/object/uuid:d45aea59-a449-46ad-ace1-3254529c05f4/datastream/OBJ/download (accessed 06 December 2016).

10

Robust Cyber Infrastructure for Cyber Attack Enabling Resilient Distribution System

Hyung-Seung Kim[1]*, Junho Hong*[2]*, and Seung-Jae Lee*[1]

[1] *Myongji University, Department of Electrical Engineering, Yongin, South Korea*
[2] *University of Michigan-Dearborn, Department of Electrical and Computer Engineering, Dearborn, MI, USA*

10.1 Introduction

The term "power distribution system" refers to a medium/low-voltage power system from the substation feeder to the end consumer who consumes the power. A power distribution system consists of the distribution line, pole, pole-mounted transformer, circuit breaker (CB), recloser, automatic switch (AS), and manual switch (MS). Since the main priority of the power distribution system is to supply electricity to the customer, it should maintain a stable and reliable power supply. In order to maintain reliability, it is necessary to reduce the power outage time during a fault and to minimize the outage area by optimized protection coordination among protective devices [1–4]. Therefore, the distribution automation system (DAS) has been introduced to perform automated outage restoration and stable system operation. DAS is a system to collect the operation data (e.g., voltage, current, and switch status) from feeder remote terminal units (FRTUs) using a communication network and then to operate the distribution system through monitoring and control. Algorithms in the operation center need to be accurate since the applications for DAS operation are installed and performed at the operation center, and the operation center sends control commands to FRTUs. Therefore, many research studies have been presented about rapid restoration algorithms for distribution operations based on reliability (e.g., restoration strategy) using expert knowledge [5, 6], restoration solutions using fuzzy theory [7], restoration methodologies in multiple faults [8], and restoration methods based on information gap decision theory [9]. In order to solve the problem of the communication load of a centralized DAS, research has been carried out on restoration and protection coordination based on the operation method of the distributed multi-agent system (MAS) [10–12]. In some cases, DAS cannot maintain resiliency due to communication failure between the front-end-process (FEP) and FRTUs, cyber attack, or misoperation, even if the above algorithm is correctly performed. In particular, it is necessary to detect and deal with cyber attacks in distribution systems because distribution systems have vulnerabilities to cyber attack. Research on cyber security

Resiliency of Power Distribution Systems, First Edition.
Edited by Anurag K. Srivastava, Chen-Ching Liu, and Sayonsom Chanda.

of power distribution systems have been presented, such as key-based protocol research [13], an attack detection method using the electric power and communication synchronizing simulator (EPOCHS) [14], and countermeasures against false data injection attack [15, 16]. Hong et al. [17, 18] proposed a host- and network-based anomaly detection system. The above-mentioned mitigation methods can be used for the existing centralized power distribution operation method to prevent data injection attack by message authentication or by detecting abnormal activities in the communication system. However, it is difficult to apply the above methods for protection settings change attacks, direct circuit breaker control attacks, and wrong switching command attacks during power restoration since the attack commands are coming from a trusted party. In other words, cyber attacks should be detected when DAS operates in a normal state, so that DAS should not be affected by cyber attack while the outage restoration or protection coordination is being performed. Therefore, this paper explains how to maintain resiliency against cyber attack as well as rapid outage restoration and accurate protection coordination for reliable operation of DAS. The concept of electric power characteristics, for example, power restoration solution [7], power flow calculation [19–21], and communication characteristics (polling, request/response), is used in the proposed method. It also describes detection and mitigation strategies for each cyber attack based on the operation of distributed MAS. In summary, Section 10.1 describes the vulnerability of DAS. Section 10.2 explains the cyber threat or attack scenarios for the power distribution system. Section 10.3 provides the operation method for the fault detection, isolation, and restoration (FDIR) process and protection coordination in multi-agent-based distributed operation compared with the existing centralized operation. Finally, Section 10.4 shows the proposed mitigation method using multi-agent-based DAS.

10.2 Cyber Security Analysis of Distribution System

10.2.1 Vulnerability of Distribution Automation System

The DAS provides capabilities for the distribution operation center to acquire information (voltage, current, fault, and switch status) from field devices, for example, FRTU, and to monitor the distribution line in real-time. In particular, the DAS is contributing to increase the reliability of distribution system operation by detecting the fault area, reducing power outage time through remote control functions. Since the distribution operation center and FRTUs are communicating via standard communication protocols, for example, Distributed Network Protocol (DNP) 3.0, the role of information and communication technology (ICT) becomes more critical for system operation and control [13]. The cyber security of distribution systems needs to be recognized as a critical issue since it is connected to the customers directly and may have critical loads, for example, hospitals, government agencies, and police stations. However, there are many potential cyber security issues, such as (i) standardized communication protocols that allow intruders to understand the power distribution system architecture, (ii) exposed field devices that allow intruders to access the communication backbone of the power distribution system, (iii) well-trained intruder(s) who compromise the distribution operation center using malware viruses, and (iv) protective devices with default or weak passwords.

10.2.1.1 Unsecured Standardized Communication Protocols

The communication protocol is an important element for the operation of a distribution system since it contains critical information, for example, measurements and controls. Integrity is one of the important cyber security measures since any modification of the communication protocols may cause a huge impact to the distribution system's operation. Despite their importance, most of the electrical distribution system's protocols are not designed for cyber security. This is because cyber security was not a major concern when industrial communication protocols were published, for example, DNP 3.0 and International Electrotechnical Commission (IEC) 60870-5. Therefore, IEC Technical Committee (TC) 57 Working Group (WG) 15 established the IEC 62351 standard, which describes the cyber security measures of the DNP 3.0 and IEC 60870-5 communication protocols. A major vulnerability of the standardized communication protocol is that once attackers learn the standard of the communication protocols, they can analyze, modify, and inject fabricated packets into the communication system.

10.2.1.2 Exposed Field Devices

Most of the control devices in a DAS are installed in a public space (e.g., utility pole, house, or building) and anyone can easily access the devices, as shown in Figure 10.1. For instance, the gas insulated load break switch (GS) and FRTU are normally installed on a utility pole, and anyone can access the control box. If well-trained attackers access the control box of the FRTU, they can easily access the communication backbone system in the DAS. Sometimes the communication system of the distribution operation center and the distribution substation are physically connected from/to the distribution system's communication backbone system, so the exposed field devices could be an entry point for cyber intrusion.

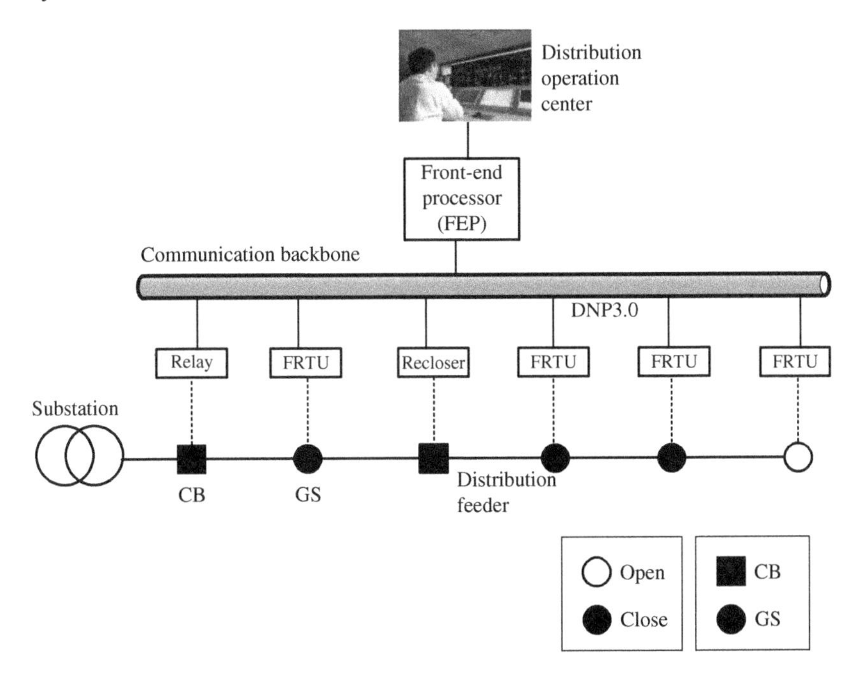

Figure 10.1 The cyber-physical system structure of a typical distribution automation system.

10.2.1.3 Malware Viruses

The United States Industrial Control Systems Cyber Emergency Response Team (US ICS-CERT) in the Department of Homeland Security (DHS) issued an alert about a coordinated cyber attack on the Ukrainian power grid [22, 23]. The first cyber attack was executed in December 2015 and caused outages to approximately 225,000 customers after disconnecting seven 110 kV and twenty three 35 kV distribution substations from the grid. Attackers successfully compromised the utility's ICS via a virtual private network (VPN) and malware viruses called "KillDisk" and "BlackEnergy." The second cyber attack was executed in December 2016 and focused on industrial communication protocols, for example, IEC 60870-5-101, -104 and IEC 61850 using a malware virus called "CrashOverride." This malware virus issues valid commands directly to remote terminal units (RTUs) over industrial communication protocols and executes abnormal circuit breaker control commands in a rapid open–close–open–close pattern. These cyber incidents clearly show the need for reliable cyber security measures in distribution systems.

10.2.1.4 Default Password and Built-In Web Server

A typical distribution feeder may have a number of protective devices and FRTUs, and it is difficult to manage the different passwords for each device. Therefore, distribution operators may use the default or same password for all devices. In addition, some field devices and user interfaces have a built-in web server and hence may be vulnerable to cyber intrusions, for example, remote configuration change and control with default passwords. Distribution system security managers have to check the security and system logs of field devices and user interfaces to detect unauthorized access.

10.3 Cyber Attack Scenarios for Distribution System

Security threats to the DAS can be divided into two parts based on the physical and cyber assets. The physical assets are the hardware components, for example, FRTU, recloser, relay, transformer, bus bar, and feeder, whereas cyber assets include physical and cyber resources, for example, firewall, communication network, and software applications in the distribution operation center, as illustrated in Figure 10.2.

As shown in Figure 10.2, potential cyber security threats and locations of intruders in a distribution automation network include:

A1: compromise protection and control field devices on feeders (e.g., relay, FRTU, and recloser)
A2: gain access to distribution system communication bus
A3: compromise the user interface of distribution operation center
A4: bypass or compromise remote access firewall
A5: gain access to substation communication bus
A6: compromise protection and control field devices in substation (e.g., relay and RTU)
A7: gain ID and password of remote access point
A8: compromise gateway
A9: bypass or compromise substation firewall
A10: bypass or compromise distribution system firewall

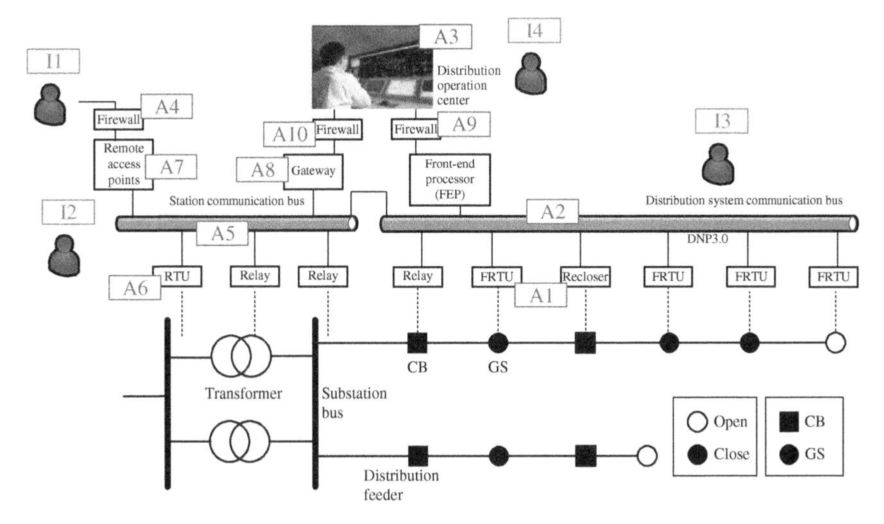

Figure 10.2 Overview of ICT network diagram and security threats for distribution substation and distribution automation system.

I1: intruder from outside of substation network via remote access points
I2: intruder from inside of substation network
I3: intruder from outside of distribution system network via field devices
I4: intruder from inside of operation center network

As depicted in Figure 10.2, possible intrusions into the distribution system local area network can originate from outside or inside a network. For instance, the intruder in (I3) may gain access to the distribution system communication network (A2), and then they can analyze the communication protocols and identify the most critical feeder or load. After that, they may generate a fabricated control communication packet and open the switch or breaker by injection attack.

10.3.1 Remote Access to Open Switch Attack

Most distribution substations are located in widespread and remote sites. Remote access to substation networks using VPN, dial-up, or wireless is a common way to monitor and maintain the distribution substation. The main problem of the remote access point is that remote access points may not be installed with adequate security features, for example, a poorly configured firewall, a weak ID and password policy, bad key management for cryptography, and use of unsecured external memory (e.g., USB flash drive). Furthermore, some of the communication networks of distribution substations and DASs are physically connected without proper cyber security measures (firewall). Hence, any successful cyber intrusions into a distribution substation's network could be an entry point to the DAS. Therefore, substation security managers have to consider the following actions in order to enhance the cyber security: (i) check firewall policies and logs periodically to identify security breaches, (ii) change IDs and passwords frequently and enhance the password policy (e.g., including numerical digits and special characters), (iii) enhance security of the key server against attackers, and (iv) provide education on security practices for operators.

Cyber attacks via remote access to an open switch can originate from outside intruders (I1). Once they compromise and bypass the firewall (A4), they can easily access the remote access point (A7) using a cracked ID and password. After that, they can access the station communication bus (A5) and execute a scanning attack to identify a vulnerability of the ICT system and critical protective and control devices. After identifying the most critical relay in the system, they may access the relay and issue open/close commands to the circuit breaker to create outages [24].

10.3.2 Man-in-the-Middle Attack

A man-in-the-middle attack is one of the cyber attacks, in which an intruder plays the role of a malicious actor between multiple trusted parties and gains access to information that the trusted parities are exchanging with each other. This attack allows an intruder to monitor, intercept, modify, send, and receive data as a trusted parties without the group of party knowing the activities, as shown in Figure 10.3a and b. The intruder is located at the middle of the communication flow, as illustrated in Figure 10.3b, and other trusted parties cannot recognize the problem. Once they intrud into the communication between trusted parties, they can inject abnormal fabricated data or replay the switching control messages in the future. For instance, the intruder learns the semantics of switching control messages from the distribution operation center to the recloser, and (s)he creates a fabricated switch open command and injects it into communication network. Then the recloser will be opened by the injected control message; however, the intruder intercepts the status signal and injects a fabricated signal again (from open status to close status), and sends it to the distribution operation center. In this scenario, operators in the operation center cannot know the actual status of the recloser since the intruder is in the middle of the communication and injects a fabricated message. Another example is the replay attack. Replay attack is to capture the actual control message and reuse it for a future cyber attack. For instance, when a fault occurs at the distribution line, the operation center sends switch control commands for fault isolation and restoration. At this moment, the intruder can capture the open or close switching commands and reinject them into the communication system to generate system failure during normal operation.

10.3.3 Configuration Change Attack

Protection coordination between the overcurrent relay (OCR), the recloser, and the fuse is one of the well-known protection schemes in DAS. In order to minimize the outage area, proper setting of protection coordination is crucial. However, these setting values can be changed by an attacker and lead to an unwanted outage area during normal operation or when a fault occurs at a feeder. Typically, there are two different ways to change the protection settings of field devices: (i) the vendor's engineering tool or (ii) the front panel of the protective device. If the attacker knows the ID and password of the user interface in the distribution operation center, they can execute the engineering tool for the protective devices. Then they can change the protection settings of devices with the default password. If the changed setting pick-up value is lower than the normal load current, the protective device immediately trips the circuit breaker, whereas if the changed setting time-delay value of the backup device is smaller than primary protection's time delay value, it will trip the entire feeder and create an unwanted outage area during the power system fault.

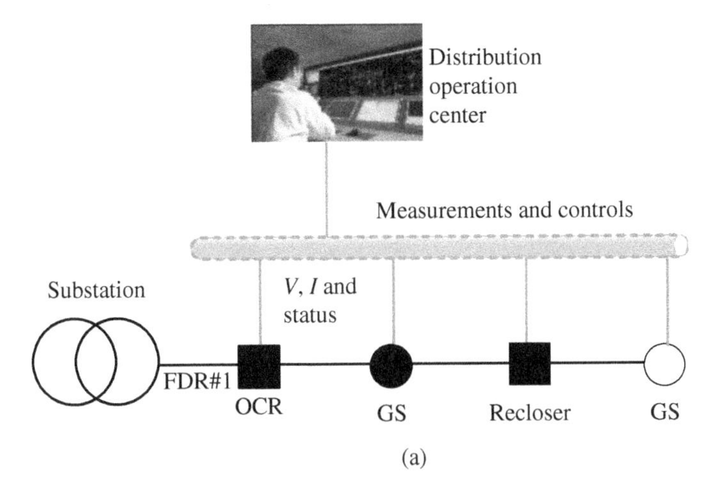

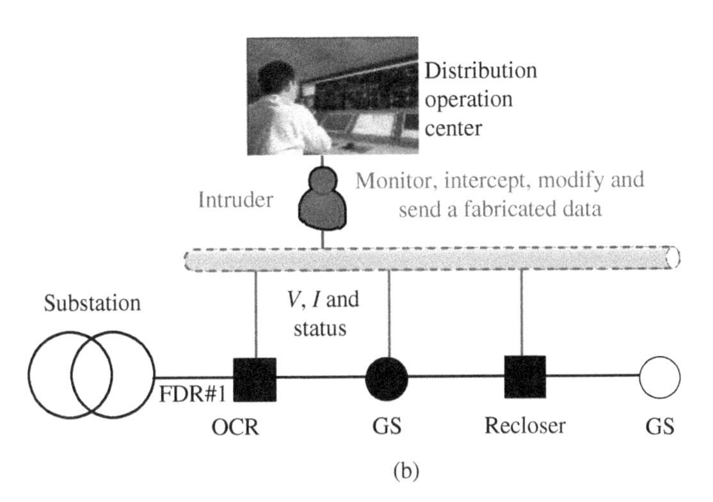

Figure 10.3 Man-in-the-middle attack in distribution automation system. (a) Normal operation. (b) Man-in-the-middle attack.

Figure 10.4 shows an example of protection coordination between primary (recloser) and backup (OCR) protective devices. In order to minimize the outage area, the primary protective device's time delay has to be faster than the backup protective device. Hence, recloser will be locked out after the reclose operation and the fault will be cleared (area 3–4) without outage of the upside of the feeder (area 1–3). However, if the intruder compromises the engineering tool and changes the time-delay setting of the OCR from 20 cycles to 2 cycles, area 1–3 will have an unwanted outage during the fault. This is because OCR will trip the circuit breaker faster than the recloser.

10.3.4 Denial-of-Service Attack

The denial-of-service (DoS) attack is one of the typical network-related cyber attacks on the ICS. The main purpose of a DoS attack is to make a network or device resource unavailable,

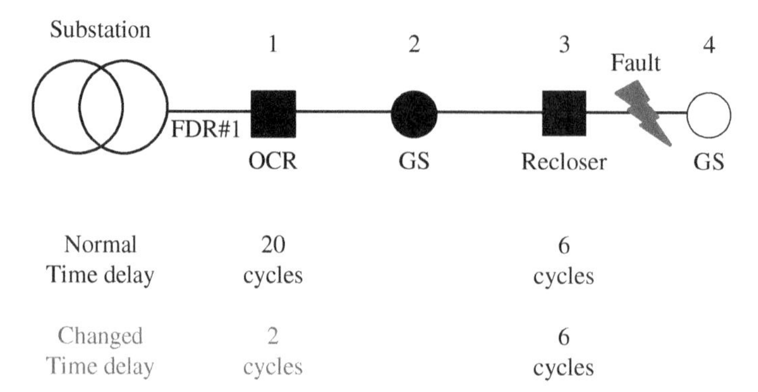

Figure 10.4 Protection coordination between OCR and recloser.

thereby disrupting normal operation or service of the target system. A DoS attack is accomplished by flooding the targeted device or system with unnecessary requests and making it overloaded. Then it can prevent the trusted users from accessing or operating the system in a timely manner. If the origin of the DoS attack is from outside of the system, it can be blocked by proper firewall settings; however, an inside DoS attack is hard to prevent. The impact of a successful DoS attack on protective devices will be huge when a fault occurs at the distribution system. For instance, during the fault isolation process, if the operation center cannot reach the proper FRTUs and is not able to open/close the switches, the outage time and economic losses will be increased.

10.4 Designing Cyber Attack Resilient Distribution System

10.4.1 Centralized Controls and Protections

The DAS consists of three main functions: (i) monitoring and control of remote breakers and switches, (ii) polling the current and voltage values from FRTUs to FEP via wired/wireless communications, and (iii) minimizing the outage areas by detecting a fault location automatically. The DAS covers the system area from a substation feeder to a customer (e.g., factories, hospitals, and houses). Figure 10.5 shows an example of a centralized DAS. The centralized DAS executes its own automation algorithms and makes a control decision (e.g., open/close breakers or switches) using the polled data from FRTUs to the operation center. Once it finishes calculating the restoration solutions, it sends corresponding control signals in a sequence.

When a fault occurs at a distribution line, the primary protective device (e.g., OCR or recloser) detects the fault and then will open a breaker to clear the fault. Once a fault is detected by FRTUs, they send the fault indicator (FI) to the operation center. The FIs can be divided into permanent and temporal fault. A temporal fault indicates that FRTUs experienced a fault current more than one time, whereas a permanent fault specifies that FRTUs still detect a fault after reclosing actions. The FI information is very crucial for detecting the location of a fault. For instance, when a fault occurs between FRTU A and FRTU B, the fault current flows toward the fault location, as described in Figure 10.6. The principle of

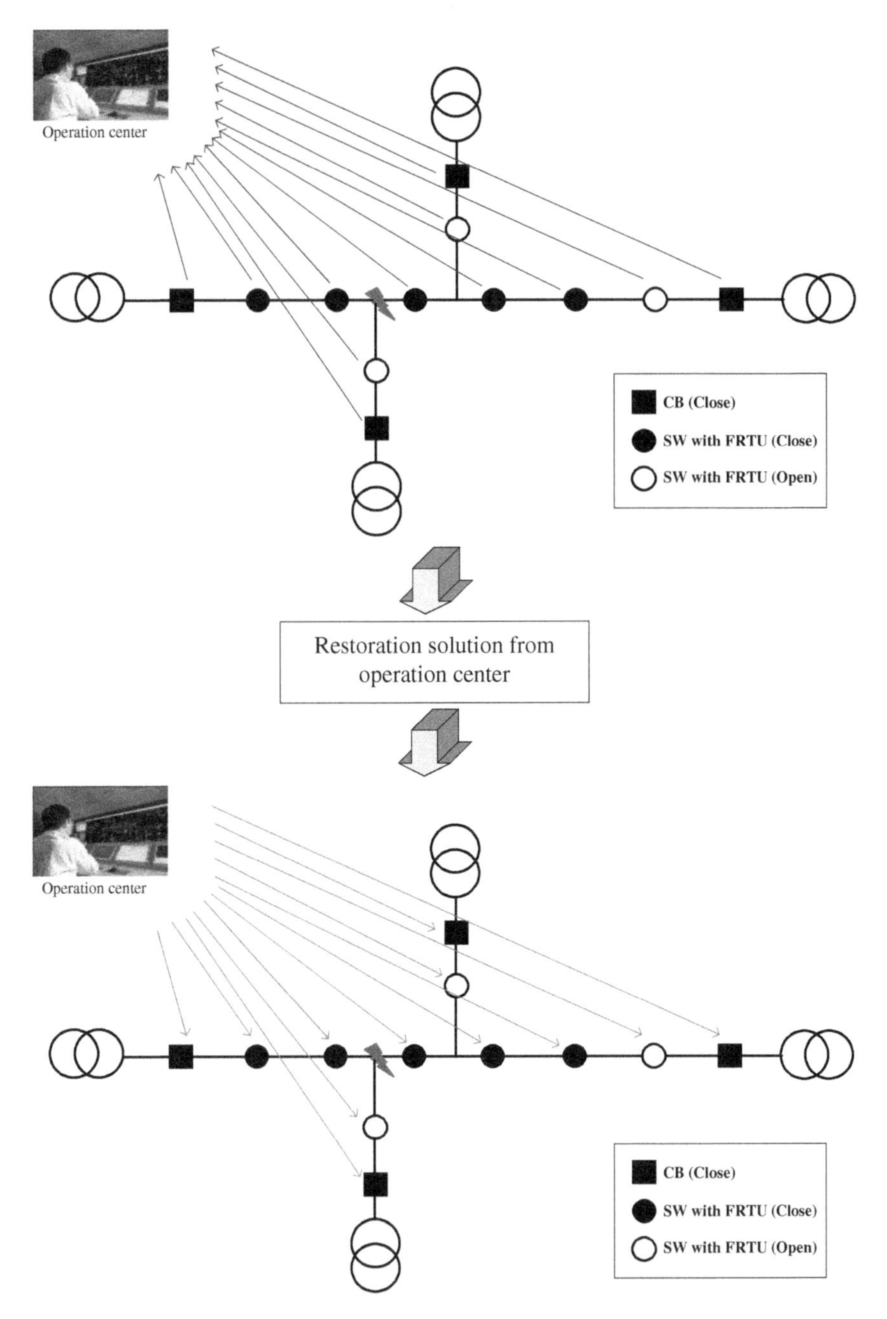

Figure 10.5 A structure of centralized distribution automation system.

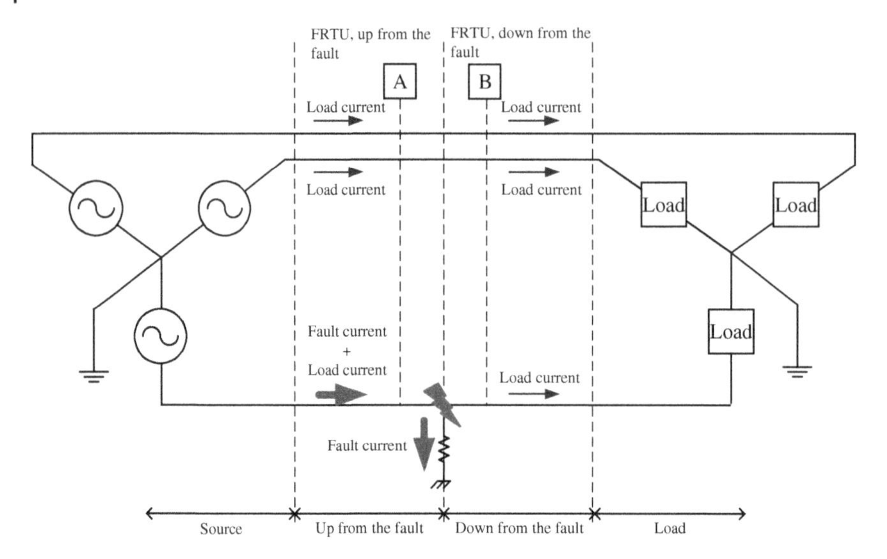

Figure 10.6 Detecting the fault location using fault indicator information.

FI information is that FRTU A can detect a fault, but FRTU B cannot detect a fault. Hence, the location of the fault can be calculated as between FRTU A and FRTU B. At the same time, OCR or overcurrent ground relay (OCGR) send the detected fault information to the operation center. An automatic switch cannot interrupt the fault current. However, it can detect and send the fault information (phase or line-to-ground fault) to the operation center if the current is higher than the predefined threshold value.

When a fault occurs, the main role of the operation center is to calculate the fault location, isolate the fault, and then execute the restoration actions in order to minimize the fault areas and outage time using transmitted FI information from FRTUs. For the restoration process, the operation center generates possible restoration scenarios and then chooses the optimal restoration plan [6, 7]. The restoration plan contains the sequence of switching actions (open or close) that can minimize the outage areas and time and maximize the system capacities. There could be multiple restoration scenarios, and it is crucial to choose the optimal restoration plan and execute the commands to each breaker or switch. Many research studies have been conducted to calculate restoration plans, and the main purpose is to prevent additional outages, minimize the switching actions, and maximize the system's capacity for stable operation.

The restoration plan can be calculated for two different systems: (i) a feeder without a tie breaker and (ii) a feeder with a tie breaker. A feeder without a tie breaker indicates that the feeder does not have any connection to other feeders, whereas a feeder with a tie breaker specifies that the feeder has a tie breaker and can be connected to another feeder. When a fault occurs at the feeder without a tie breaker, the restoration plan can be calculated as a small number of switching actions. For instance, if a fault has occurred between GS2 and GS3 in Figure 10.7, the recloser will be opened and clear the fault. Then there will be an outage area from the recloser to the end of the feeder (GS4). In this scenario, the restorable area is from the recloser to GS2 since the example system does not have a tie breaker to the other feeder so it cannot transfer the load between GS3 and GS4. Once the operation center

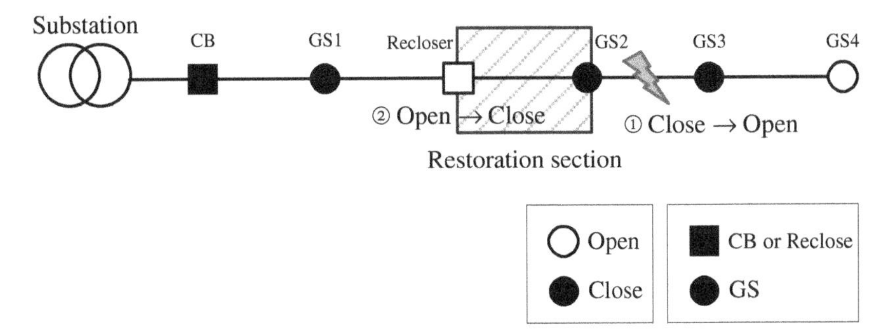

Figure 10.7 Fault detection and clearance by recloser.

receives FI information, it sends a sequence of commands to open GS2 and GS3 and then closes the recloser. Finally, the fault location is isolated by GS2 and GS3, and other areas are energized again.

As explained above, the system without a tie breaker has a limited restoration plan. However, if the system has a tie breaker, a number of restoration plans can be calculated (it depends on how many tie breakers exist in the system). One of the well-known restoration algorithms is to use fuzzy logic [7]. The fuzzy-based restoration algorithms use various types of data (e.g., restoration speed, restoration percentage, number of switching actions, load balance, voltage drop, load shedding, loss, protection coordination) and evaluate the priorities of data to calculate the restoration plans (Figure 10.8).

Once the system finishes calculating the restoration plan, the operator will choose the optimal plan to restore the faulted system. If the faulted system has a tie breaker and the connected feeder has enough capacity to pick up the faulted system, the system can be restored with one switching action by closing the tie breaker (i.e., transfer loads). After finishing the initial restoration process, the system should have a minimized isolated area (where the fault occurred), and then the dispatch center will send a field crew to find the actual fault location. Typical examples of distribution system faults include natural disaster,

Outage load / Total margin	Small (S)	Medium (M)	Big (B)
Small (S)	Normal (N)	Low (L)	Low (L)
Medium (M)	High (H)	Normal (N)	Low (L)
Big (B)	High (H)	Normal (N)	Low (L)

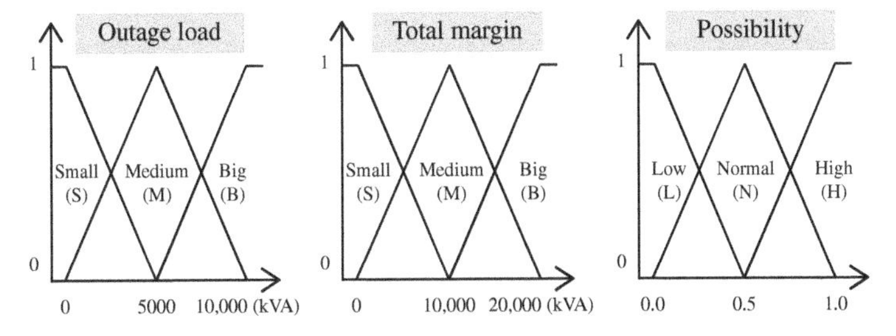

Figure 10.8 Fuzzy-based restoration for distribution automation system.

tree or animal contact, lightning strike, and aging of equipment. Once the dispatched field crew finds the fault location and restores the faulted area, the system operators will bring system back to normal.

10.4.2 Multi-Agent-Based Controls and Protections

10.4.2.1 Multi-Agent System

An agent is an intelligent processor that is composed of hardware or software within an environment that conducts a special task. A multi-agent system (MAS) is the system in which agents are all connected. The agents in an MAS are not processing system tasks for the final goal of the system but are processing their individual goals to achieve the system goal. Hence, the purpose of the MAS is to have an active response for the environment changes and to finish the final goal under special cases by cooperation between agents.

The characteristics of the MAS include autonomy, local view, and decentralization. The individual agent has an autonomy that can react to the environment changes automatically. When there are many terminal devices in the power distribution system and they are structured to cooperate with each other as an MAS, it is difficult to manage the entire system. Hence, they need to be clustered and operated as local-view agents. All agents are independent and have a characteristic to cooperate when there is an environment change so they have decentralized characteristics in the control aspect. Furthermore, the basic functions that are required from MAS are system operation, inspection, diagnosis for system reliability, and simulation functions for decentralized control and connection of the previous and future system. A system platform, tool kit, agent design, communication protocol, data standard, and cyber security are needed and considered. MAS should have a decentralized structure, peer-to-peer communication between agents, and a flexible platform for expandability. Since agents communicate via standardized communication protocols, cyber security measures have to be considered.

Due to the flexibility and expandability of the DAS, the MAS can be implemented and enhance the reliability of DAS. There are requirements for implementing the MAS to DAS: (i) observation and diagnosis, (ii) decentralization, (iii) distribution system modeling, and (iv) system protection. The DAS requires observation and diagnosis of measured data from FRTUs. Hence, if an FRTU has MAS functions that include local observation and diagnosis, it can enhance the reliability and resiliency of the DAS. A single communication failure of the centralized system may cause huge system problems. However, a single communication failure in MAS can be resolved by rerouting the communication via other agents, so decentralization is required in MAS. A study and simulation of DAS modeling with MAS is required, and it has to be tested whether the performance of MAS in DAS is reliable and performing its functionalities [25, 26]. Protection is one of the most important functions in DAS and requires a complex structure of protection coordination between devices. Even though the efficiency of MAS is higher than that of a centralized structure, the reliability, stability, and availability of protection has to be tested and simulated before deploying the MAS-based DAS.

As shown in Figure 10.9, the structure of MAS is a meshed network in which all agents are connected via a communication network (peer-to-peer) for cooperation, and each agent has its own local goal. In DAS, MAS is implemented in FRTUs that can communicate with each

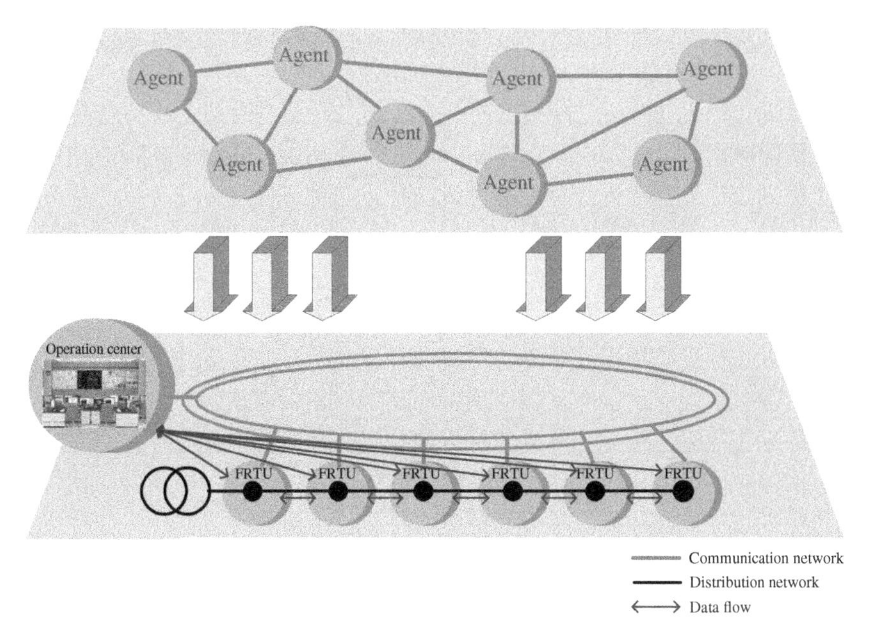

Figure 10.9 Multi-agent-based distribution automation system.

other and also can communicate with the operation system. FRTUs are installed at each circuit breaker, automated switch, and recloser so each node is represented as an FRTU. As explained above, MAS is the connected agent system, and they mutually cooperate to achieve the final goal. Hence, DAS can benefit from adopting the MAS.

10.4.2.2 Multi-Agent-Based Message Authentication System

In order to secure the communications between agents, message authentication code (MAC) algorithms need to be applied. The overall authentication process of the proposed MAC scheme is shown in Figure 10.10. The sender and receiver share the same private key, KEY (K). Although there is no requirement for the length of the shared private key, it is recommended to have a key length higher than 128 bits in order to enhance the security strength of MAC-based authentication. The message synchronization number (Sync) is introduced to check the freshness of the message. Sync will be increased by "1" for every transmission, so the new Sync should be higher than the Sync number previously received. By comparing these two numbers, the receiver knows whether the received messages are reused by an attacker. The sender will generate a MAC tag using the MAC algorithm together with the shared key, time, synchronization number, and message body. Then the generated MAC tag is appended to the main body of the message and sent to the receiver.

Once the receiver receives the packet with the MAC tag, it will calculate the MAC tag again using the shared key, synchronization number, time, and delivered message body. In order to verify the authentication, three conditions need to be checked as follows: (i) if the calculated MAC tag and delivered MAC tag are matched, (ii) the delivered Sync number is the same as the expected Sync number ("1" higher than the previous Sync number), and (iii) the delivered time is faster than the old time, then the delivered communication message is verified. Another additional verification could be timeout. For instance, operators

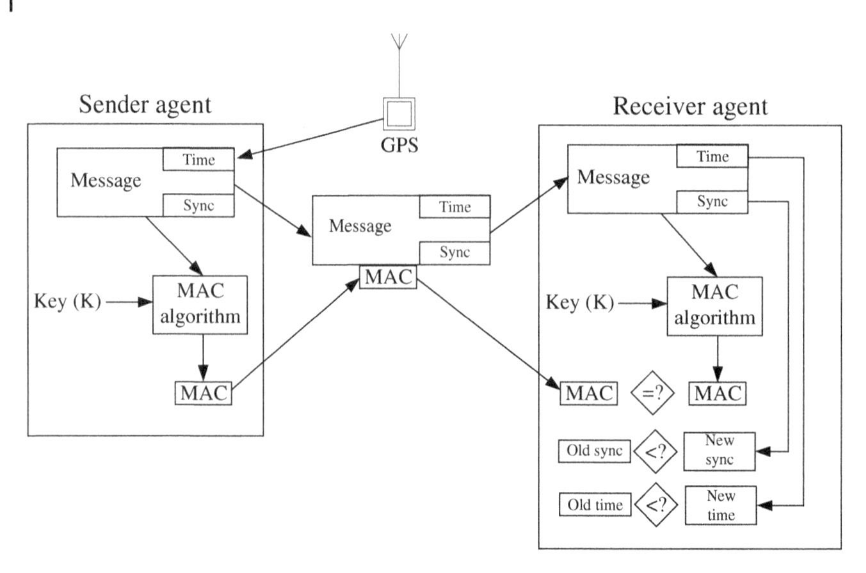

Figure 10.10 Authentication process for the communication between agents.

can set the timeout threshold as 3 (ms), and if the delay of the delivered message is higher than 3 (ms), receiver may discard the delivered message and request a new message from the sender. Therefore, it can make it more difficult for attackers to decode and generate the messages during normal communication.

10.4.2.3 Multi-Agent-Based Distributed Restoration System

An electric power distribution system is the final stage in the delivery of electric power from generation to customers. Therefore, the main purpose of the distribution system is to provide reliable electric power to the customers and enhance the reliability and quality of electricity. In order to maintain the above-mentioned requirements and purposes, outage restoration and protection coordination are crucial. The definition of outage restoration is to calculate and isolate the faulted area, minimize the outage area, and then energize the outage area by load transfer to the adjacent feeder. The traditional operation of DAS is centralized FDIR, whereas MAS-based DAS is decentralized FDIR. The difference between centralized and decentralized FDIR is described in Figure 10.11. Figure 10.11a shows the centralized FDIR and Figure 10.11b describes decentralized FDIR.

The centralized FDIR system gets data (e.g., voltage, current, switch status, and FI) from FRTUs in DAS. Then, the centralized operation center calculates restoration plans and sends FDIR commands (e.g., switch open or close) to the corresponding FRTUs. In contrast, in a decentralized system, the central agent (operation center), and the decentralized agents (FRTUs) collect data and operate FDIR processes by cooperation. The purpose of FDIR is to minimize the isolation area and reduce the time of outage restoration. However, the centralized FDIR system requires more steps and time for polling data, processing FDIR algorithms, and sending sequential switching control commands whereas decentralized FDIR based on MAS requires less time and decision processing compared with the centralized FDIR system. Agents can detect abnormal system status (e.g., fault), and then communicate with each other to find the fault location. Once they detect the fault location, they can make an optimal decision to isolate the fault and do restoration processes. In a decentralized FDIR system, the role of the operation center can be performed as one of the agents, so the size of the function is smaller as compared to centralized FDIR. It supports other agents (FRTUs) to make a better decision and reduce the FDIR process as follows:

(a) Monitoring of Other Agents (FRTUs): The operation center monitors all agents and checks their operation, communication, switch status, and device health. So, if there is any abnormal activity or log from agents, it will report to operators for further investigation.
(b) Update Communication Structure: In a multi-agent system, each agent should have the other agents' communication information for exchanging data so they can make a decision when a fault occurs in the system. For instance, each agent needs to have communication information of upstream and downstream agents. If the location of the tie breaker is changed for load balance, the system structure will be changed as well. In this case, the operation center sends the updated communication structure to all agents so they can make optimal decisions for future events.

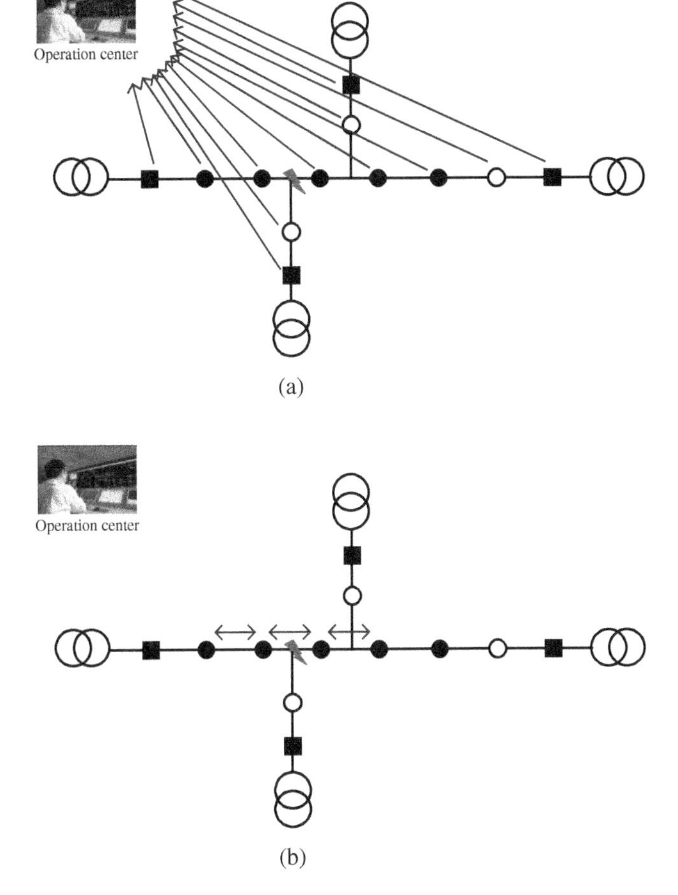

Figure 10.11 An operation of centralized and distributed (decentralized) distribution automation system. (a) Centralized FDIR. (b) Decentralized FDIR.

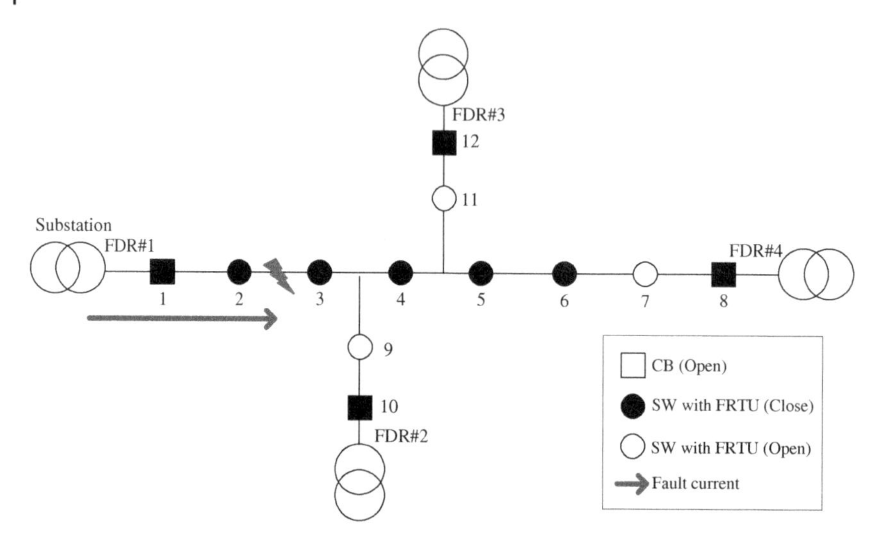

Figure 10.12 Fault occurring in a decentralized distribution automation system.

(c) Update FDIR Process: In a multi-agent-based FDIR system, the FDIR process is executed based on cooperation and precalculated solutions. The precalculated solutions are transferred from the operation center to each agent. For instance, the operation center calculates all FDIR processes and steps for each fault scenario (all segments between switches) and then sends this precalculated information to all agents. Therefore, when a fault occurs in the system, each agent can send a command and process the optimal FDIR.

Figure 10.12 shows one of examples of a decentralized FDIR process using multi-agents. If the fault occurs between SW2 and SW3, the protection device sees the fault and opens CB1. The agents that experience the fault generate FI in the multi-agent-based FDIR system. First CB1 asks the SW2 agent whether it generated FI. SW2 responds to CB1 that it generated FI, and then SW2 asks SW3 whether it generated FI. Since the SW3 agent did not generate FI, it responds to the SW2 agent that it did not generate FI. Therefore, agents can know that the fault occurred between SW2 and SW3, and they will open its switch to isolate the fault.

Since the area between CB1 and SW2 can be energized from the substation, SW2 sends a close command to CB1, as shown in Figure 10.13. Once it restores the upstream area, it executes the downstream area restoration. It is assumed that the precalculated FDIR process is to transfer the load between SW3-SW4-SW9 to FDR#2, the load between SW4-SW5-SW9 to FDR#3, and the load between SW5-SW6-SW7 to FDR#4, as described in Figure 10.14. In this FDIR scenario, SW4 and SW5 need to be opened, and SW7, SW9, SW11 need to be closed. The FDIR process starts from agent SW3 and will find the switches to be opened. SW3 finds that SW4 and SW5 need to be open switches and then sends open commands 3-4-O and 3-5-O. If there are no more switches to be opened, it will find the switches to be closed. During the closing process, the controlled SW can issue the close command to its downstream SWs. Hence, SW3 issues close command 3-9-C, SW4 issues close command 4-11-C, and SW5 issues close command 5-7-C. Finally, the following FDIR plans are transferred, stored, and executed at SW3 (agent) when a fault occurs between SW2 and SW3.

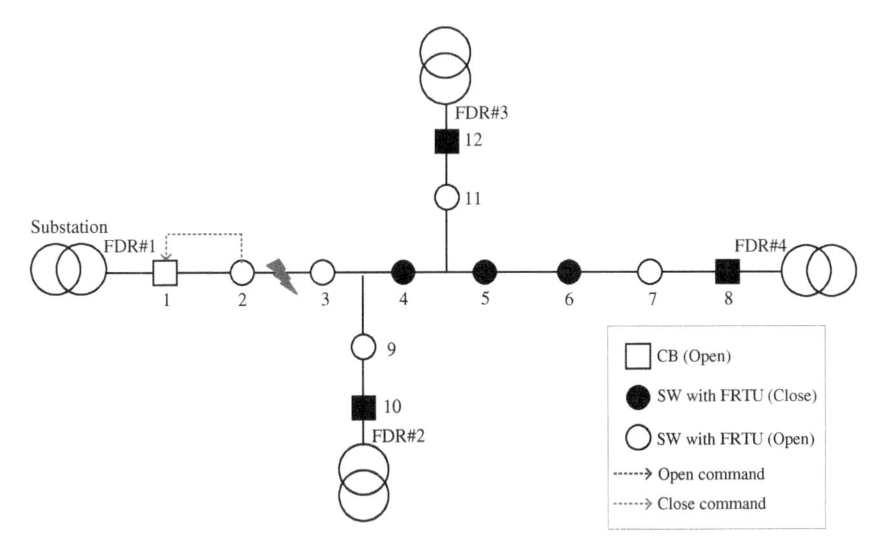

Figure 10.13 Fault isolation and initial restoration process using multi-agents.

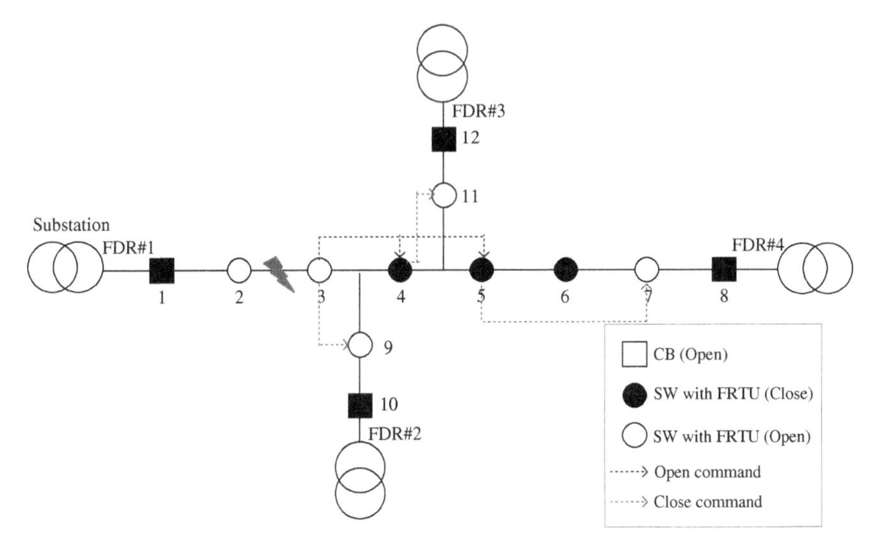

Figure 10.14 Restoration process using multi-agents.

3-4-O: SW3 issues open command to SW4
3-5-O: SW3 issues open command to SW5
3-9-C: SW3 issues close command to SW9
4-11-C: SW4 issues close command SW11
5-7-C: SW5 issues close command to SW7

After finishing the FDIR process, the faulted area needs to be restored, so the operation center dispatches the field crews and also needs to update the changed system information to all agents. In this scenario, the FDIR process changes all system structures and locations of the tie breakers. Therefore, each agent needs to know their new upstream and downstream agent information. Also, the operation center needs to calculate new FDIR plans and to send the update to all the agents. This process needs to be executed when there is a faulted agent that cannot be used for the FDIR process.

10.4.2.4 Multi-Agent-Based Protection Coordination System

Protection coordination refers to a time sequence of operations between protective devices. The purpose of protection coordination is to minimize the outage area by collaboration between protection devices (relay, recloser, and fuse) and to operate the CB or switch nearest to the fault. This can be configured by the time delay settings of each protective device, and the settings need to be calculated and verified before the protective devices are deployed. After reconfiguration of the distribution system due to the FDIR process, these setting values are not changed. Because of this problem, if there is a fault after the first FDIR process, the protective devices may not perform correctly. However, a multi-agent-based decentralized FDIR system can adopt new calculated FDIR plans from the operation center after reconfiguration of the distribution system or if there is a faulted agent in the system. In order to distribute the newly calculated FDIR plans to each agent, the following functions are needed.

(a) A protection coordination calculation function is required at the operation center. When a system is operated as multi-agent based, each agent can store protection coordination settings (time delay settings). These settings can be dynamically adopted based on the changed or reconfigured distribution system, and this can prevent wrong protection coordination after the first FDIR process. In this case, the central agent (operation center) needs to calculate the new protection coordination and distribute it to all agents in the system. The protection coordination information contains communication information of the upstream and downstream agents, communication information of the central agent, decentralized FDIR plans, and time-delay settings of protective devices. Normally a single feeder does not have a large number of protective devices (about two to three devices), so new protection coordination that includes time delay settings can be calculated in a few seconds.

(b) Both protection and time-delay settings are very crucial for protection coordination between protective devices. The time-delay settings of the protective devices have to be recalculated when a protective device is allocated to the new feeder due to the reconfiguration of the distribution system (e.g., load balance or FDIR). For instance, if the location of the open tie breaker between two feeders is changed due to power restoration or overload (load balance), as shown in Figure 10.15, the recloser in the middle of the line is transferred to the other feeder. In this case, the setting of the protective device should be set again. Before load balance, two feeders are connected via a normally open tie breaker (SW4). However, if the normally open switch is moved to SW7 due to fault or overload, Recloser 6 belonging to existing Feeder 2 will be moved to Feeder 1. In this case, it is crucial to modify the setting values of protection coordination. Therefore, when the location of the main protective device is changed by comparing the protection information (i.e., multi-agent-based protection coordination) that switches (FRTUs) have, reconfiguration of protection coordination needs to be recalculated and updated.

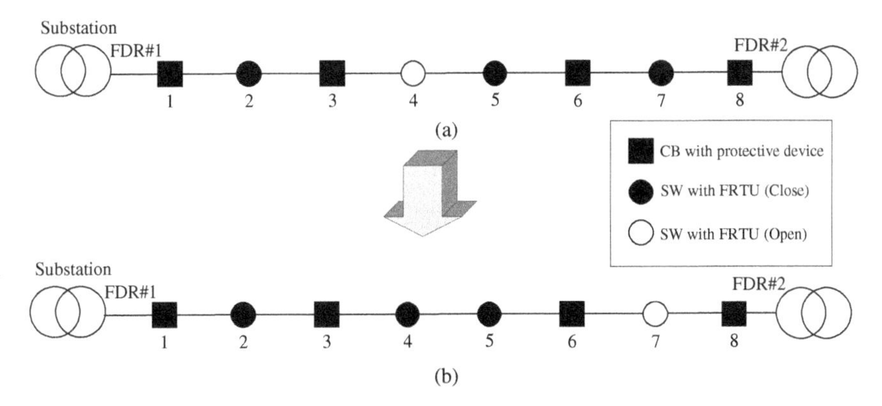

Figure 10.15 Tie breaker reallocation due to load balancing. (a) Before. (b) After.

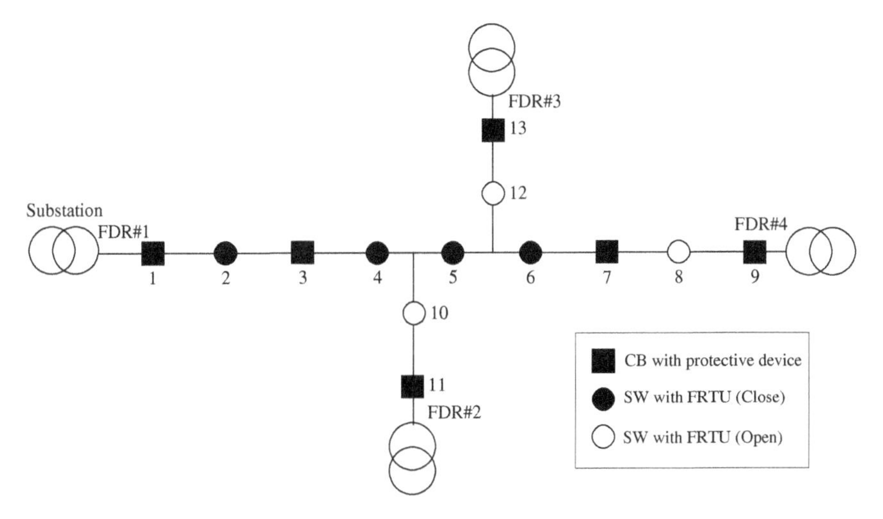

Figure 10.16 Multi-agent-based protection coordination.

The multi-agent-based protection coordination process is as follows. As shown in Figure 10.16, Feeder 1 includes three protective devices, and these three protective devices need to have correct settings of protection coordination. First, the central agent needs to transmit the information (e.g., communication address of the central agent, the upper and lower agent communication addresses, the distributed restoration solutions, and the main protective device communication information for protection coordination) to each terminal agent (FRTUs). In addition, the terminal agents that have a protective device receive the upper and lower protective device communication information and the setting values of protection coordination. If the power distribution system is operating in the normal state, CB1 operates as the primary protection device for the area 1 to 3, CB3 (recloser) operates as the primary protection device for the area 3 to 7, and then CB7 (recloser) operates as the primary protection device for the area 7 to 8.

If there is a fault between SW4 and SW5, Recloser 3 is opened by the protection coordination and the fault current is interrupted. Then FRTU3, 4, and 5 will consecutively

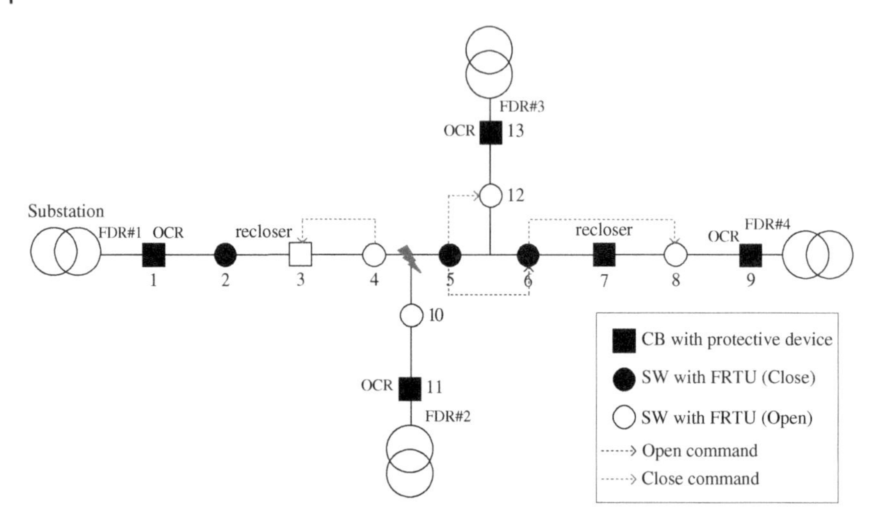

Figure 10.17 Multi-agent-based protection coordination and FDIR process.

inquire whether they have FI information to find the fault location. Once they find the fault location, they isolate the fault by opening switches. After that, according to the multi-agent-based FDIR process, the close command (4-3-C) will be executed at the upper side of the fault. At the downside of the fault, the open command (5-6-O) will be executed first, and then the close command (5-12-C and 6-8-C) will be executed to finish the FDIR process, as illustrated in Figure 10.17.

Once the FDIR process is completed, the setting values of protection coordination should be set again for the newly changed system. For instance, the range of Recloser 3, which is the primary protective device of Sections 3–7, is reduced. Since the range of protection of OCR 1 has been reduced, FRTU 4 transmits the fault location information to Recloser 3, and then Recloser 3 forwards the received information to OCR 1. Then, the terminal agents that are connected with protective devices (1 and 3) find the corresponding setting values of protection coordination that were previously stored and set the protection coordination again for the changed system. Likewise, in the downside of the fault, both switching command and fault location information is transmitted from SW5 to SW12. SW12 sends the fault zone information to OCR13, which is the main protection of Feeder 3, to recognize that the range of the protection zone is changed and then sets the correct setting values again. For Feeder 4, SW5 sends an open command and fault location information to SW6, and SW6 sends open command and fault location information to SW8. At this time, Recloser 7 (which is a primary protective device of SW8) recognizes the change of its feeder through the fault information, and SW8 sends fault location information to Recloser 7 and OCR 9. After that, Recloser 7 and OCR 9 load the correct saved setting values of protection coordination for the changed distribution system, as described in Figure 10.18.

10.4.2.5 Multi-Agent-Based Communication Failure Resilient System

Reliability is one of the most important factors in distribution system operation. The system may have an interruption and normal system operation may not be possible due to various causes, e.g., malfunction, cyber attack, and misoperation. In such a case, the system should enhance its reliability by improving resiliency in order to operate the system normally.

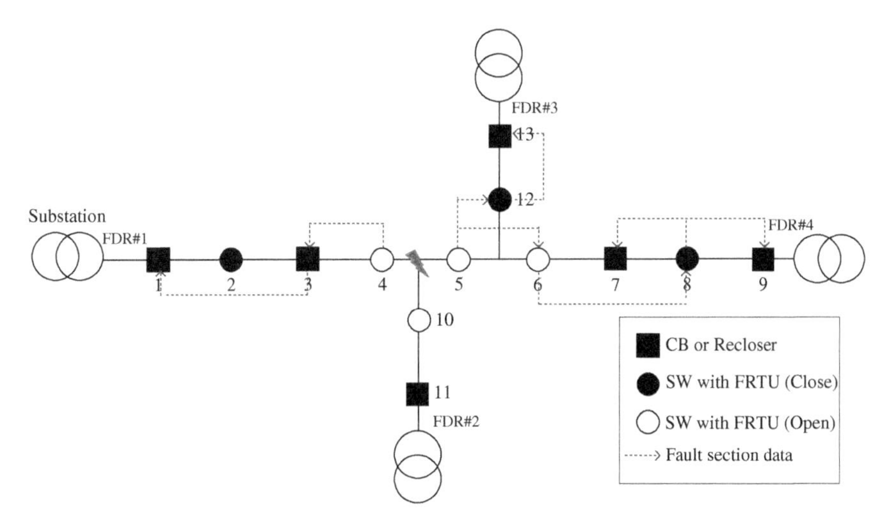

Figure 10.18 Protection coordination update for the newly changed system.

In the multi-agent-based DAS, mutual cooperation among the terminal agents is crucial. Therefore, any application problems or communication failures from an agent can cause significant delay during FDIR processes, and the damage of power outage may increase. In terms of resiliency, a mitigation strategy is required against potential single-agent communication faults or cyber attack, for example, DoS.

First, a communication failure problem during an FDIR process needs to be considered. When the agent that needs to issue a switching operation during an FDIR process has a communication failure problem, adjacent agents can monitor the communication status of the agent. However, in order to apply this method, the central agent should recalculate the FDIR solution without the faulted agent and then transmit the new solution to all terminal agents. Although the above process can complete the FDIR process with a faulted agent, the time delay in recalculating the FDIR solution needs to be considered. In order to prevent delay in recalculating the FDIR solution during the fault, periodical monitoring of communication health for all agents should be performed during normal operation.

Figure 10.19 shows the mitigation method when a communication fault occurs at an agent.

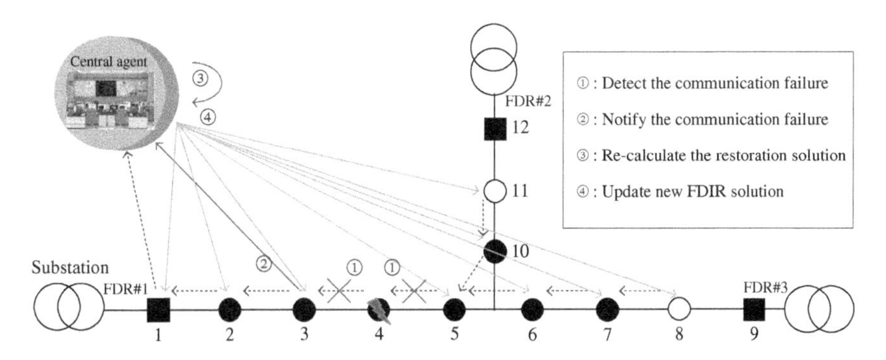

Figure 10.19 Mitigation method for faulted agent in the multi-agent-based DAS.

In DAS, each agent checks the communication health by monitoring adjacent agents during normal operation. The communication status is monitored by a polling method at every periodic time. Each agent periodically receives the status of the downstream agent, and it can detect a communication failure if there is no polled data from the downstream agent. Then the central agent calculates a new FDIR solution except for the faulted agent and distributes the new FDIR solution to all agents in the system. For instance, if Agent 3 cannot get a monitoring signal from Agent 4, it can detect the communication problem of Agent 4 and send the notification to the central agent. Then the central agent calculates a new FDIR solution except for Agent 4 and distributes the new FDIR solution to all agents in the system. At the same time, operators dispatch the field crews to Agent 4 and solve the communication problem. Once field crews finish the repair, the central agent distributes the original FDIR solution to all agents in the system. Therefore, the FDIR process can be executed and complete the distribution system restoration under the communication failure scenario.

10.5 Mitigation Methods Against Cyber Attacks

10.5.1 Remote Access to Open/Close Switch Attack and Mitigation

One of the attack methods in [22, 23] was to inject a number of unauthorized switching open/close commands. For instance, intruder(s) bypass the firewall and gain access to the communication network of the operation center. Then they do sniffing attacks to find all communication-enabled devices and successfully crack the ID and password of the user interface. Once they gain access to the user interface, they can issue switching control commands to any breaker or switch in the distribution system. Unauthorized switching commands may create voltage instability problems and could lead to a feeder outage problem. Therefore, mitigation methods against unauthorized switching commands need to be studied. This paper proposes a method to prevent unauthorized switching command attacks by estimating the system condition before operating the switch.

Figure 10.20 shows an example of power flow results under normal operation. It consists of four feeders with substations in a radial structure. Switches 4, 9, and 11 are tie breakers that are normally open to separate the feeders from each other.

It is assumed that Agents 7 and 4 receive open and close commands, respectively, from the operation center (Agents 7 and 4 do not know whether this command is from an attacker or the operator). Before accepting and executing these commands, Agents 7 and 4 will then run the local power flow analysis to determine whether there is any operational violation (e.g., voltage violation). As shown in Figure 10.21, Agent 7 can know that the section between 4-5-6-7 will have an outage if it accepts the switching command from the operation center. Similarly, Agent 4 can know that there will be a voltage instability problem (less than 0.98 p.u.) once it closes the switch of Agent 4 after the switch of Agent 7 is opened. As a combined result of these two estimated conditions of the system, Agents 4 and 7 can make a decision on whether they need to accept and execute the delivered commands from the operation center. If the mitigation setting is high, agents can block all these commands, whereas a low mitigation setting will allow Agents 4 and 7 to accept the command but issue an alarm to the operator (Figure 10.22).

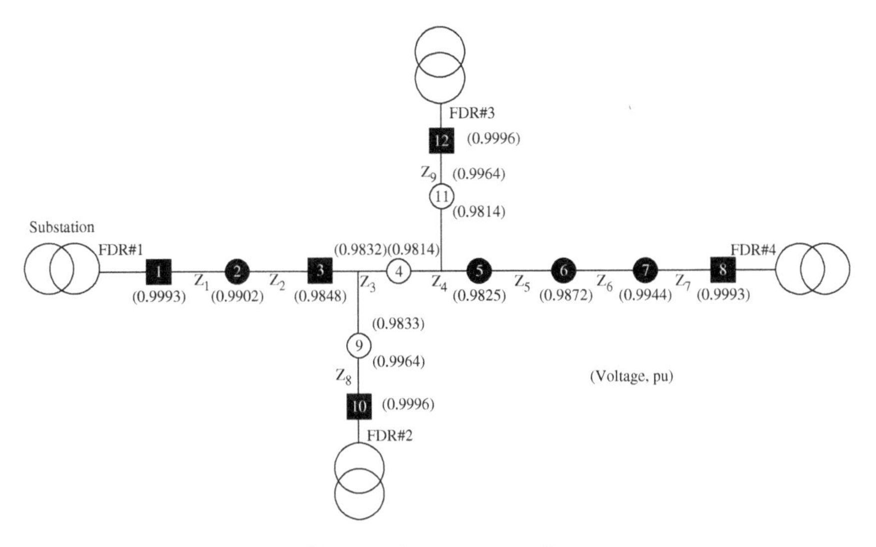

Figure 10.20 Bus voltages with normal system operation.

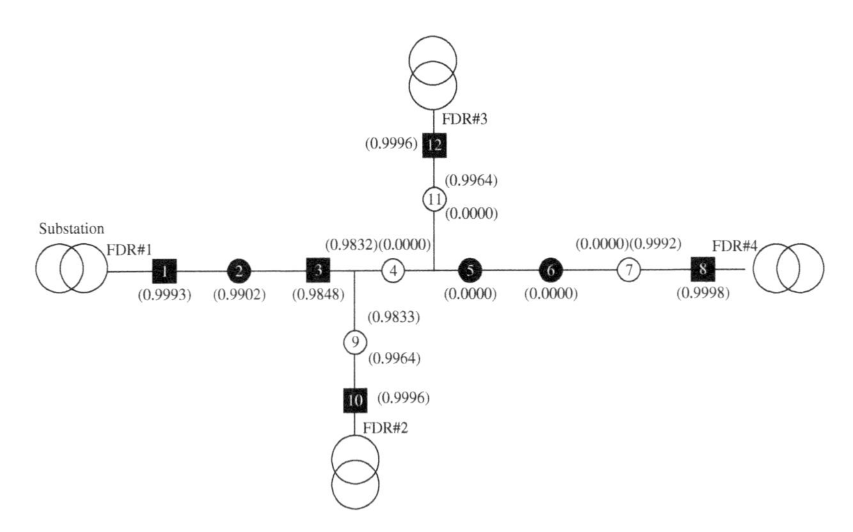

Figure 10.21 System condition estimation method for control command (Switch 7).

10.5.2 Man-in-the-Middle Attack and Mitigation

In this scenario, it is assumed that attackers access one of the remote FRTU devices and physically install a communication wiretap for a man-in-the-middle attack. Once they install the wiretap, they can connect to the distribution system's communication backbone network and gain access to all the unencrypted communication packets. Then they can capture one of the control commands, modify it, and then reinject it into the communication network for an open/close switching attack. In addition, they can also capture the measurements (e.g., voltage, current, or status of switch) and inject abnormal

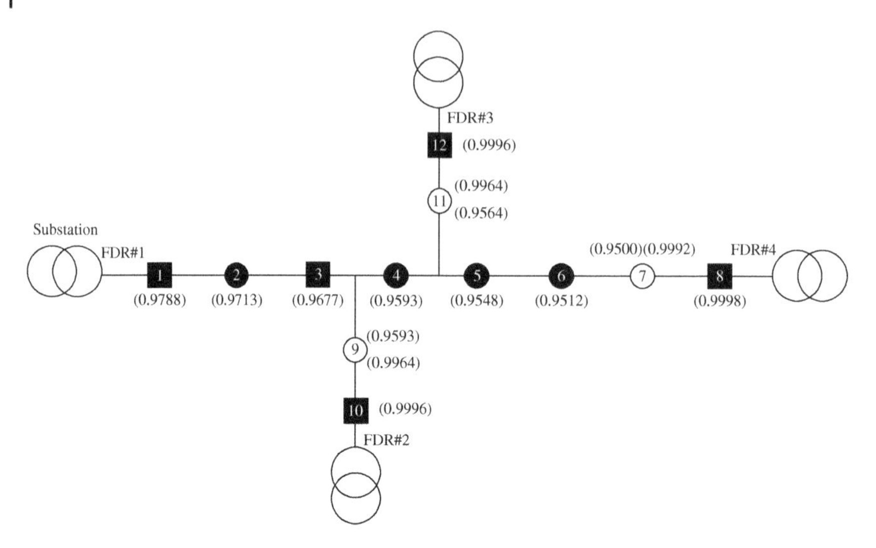

Figure 10.22 System condition estimation method for control command (Switch 4).

measurements. The operators in the operation center cannot know of the ongoing man-in-the-middle attack, and they may make a wrong decision. For instance, if the attackers inject high voltage measurements, operators may change the transformer tap position to decrease the feeder voltage. This action will decrease the feeder voltage smaller than the normal voltage and may create voltage instability problems. Figure 10.23 describes the proposed mitigation method and shows how it can detect and block the injected abnormal packet. In this case, the attacking agent captures the original sender agent's message and modifies it. When the attacking agent modifies the message, they will append the delivered MAC without any modification, since they does not have the shared private key. Once the receiver agent receives the modified packet, it will calculate MAC from the packet. However, the calculated MAC is not same as the delivered MAC. Furthermore, the

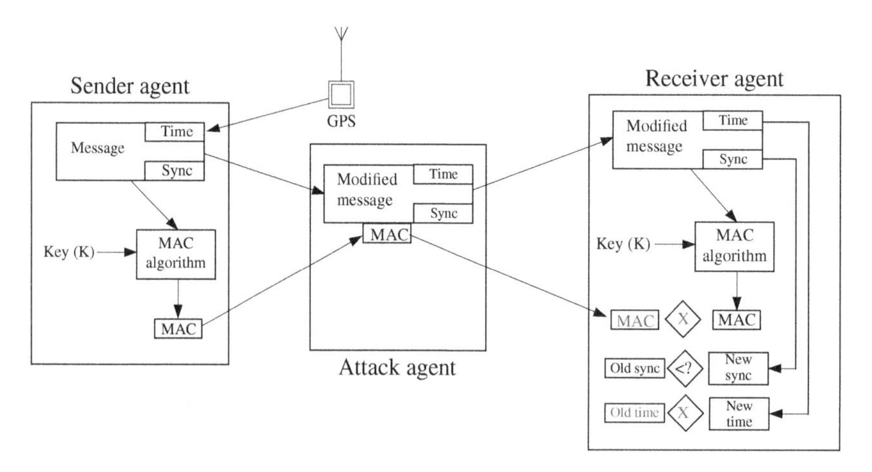

Figure 10.23 Mitigation method for the man-in-the-middle attack.

delivered time delay is higher than the threshold value since the attacking agent spends some time to capture and modify the original message. So, the receiver agent can know that the delivered message has been modified during transmission, and therefore it will discard the message and issue an alarm to the operator.

10.5.3 Configuration Change Attack and Mitigation

A proper protection coordination setting between protective devices is very crucial during a power system fault. However, most of the existing protective devices allow operators to change the setting when the device is online. Attackers may use this vulnerability and set up an invisible cyber attack. For instance, attackers send spam email to distribution system operators, and the operators open the infected email. After infecting an operator's laptop, the worm virus will wait until the operator connects the laptop inside the distribution operation center. Once the operator connects the infected laptop, the worm virus will generate other worm viruses and will spread out to the entire distribution system communication network. Then it will create an open port at the firewall so attackers can gain access to the distribution operation center. After cracking the ID and password of the engineering machine, attackers can change the settings of the protective devices. For instance, if they change the setting value of the primary protective device slower than the backup protective device, it will create an unwanted outage area when a fault occurs at the distribution feeder. The proposed mitigation method against configuration change attack will check the protection coordination between protective devices before they accept the changed setting value. Figures 10.24a and b illustrate an example of a two-feeder distribution system for configuration change attack mitigation. As shown in Figure 10.24a, Recloser 1 is the primary protective device and OCR 1 is the backup protective device. So, when a fault occurs at the section between 3 and 4, Recloser 1 will open the switch first and check whether it is a temporal fault or not. However, if the attacker changes the setting of Recloser 1 from 0.5 to 1.2, OCR 1 will open the circuit breaker faster than Recloser 1, and this will create unwanted outages (Sections 1-2-3-4). Therefore, the proposed mitigation method is to check the protection coordination before acceptance and apply it to the protective devices. For instance, if Recloser 1 receives a change setting value from 0.5 to 1.2, it will check the protection coordination and see whether there is any violation. If it can find any violation, it will discard the changed setting value and issue an alarm to the operators.

10.5.4 Denial-of-Service Attack and Mitigation

Installation of the intrusion detection system not only increases the cost of operation of the distribution system but also requires maintenance costs. However, it is not easy to detect a DoS attack without an intrusion detection system. If a DoS attack is initiated before the fault in the power distribution system, it can be detected efficiently by the proposed state monitoring between the agents, but if the DoS attack targets the command to operate the switch, it will be difficult to detect the attack before the switch is activated. Therefore, there is a need for a mitigation method to cope with a DoS attack against the switching command during the FDIR process.

Figure 10.25 shows a case in which a fault occurs in a feeder (section between 2 and 3) where a DoS attack is targeted at Agent 5 of the distribution system. If a fault occurs between

<TC curve formula>

$$T = \left(\frac{K}{\left(\frac{I}{I_{pu}}\right)^{a} - 1} + C \right) * \frac{M}{10}$$

T : operation time in seconds
M : multiplier setting(lever)
I : input current
I_{pu} : setting current(Pick up)
K, a, C : defined constant based on standard (K : 0.11, a : 0.02, C : 0.42)

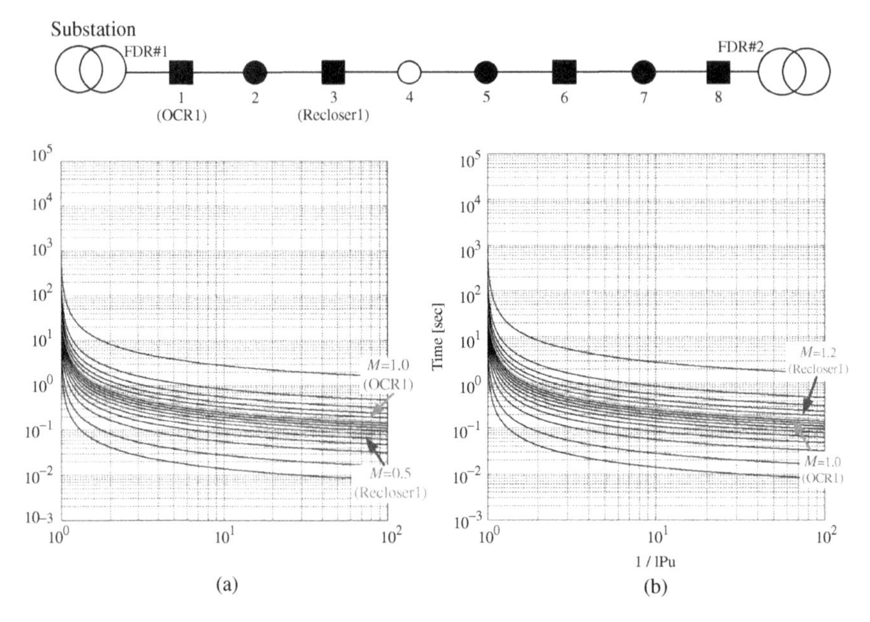

Figure 10.24 Mitigation method for configuration change attack. (a) Before the configuration attack. (b) After the configuration attack.

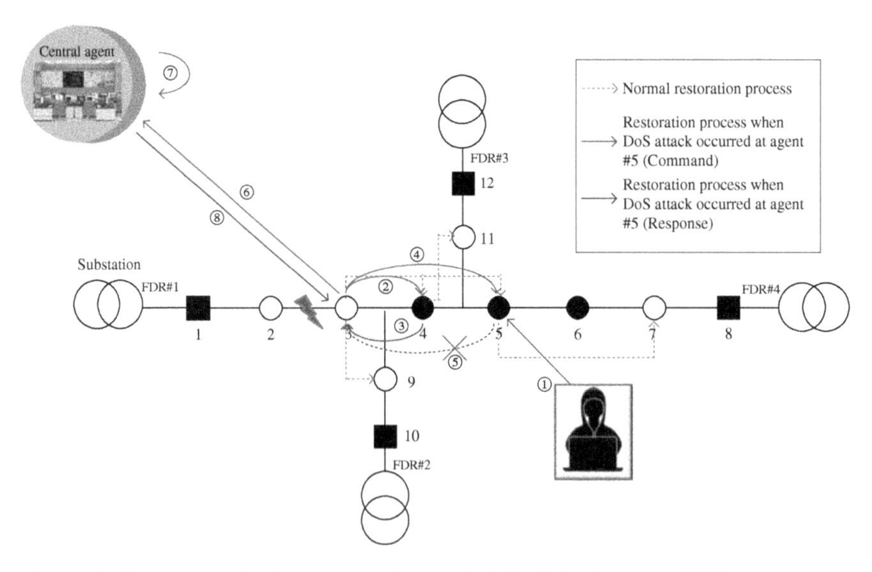

Figure 10.25 Proposed mitigation method against DoS attack during FDIR process.

GS 2 and 3, the downstream of the fault area is restored according to a predetermined FDIR solution. If the predetermined FDIR solutions are 3-4-C, 3-5-O, 3-9-C, 4-11-C, and 5-7-C, Agent 3 executes an open command to Agent 4 and Agent 4 forwards a response to Agent 3 that includes a switching action after opening the switch. After that, Agent 3 executes the open command to Agent 5; however, Agent 5 cannot execute the switching operation and send the command response due to the DoS attack by the intruder. Therefore, Agent 3 will detect a DoS attack on Agent 5 and inform the central agent of it. The central agent recalculates the FDIR solution after excluding Agent 5 and sends the newly calculated FDIR solution to Agent 3. In this case, the reason for not resetting the newly calculated FDIR solution to all the agents is to prevent unnecessary delays (since FDIR is in progress) and to speed up the FDIR process. Agent 3, which has received the newly calculated FDIR solution, performs its own tasks for the next FDIR process, and transmits the newly calculated FDIR information to other agents that need to perform the next FDIR process. After the FDIR process is completed, the DAS can perform normal FDIR process by calculating the FDIR solution excluding Agent 5 and distributing it to all other agents until the DoS problem is mitigated.

10.6 Summary

The DAS collects data (e.g., voltage, current, and status of switch) from the FRTUs, which are terminal devices installed at remote places, to the control center where the operation is performed and executes the application for the operation. In addition, it is an automation system that enables the system to be operated efficiently by monitoring and controlling the condition of switchgears. For operation of DAS, the information and control commands are transmitted by communication. Due to the structure of the power distribution system and the operational characteristics, there is vulnerability in terms of communication security.

Since DAS is using a standardized communication protocol such as DNP 3.0, it is easy to understand the ICT structure of the power distribution system, and the terminal devices are installed in exposed environments such as a poles, so the communication backbone can be easily accessed by adversaries. Attackers can also use a malware virus to access the operation centers, and some of the protective devices are set to default passwords, which are easy to access. Therefore, cyber attacks may occur using the above-mentioned communication vulnerabilities in the DAS. Potential threat models that can cause critical problems include remote access to open/close switching attacks, man-in-the middle attacks, configuration change attacks, and DoS attacks. In particular, the power distribution system directly supplies power to the customer, so the outage time (duration) and area must be minimized by accurate and prompt switch operations. However, such attacks may cause undesired results in the control and the information acquisition of the switchgear, which may reduce the reliability of the operation of the distribution system and may even cause blackout of the entire system. Therefore, it is necessary for the power distribution system to maintain resiliency and operate normally even if the above cyber attack occurs.

In this paper, we have presented an analysis of the operation method of the DAS based on the existing centralized processing method and the multi-agent-based distributed operation method for maintaining resiliency and efficient operation against cyber attack. In addition, a detection and mitigation strategy against cyber attack, resilient power restoration, and protection coordination algorithms are described for the multi-agent-based distributed system. The proposed mitigation method utilizes the characteristics of the power distribution system, for example, power flow analysis, and protection coordination. As a result, this paper describes a detection and mitigation strategy for a distribution system cyber attack through the combined use of physical (power system) and cyber (ICT) characteristics for reliable operation and resiliency of a power distribution system. By applying the proposed mitigation methods, it will be possible to maintain reliability and stable system operation as well as communication in the power distribution system.

References

1 Wang, P. and Li, W. (2007). Reliability evaluation of distribution systems considering optimal restoration sequence and variable restoration times. *IET Generation, Transmission and Distribution* 1 (4): 688–695.

2 Li, W., Wang, P., and Li, Z. (2004). Reliability evaluation of complex radial distribution systems considering restoration sequence and network constraints. *IEEE Transactions on Power Delivery* 19 (2): 753–758.

3 Chen, C., Wu, W., Zhang, B., and Singh, C. (2015). An analytical adequacy evaluation method for distribution networks considering protection strategies and distributed generators. *IEEE Transactions on Power Delivery* 30 (3): 1392–1400.

4 Graziano, R.P., Kruse, V.J., and Rankin, G.L. (2015). Systems analysis of protection system coordination: a strategic problem for transmission and distribution reliability. *IEEE Transactions on Power Delivery* 30 (3): 1392–1400.

5 Curcic, S., Ozveren, C.S., and Lo, K.L. (1997). Computer-based strategy for the restoration problem in electric power distribution systems. *IEEE Proceedings of Generation, Transmission and Distribution* 144 (5): 389–398.

6 Liu, C.-C., Lee, S.-J., and Venkata, S.S. (1988). An expert system operational aid for restoration and loss reduction of distribution systems. *IEEE Transactions on Power Systems* 3 (2): 619–626.

7 Lee, S.-J., Lim, S.-I., and Ahn, B.-S. (1998). Service restoration of primary distribution systems based on fuzzy evaluation of multi-criteria. *IEEE Transactions on Power Systems* 13 (3): 1156–1163.

8 Lim, S.-I., Lee, S.-J., Choi, M.-S. et al. (2006). Service restoration methodology for multiple fault case in distribution systems. *IEEE Transactions on Power Systems* 21 (4): 1638–1644.

9 Chen, K., Wu, W., Zhang, B., and Sun, H. (2015). Robust restoration decision-making model for distribution networks based on information gap decision theory. *IEEE Transactions on Smart Grid* 6 (2): 587–597.

10 Lim, I.-H., Sidhu, T.S., Choi, M.-S. et al. (2013). Design and implementation of multiagent-based distributed restoration system in DAS. *IEEE Transactions on Power Delivery* 28 (2): 585–593.

11 Elmitwally, A., Elsaid, M., Elgamal, M., and Chen, Z. (2015). A fuzzy-multiagent service restoration scheme for distribution system with distributed generation. *IEEE Transactions on Sustainable Energy* 6 (3): 810–821.

12 Eriksson, M., Armendariz, M., Vasilenko, O.O. et al. (2015). Multiagent-based distribution automation solution for self-healing grids. *IEEE Transactions on Industrial Electronics* 62 (4): 2620–2628.

13 Lim, I.-H., Hong, S., Choi, M.-S. et al. (2010). Security protocols against cyber attacks in the distribution automation system. *IEEE Transactions on Power Delivery* 25 (1): 448–455.

14 Ross, K.J., Hopkinson, K.M., and Pachter, M. (2013). Using a distributed agent-based communication enabled special protection system to enhance smart grid security. *IEEE Transactions on Smart Grid* 4 (2): 1216–1224.

15 Xiang, Y., Ding, Z., Zhang, Y., and Wang, L. (2017). Power system reliability evaluation considering load redistribution attacks. *IEEE Transactions on Smart Grid* 8 (2): 889–901.

16 Isozaki, Y., Yoshizawa, S., Fujimoto, Y. et al. (2016). Detection of cyber attacks against voltage control in distribution power grid with PVs. *IEEE Transactions on Smart Grid* 7 (4): 1824–1835.

17 Hong, J., Liu, C.-C., and Govindarasu, M. (2014). Integrated anomaly detection for cyber security of the substations. *IEEE Transactions on Smart Grid* 5 (4): 1643–1653.

18 Hong, J., Liu, C.-C., and Govindarasu, M. (2014). Detection of cyber intrusions using network-based multicast messages for substation automation. In: *IEEE Innovative Smart Grid Technologies (ISGT) Conference.*

19 Muhammad, A. (2017). A practical state estimation for distribution automation system under distribution automation environment. M. S. thesis, Department of Electrical Engineering, Myongji University, Yongin, Korea.

20 Vempati, N., Shoults, R.R., Chen, M.S., and Schwobel, L. (1987). Simplified feeder modeling for loadflow calculations. *IEEE Transactions on Power Systems* 2 (1): 168–174.

21 Cheng, C.S. and Shirmohammadi, D. (1995). A three-phase power flow method for real-time distribution system analysis. *IEEE Transactions on Power Systems* 10 (2): 671–679.

22 Industrial Control Systems Cyber Emergency Response Team (ICS-CERT) (2016). Cyber-attack against Ukrainian critical infrastructure. https://ics-cert.us-cert.gov/alerts/IR-ALERT-H-16-056-01.

23 United States Computer Emergency Readiness Team (US-CERT) (2017). CrashOverride malware. https://www.us-cert.gov/ncas/alerts/TA17-163A.

24 National Electric Sector Cybersecurity Organization Resource (NESCOR), Electric Power Research institute (EPRI) (2015). Analysis of selected electric sector high risk failure scenarios. https://smartgrid.epri.com/doc/NESCOR%20Failure%20Scenarios%20v3%2012-11-15.pdf.

25 McArthur, S.D.J., Davidson, E.M., Catterson, V.M. et al. (2007). Multi-agent systems for power engineering applications – Part I: concepts, approaches, and technical challenges. *IEEE Transactions on Power Systems* 22 (4): 1743–1752.

26 McArthur, S.D.J., Davidson, E.M., Catterson, V.M. et al. (2007). Multi-agent systems for power engineering applications – Part II: Technologies, standards, and tools for building multi-agent systems. *IEEE Transactions on Power Systems* 22 (4): 1753–1759.

11

A Hierarchical Control Architecture for Resilient Operation of Distribution Grids

Ahmad R. Malekpour[1], Anuradha M. Annaswamy[1], and Jalpa Shah[2]

[1] *Department of Mechanical Engineering, Massachusetts Institute of Technology, Cambridge, MA, USA*
[2] *Sensata Technologies, Eaton Corporation Inc., Eden Prairie, MN, USA*

11.1 Resilient Control Theory

Cyber-physical Energy Systems (CPESs), a specific example of which are smart distribution grids, are not only vulnerable to security threats but also physical outages which may occur due to natural disasters such as hurricanes, earthquakes, and other unforeseen anomalies. A physical outage can be also caused by malicious human physical actions. An example is the "Metcalf sniper attack" on PG&E Corp's Metcalf transmission substation in San Jose, California that happened in 2013 and caused a damage worth over $15 million [1]. Given that the end-goal of CPES is reliable delivery of power to its end-user at all times, cyber-physical resilience of CPES is a necessary requirement. This corresponds to the ability of the CPES to withstand high-impact disturbances, which may occur due to either physical or cyber causes, and continue to operate with acceptable performance. In this chapter, we present a framework toward such cyber-physical resilience of a distribution grid in the presence of outages.

Similar to the classification of power systems, we classify a CPES into five modes, normal, alert, emergency, in-extremis, and restorative [2] (see Figure 11.1). Depending on the mode that the energy system is in, a corresponding set of decisions and actions is pursued to ensure safety and desirable performance. Two points are worth making regarding these five modes. The first is that these CPES are typically designed to be robust with their control and protection measures ensuring that they lie in the nominal mode for the most part, and for any perturbations that may cause a transition to the alert stage that a suitable robust action brings the system back to a nominal state. And those disturbances that trigger a transition to the emergency mode, or worse still, the in-extremis mode, are assumed to be a high impact, but low probability event. In the absence of cyber-attacks, while such an assumption is valid, a combined presence of both physical and cyber anomalies can significantly increase the probability of this transition, as well as reduce the time interval

This work was supported by Eaton Corporation.

Resiliency of Power Distribution Systems, First Edition.
Edited by Anurag K. Srivastava, Chen-Ching Liu, and Sayonsom Chanda.

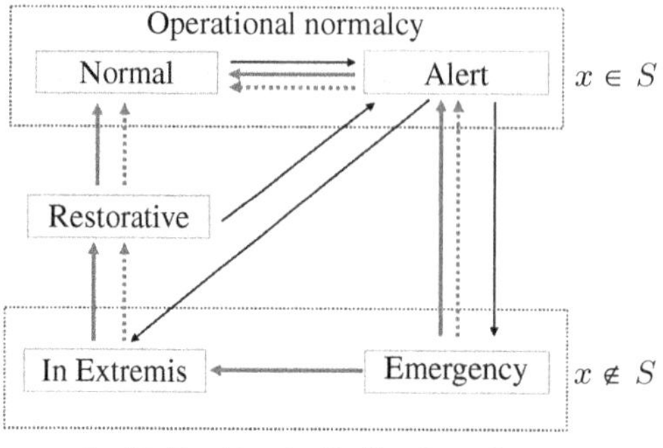

Figure 11.1 Feasible and nonfeasible transitions of a resilient and a nonresilient system.

of this transition. The second is that the schematic proposed in Figure 11.1 also sets the stage for the notion of resilience and its distinction from robustness. CPES can be characterized as robust when they are able to operate normally (which can correspond to the case when the total electricity demand is fully served) under disturbances that only cause transitions between normal and alert states [3, 4]. In contrast, in the presence of high-impact disturbances, if the CPES transition to an emergency state and further to an extremis state they can be characterized as being resilient when they can return to an alert or normal state within an acceptable time. A CPES system must therefore be designed to be both robust and resilient, as disturbances can have a range of impact, making the transition to any of the five states mentioned above equally likely. Such a design principle is applicable not only to CPES but any critical infrastructures such as transportation, water, and health care [5–7].

Robustness of systems has been investigated extensively over the past two to three decades including several papers, textbooks, conferences, and journals by the controls community. And more recently, robustness of CPES has been examined in [8–10]. In contrast, a formal definition of resilience, either in the context of power systems, or cyber physical system (CPS) in general, is yet to emerge. Broadly speaking, it is widely accepted that resilience connotes the ability of a CPS to sustain and recover from extreme and severe disturbances that can drive the system to its physical operational limits [2]. In contrast, robustness is a precisely defined notion in control theory that denotes a property that characterizes the system's ability to retain normal operation after being subjected to a range of bounded, and small disturbances or uncertainties [11]. Clearly new tools for analysis and synthesis of resilient control methods for CPES are needed and are currently lacking.

11.1.1 Definition of Resilience

The term resilience is being discussed increasingly of late in the context of CPS, ranging from transportation [7], power [3, 12], control systems [3, 13, 14] as well as other types of systems such as ecological [15, 16] and biological [5]. Resilience is often discussed concomitantly with other system-oriented notions such as robustness, reliability, and stability [5] and quite often used interchangeably with the term robustness. We argue, however, that these two terms are distinct. The reason is that resilience and robustness characterize fundamentally different system properties. As mentioned earlier, the term robustness applies in the context of small-bounded disturbances while resilience, in the context of extreme high-impact disturbances. We offer the following conceptual definition of resilience: *resilience of a CPS with respect to a class of extreme and high-impact disturbances, is the property that characterizes its ability to withstand and recover from this particular class of disturbances by being allowed to temporarily transit to a state where its performance is significantly degraded and returning within acceptable time to a state where certain minimal but critical performance criteria are met.*

In the context of power systems, the feasible and nonfeasible transitions of a resilient and a nonresilient system are conceptually depicted in Figure 11.1 with solid and dotted arrows, respectively. Suppose that the system descriptions in these various states are described as follows:

$$\text{nominal: } \dot{x} = f_1(x), x \in R^n, 0 \leq t \leq t_1 \tag{11.1}$$

$$\text{during outage: } \dot{x} = f_2(x, \alpha), t_1 \leq t \leq t_2 \tag{11.2}$$

$$\text{recovery to normalcy: } \dot{x} = f_3(x, \alpha, d), t \geq t_2 \tag{11.3}$$

where $\alpha \in R^p$ is an exogenous outage with respect to which we need to be resilient, and $d \in R^q$ is a resilient control input. We define the set $S \in R^n$ to be the region in state-space that corresponds to operational normalcy. With the above definitions, we can define resilience in a quantitative manner as the ability of the system to transition from the emergency or the in-extremis mode where $x \notin S$, to either the normal or the alert mode, where $x \in S$, that is, the system still maintains operational normalcy, following a high-impact disturbance such as an outage. This is quantified in the following definition:

Definition 11.1 ***(Resilience)*** *Suppose that a system is in state (11.1) for $0 \leq t \leq t_1$, and in state (11.2) for $t_1 \leq t \leq t_2$. Then the system (11.1) is resilient if it enters state (11.3) with $x(t) \in S$ for $t \geq t_2$. The resilience metric of the system is given by $t_2 - t_1$, the time taken for the system to return to normalcy.*

11.1.2 Smart Distribution Grids

The U.S. power grid is rapidly transforming from a centralized and unidirectional power distribution network to a decentralized, bidirectional network supporting large penetrations of distributed generation resources. As distributed generators (DGs) introduce distributed reactive power injection, the spatio-temporal management of voltage fluctuations and therefore volt/var control becomes highly nontrivial. For example, with increasing penetration of renewable sources such as solar power on a distribution grid can exacerbate problems associated with power quality such as voltage rise [17], line overloading, reverse flow [18], voltage fluctuation [19], harmonics, and flickers [20]. While power technology to address voltage regulation concern exists in legacy devices such as load-tap changers (LTCs), voltage-regulators (VRs), and capacitor banks (CAP) [21, 22], inverter technology has drastically improved, and smart inverters can now provide a suite of advanced functionalities under new standards and regulations such as reactive power support to improve grid voltage stability and control [23]. Such recent developments have significantly enhanced the level of visibility and automation in the distribution grids, which can be utilized to enhance system performance. For example, with the high-level deployment of digital control, DGs can be utilized efficiently after power outages (see [1, 24–31]).

When the renewable penetration becomes significant, the use of legacy devices alone may not be sufficient, especially for resilience in the presence of outages. A standard practice for ensuring volt-var control with DGs is the use of centralized decision-making for optimal reactive power injection [32–39]. This enables optimization of power losses, which is a major factor in distribution grids. Our proposed resilient control architecture proposes the use of such a centralized optimization as the top layer. An important additional layer of coordination of active and reactive power is needed to ensure resilience, as the use of centralized decision-making alone is too global, as it does not accommodate intermittencies in DGs and real-time changes in the operating conditions. Therefore we introduce a second layer of decision-making, and use the concept of consensus, studied extensively in several disciplines including computer science [40], physics [41], operations research [42], and control theory [43]. The return to normalcy, referenced in Definition 11.1 is defined as the time taken for critical loads to be picked up and fully accommodated following an outage. We will use the corresponding resilience metric defined as in Definition 11.1 for the purposes of this paper. Together with the top layer, the resulting hierarchical resilient control architecture, with the top layer providing centralized optimization and the bottom layer providing distributed coordination, we show that we can significantly improve the resilience metric of a distribution grid. The hierarchical control architecture is described in Section 11.2, and the results obtained using this architecture in a modified IEEE-34 bus are presented in Section 11.3.

11.2 A Hierarchical Control Strategy

The proposed resilient controller has a hierarchical architecture with two layers, where the top layer carries out centralized optimization, and the bottom layer carries out distributed, real-time control using a consensus algorithm. Figure 11.2 shows the overall control architecture, including both physical layer, the control layer, and real-time measurements and exogenous inputs.

Layer one of the control architecture is assumed to receive a day-ahead 15-minute load and generation forecasted data and calculates the on/off status of capacitor banks, tap operation of LTC/VRs, and reactive power provisioning from DGs for the next 24 hours. As shown in Figure 11.2, these setting will be communicated to layer two, where each DG receives the setting from the top layer, measures the voltage at its terminal voltage, and determines the required active and reactive power for better voltage regulation. If the voltage is higher or lower than predefined upper/lower critical voltages, DG requests for active and reactive power from its neighboring DGs that have additional capacity. Each DG calculates its share of contribution to meet the requested reactive power via a distributed algorithm that requires communication network (e.g. Wi-Fi, ZigBee, or power line communication) to exchange information among neighboring DGs. Each of these two layers is described below.

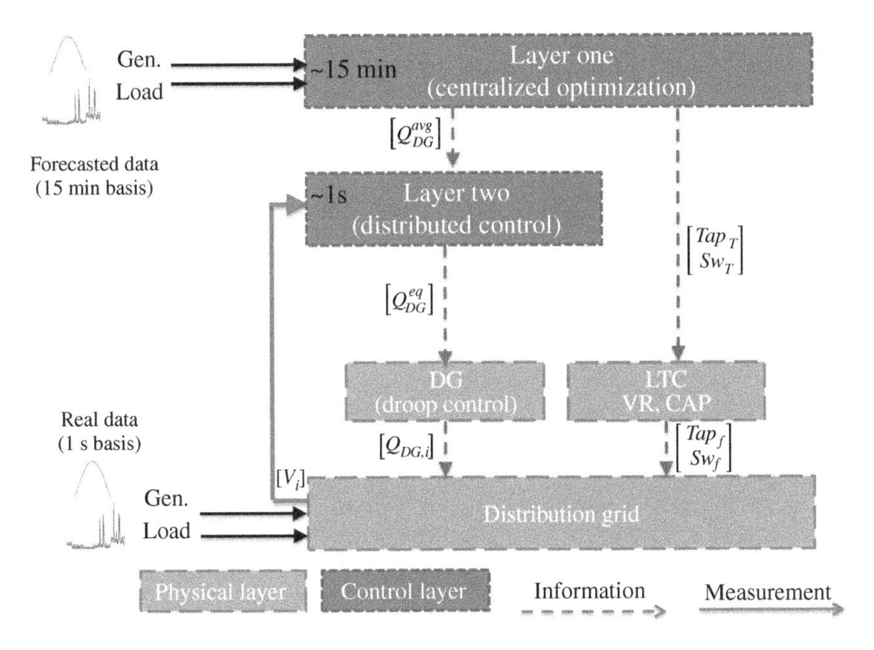

Figure 11.2 A hierarchical resilient control architecture with the top layer providing centralized optimization and the bottom layer providing distributed coordination.

11.2.1 Centralized Optimization Formulation: Top Layer

Let θ_b indicates the set of N nodes in the on-outage portion of the distribution grid and θ_{DG}, θ_{C}, and θ_{VR} denote the set of nodes in θ_b that include DGs, CAP, and LTC/VR, respectively. Layer one maximizes the out-of-service loads to be picked up as:

$$\max_{p_i^D, p_i^G, q_i^G, S_i, T_{ij}} F = \sum_{k=1}^{H} \sum_{i=1}^{N} w_i p_i^D[K] \tag{11.4}$$

s.t. the following constraints are all satisfied.

Distribution power flow and voltage constraints, $\forall i, j \in \theta_b$:

$$\sum_{m,i,j \in \varphi_b} P_{mi}[K] = P_{ij}[K] + p_i^D[K] - p_i^G[K] \tag{11.5}$$

$$\sum_{m,i,j \in \varphi_b} Q_{mi}[K] = Q_{ij}[K] + q_i^D[K] - q_i^G[K] \tag{11.6}$$

$$V_j[K] = V_i[K] - (r_{ij}P_{ij}[K] + x_{ij}Q_{ij}[K])/V_1 \tag{11.7}$$

$$l_{ij}[K] = \left((P_{ij}[K])^2 + (Q_{ij}[K])^2\right)/V_1 \tag{11.8}$$

$$\underline{V}_i \leq V_i[K] \leq \overline{V_i} \tag{11.9}$$

$$(P_{ij}[K])^2 + (Q_{ij}[K])^2 \leq (\overline{S_{ij}})^2 \tag{11.10}$$

Constraints on DG operation, $\forall i \in \theta_{\text{DER}}$:

$$\underline{p}_i < p_i^G[\text{K}] < \overline{p_i} \tag{11.11}$$

$$\left(p_i^G[K]\right)^2 + \left(q_i^G[K]\right)^2 \le \left(\overline{S_i^{DG}}\right)^2 \tag{11.12}$$

$$-p_i^G \tan\left(\cos^{-1}\mathrm{PF}_i^G\right) < q_i^G[\mathrm{K}] < p_i^G \tan\left(\cos^{-1}\mathrm{PF}_i^G\right) \tag{11.13}$$

Constraints on CAP operation, $\forall i \in \theta_C$:

$$q_i^G[K] = S_i[K]Q_i^C \tag{11.14}$$

$$S_i \in \{0, 1\} \tag{11.15}$$

Constraints on LTC/VR operation, $\forall i, j \in \theta_{VR}$:

$$T_{ij}[K]V_j[K] = V_i[K] - (r_i P_{ij}[K] + x_i Q_{ij}[K])/V_1 \tag{11.16}$$

$$\_T_{ij} < T_{ij}[K] < \overline{T_{ij}} \tag{11.17}$$

$$T_{ij} \in \{0.95, 096, \dots, 1.05\} \tag{11.18}$$

The decision variables of problem (11.4) are $\mathbf{x} = \left[p_i^{D^*} \in \left\{0, p_i^D\right\}, p_i^G, q_i^G, S_i, T_{ij}\right]$ and state variables are $\mathbf{y} = [P_{ij}, Q_{ij}, V_i]$. Due to the outage, some of the loads may need to be dropped, which corresponds to the decision variable $p_i^{D^*} = 0$.The objective function (11.4) is the total load that needs to be picked up over the fault-clearance period H. Denoting $n_D = \left\{i | p_i^{D^*} = p_i^D\right\}$ as the most critical loads that needs to be picked up, we impose their criticality by choosing the corresponding w_i in (11.4) to be appropriately large. The more the weight, the more important it is for that load to be restored. In the above formulation, (11.5)–(11.7) are the linear form of the DistFlow equations which have been extensively verified and used in the literature [44–46], (11.8) is the power loss in line segment ij at time instant K, and (11.9) and (11.10) are bus voltage magnitude and line thermal limits. Constraint (11.11) is the DGs active power generation limit. Constraint (11.12) imposes the inverter rating curve limit on DG generation capacity while constraint (11.13) enforces upper and lower bounds on the reactive power provisioning of DGs (specified by a given power factor [PF]). The on/off status of the CAP is established by Eq. (11.14). Constraint (11.15) represents the zero or full capacity reactive power injection by CAP at node i and time instant K. Equation (11.16) calculates the voltage at LTC/VR buses. Constraint (11.17) represents the maximum and minimum LTC/VR tap steps, and constraint (11.18) stablishes the discrete status of LTC and VRs. We note that all of the switching devices are completely specified by the top layer with no other edge intelligence devices employed.

11.2.2 Distributed Control Formulation: Bottom Layer

11.2.2.1 Communication Network Model

Consider a grid in which nodes interact with each other through local communication. The exchange of information nodes is represented by a weighted graph $\mathcal{G}(\mathcal{V}, E, \mathcal{A})$, where the set of active nodes (vertices) $\mathcal{V} = \{1, 2, \dots, n\}$ are communicating through a set of undirected links (edges) $E \subseteq \mathcal{V} \times \mathcal{V} \backslash \mathrm{diag}(\mathcal{V})$. The nonnegative matrix $\mathcal{A} = [a_{ij}]$ (with the a_{ij} entry at the ith row, jth column) is a weighted adjacency matrix that matches exactly the set of links in the communication graph. a_{ij} is positive if there is a communication link between nodes

i, j, that is, $(i, j) \in E$, and is zero otherwise. In order to follow the literature, we assume that the communication graph does not include any self-loops. Nodes that can send information to node j are defined as in-neighbors of node j and denoted by $\mathcal{N}_j^- = \{i \in \mathcal{V} | (j, i) \in E\}$. Similarly, nodes that receive information from node j are represented as out-neighbors of node j and denoted by $\mathcal{N}_j^+ = \{i \in \mathcal{V} | (i, j) \in E\}$.

In order to implement the proposed distributed control approach and quarantee the convergence, we make the following assumptions on the network communication graph $\mathcal{G}$ [40–43]:

(1) Strong Connectivity: there exists a path between any two vertices, that is, $\forall i, j \in \mathcal{V}, i \neq j$, there is a sequence of nodes $v_i = v_{l1}, v_{l2}, v_{l3}, \ldots, v_{lB} = v_j$ $(B \geq 2)$ such that $(I, I+1) \in E$ for all $I = 1, 2, \ldots, B-1$.
(2) Nondegeneracy: there exists a constant $\beta > 0$ such that $a_{ii} \geq \beta$, and a_{ij}, for $i \neq j$, satisfies $a_{ij} \in \{0\} \cup [\beta, 1]$.
(3) column Stochastic Adjacency Matrix:

$$\sum_{i \in \{j\} \cup \left\{\mathcal{N}_j^-\right\}} a_{ij} = 1, \forall j \in \mathcal{V} \tag{11.19}$$

In order to fulfill the nondegeneracy and column stochastic characteristics, we define matrix $\mathcal{A} = [a_{ij}]$ with the following characteristic:

$$a_{ij} = \begin{cases} 2 \Big/ \left[\left(\mathcal{N}_i^+ + \mathcal{N}_j^+\right) + \varepsilon\right], & \text{if } i \in \mathcal{N}_j \\ 1 - \sum\limits_{j \in \mathcal{N}_i} 2 \Big/ \left[\mathcal{N}_i^+ + \mathcal{N}_j^+ + \varepsilon\right], & \text{if } i = j \\ 0, \text{otherwise} \end{cases} \tag{11.20}$$

11.2.2.2 Distributed Control Algorithm

In order to ensure a resilient operation, supply-demand should be balanced within the on-outage portion of the grid using real-time information to ensure that the system operates autonomously. This requires the DGs to coordinate with each other properly. For this purpose, we use the above connectivity of the communication graph, and the following definitions.

Let the total active power demand of selected nodes from layer one be denoted as P_{L}, the total maximum available power from DGs as $\overline{P}_{\mathrm{DG}}$, and the actual available power from DGs as P_{DG}. In order to ensure that power balance occurs, that is, $P_{\mathrm{DG}} = P_{\mathrm{L}}$, we define an active power utilization index as:

$$K_G^P = \frac{P_{\mathrm{L}}}{\overline{P}_{\mathrm{DG}}} \tag{11.21}$$

which corresponds to the fraction of the utilization of the DG toward the load. Denoting the maximum available power from the jth DG as $\overline{P}_{\mathrm{DG}_j}$, it follows that:

$$\overline{P}_{\mathrm{DG}} = \sum_{j=1}^{n} \overline{P}_{\mathrm{DG}_j} \tag{11.22}$$

It is easy to see that power balance can be ensured if the contribution of the active power of the jth DG is set as:

$$P_{\mathrm{DG}_j} = K_G^P \overline{P}_{\mathrm{DG}_j} \tag{11.23}$$

The utilization index however is not known in advance due to load uncertainties. We therefore propose a distributed consensus algorithm that each node automatically determines this value through communication with its neighboring nodes. This algorithm is described in detail, and is very similar to that proposed in [47]:

We define auxiliary variables x_j and y_j, where x_j represents the active power of the load and y_j denotes maximum available active power from the DG at node j. The initial values for x_j, y_j are set accordingly, as:

$$x_j[0] = P_{\mathrm{L}_j} \tag{11.24}$$

$$y_j[0] = \overline{P}_{\mathrm{DG}_j} \tag{11.25}$$

Let n be the number of nodes in on-outage portion of the grid. If there is no DG at node $j \in n$, $\overline{P}_{\mathrm{DG}_j}$ is zero. If node $j \in n$ is not selected from layer one for load restoration (i.e. $j \notin n_D$), P_{L_j} is zero. At every step t, each node j updates its information states $x_j[t]$ and $y_j[t]$ via a weighted linear combination of its own value and the information from its neighbors as:

$$x_j[t+1] = a_{jj}x_j[t] + \sum_{i \in \{\mathcal{N}_j^-\}} a_{ji}x_i[t] \tag{11.26}$$

$$y_j[t+1] = a_{jj}y_j[t] + \sum_{i \in \{\mathcal{N}_j^-\}} a_{ji}y_i[t] \tag{11.27}$$

with the coefficients specified as in (11.18). Let $x[t] = [x_1[t], \ldots, x_n[t]]^T$ and $y[t] = [y_1[t], \ldots, y_n[t]]^T$. From (11.26) and (11.27) it is easy to see that:

$$x[t+1] = \mathcal{A}^t x[0] \tag{11.28}$$

$$y[t+1] = \mathcal{A}^t y[0] \tag{11.29}$$

The speed of convergence is determined by matrix $\mathcal{A}$ with faster convergence achieved for lower values of λ_2, the second largest eigenvalue of $\mathcal{A}$. If we define:

$$\Gamma_j[t] = \frac{x_j[t]}{y_j[t]} \tag{11.30}$$

it follows that $\Gamma_j[t]$ will asymptotically converge to [47]:

$$\Gamma = \frac{\sum_{j=1}^{n} x_j[0]}{\sum_{j=1}^{n} y_j[0]} \tag{11.31}$$

From the choices in (11.24) and (11.25), it follows that $\Gamma = K_G^P$, which is the unknown utilization index. This follows since:

$$\Gamma = \frac{\frac{1}{n}\sum_{i=1}^{n} P_{\mathrm{L}_i}}{\frac{1}{n}\sum_{i=1}^{n} \overline{P}_{\mathrm{DG}_i}} = \frac{\sum_{i=1}^{n} P_{\mathrm{L}_i}}{\sum_{i=1}^{n} \overline{P}_{\mathrm{DG}_i}} = \frac{P_{\mathrm{L}}}{\overline{P}_{\mathrm{DG}}} = K_G^P \tag{11.32}$$

The active power selection for each DG in the bottom layer is simply determined using (11.23) and set to the maximum value if the converged value of Γ is greater than one. More precisely, the distributed consensus algorithm of the proposed second layer is specified as:

$$P_{\mathrm{DG}_j} = \begin{cases} K_G^P \overline{P}_{\mathrm{DG}_j}, & \text{if } \lim_{t\to\infty} \Gamma_j[t] \leq 1 \\ \overline{P}_{\mathrm{DG}_j}, & \text{if } \lim_{t\to\infty} \Gamma_j[t] > 1 \end{cases} \tag{11.33}$$

The same procedure as above is used to calculate the reactive power selection for each DG in the second layer, and is briefly described below. We define a reactive power utilization index as:

$$K_G^Q = \frac{Q_{\mathrm{L}}}{\overline{Q}_{\mathrm{DG}}} \tag{11.34}$$

where Q_{L} and $\overline{Q}_{\mathrm{DG}}$ are the total reactive power demand of selected nodes from top layer and maximum available reactive power from the DGs, respectively. By initializing auxiliary variables x_j' and y_j' as:

$$x_j'[0] = Q_{\mathrm{L}_j} \tag{11.35}$$

$$y_j'[0] = \overline{Q}_{\mathrm{DG}_j} \tag{11.36}$$

where x_j' represents the reactive power of the load and y_j' denotes maximum available reactive power from the DG at node j, we propose a distributed algorithm similar to (11.26) and (11.27) for communicating x_j' and y_j' between neighboring nodes:

$$x_j'[t+1] = a_{jj} x_j'[t] + \sum_{i \in \{\mathcal{N}_j^-\}} a_{ji} x_j'[t] \tag{11.37}$$

$$y_j'[t+1] = a_{jj} y_j'[t] + \sum_{i \in \{\mathcal{N}_j^-\}} a_{ji} y_j'[t] \tag{11.38}$$

This allows us to realize the desired Q_{DG_j} as:

$$Q_{\mathrm{DG}_j} = \begin{cases} K_G^Q \overline{Q}_{\mathrm{DG}_j}, & \text{if } \lim_{t\to\infty} \Gamma_j'[t] \leq 1 \\ \overline{Q}_{\mathrm{DG}_j}, & \text{if } \lim_{t\to\infty} \Gamma_j'[t] > 1 \end{cases} \tag{11.39}$$

The updates in (11.26) and (11.27) and (11.37) and (11.38) are assumed to occur at a sufficiently fast sampling frequency so that this convergence takes place over a desired period of interest. In practice, a terminating criterion is defined to quantify the number of iterations and reach the equilibrium as $\|x[t]-x[t-1]\|_\infty \leq \sigma$, $\|y[t]-y[t-1]\|_\infty \leq \sigma$, $\|x'[t]-x'[t-1]\|_\infty \leq \sigma$, $\|y'[t]-y'[t-1]\|_\infty \leq \sigma$ [48]. In order to ensure satisfactory resilience, it was assumed that each iteration occurs once every 10 ms. We observed that a choice of $\sigma = 0.0001$ was met for iterations no greater than a 100. That is, a total duration of one second was sufficient for Layer 2 to determine new active and reactive power injections starting from the decision variables that were provided by the optimization algorithm in Layer 1.

In summary, top layer provides an estimated active and reactive power of DGs as well as CAP switching and VR operation every 15 minutes. Bottom layer uses these values in (11.25) and (11.36), set time $t = 0$, and the real-time values of the load in (11.24) and (11.35) to determine the updated active and reactive power generation of DGs through several iterations, till time $t = \tau$. As observed above, a 10 ms update was found to be sufficient in our simulation studies, and convergence taking place for $\tau < 1$ s. The process was repeated using new load measurements every second, resetting $t = 0$, and repeating the iterations, and P_{DG_j} and Q_{DG_j} were recomputed.

11.3 Resilient Operation Using the Hierarchical Architecture

11.3.1 Network Description

A modified IEEE 34 bus distribution system is used to investigate performance of the proposed approach as shown in Figure 11.3. The network line and load data can be found in [49]. DGs are connected to buses #5, #7, #11, #15, #20, #26, #29 with the corresponding capacity of 150, 150, 150, 200, 200, 200, 200, and 200 kW, respectively. Two capacitor banks of 20 kVar each are located in buses #25, #27 with corresponding time delay of 30 and 25 seconds.

Home load data were extracted from the eGauge website [50], which provides load data with up to one-minute resolution. Typical home data is shown in Figure 11.4. Load reactive power is defined in proportion to the real load connected at the same bus with a PF of 0.9 lagging. The one-second resolution photovoltaics (PV) generation is obtained from a station near Hawaii's Honolulu International Airport on the island of Oahu as available in [51]. Global Horizontal Irradiance is shown in Figure 11.4, which includes both the clear sky and transient cloud movement days. The assumption was made that the nodes are geographically close in the network such that outputs of PV units follow the same generation pattern.

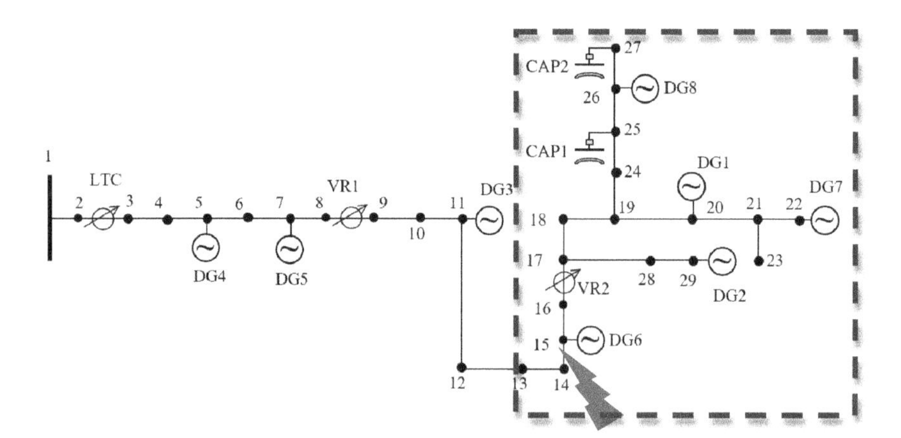

Figure 11.3 Modified IEEE 34 node test feeder.

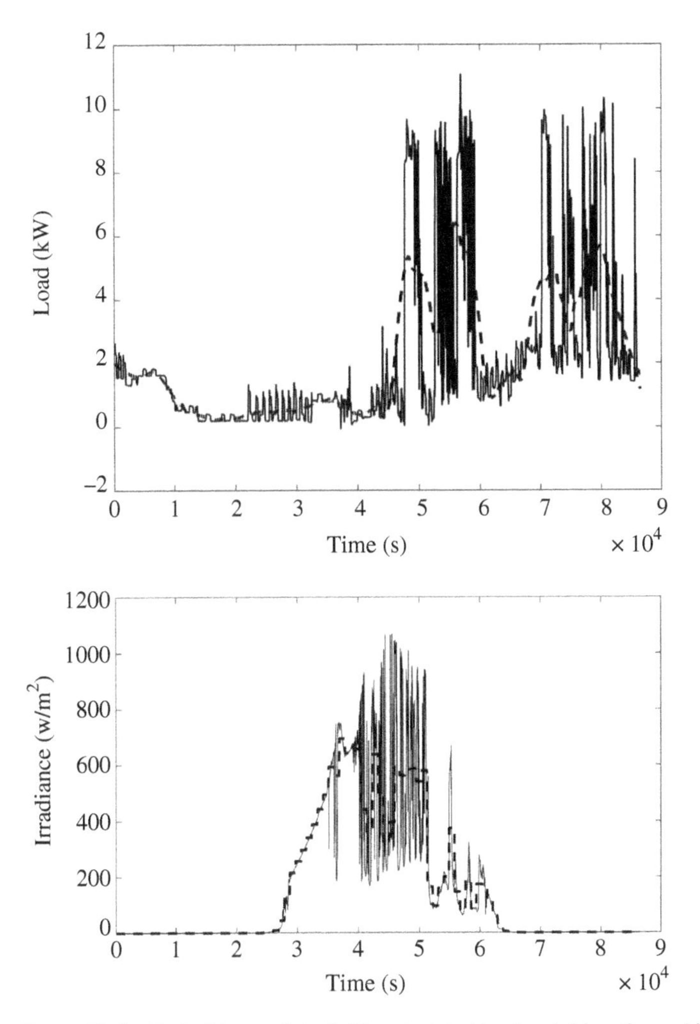

Figure 11.4 Typical home data (left) and global horizontal irradiance (right). The dashed curves correspond to the averaged load and generation data, used by top layer. The black curves correspond to the real-time load and generation data. The real-time load data were used by the bottom layer. The real-time load and generation data were used to evaluate the effectiveness of the overall resilient hierarchical architecture.

11.3.2 Results

A permanent fault was assumed to occur in line 14–15 which isolates part of the grid with the fault clearance time of one hour (see Figure 11.3). The goal of the resilient architecture is to pick up all of the critical loads based on the available generation in an intermittent cloudy day for the time period between 1:30 p.m. and 2:30 p.m. (3600 seconds). Loads at nodes 16, 20, 21, and 24 are considered to be the critical ones. This is reflected in the algorithm by choosing $w_i = 10$ for nodes 16, 20, 21, and 24 whereas $w_i = 1$ for other nodes. The proposed hierarchical architecture with both layers implemented as described in Section 11.1, and was compared to a centralized method that only employs the top layer.

Table 11.1 Picked up loads, VR, capacitor bank position during the self-healing operation via the proposed hierarchical approach.

Time	Picked up loads (node #)	CAP position		VR position
		C1	C2	
15	16, 18, 19, 20, 21, 22, 23, 24, 26	0	0	1
30	16, 18, 19, 20, 21, 22, 24, 26	1	1	1
45	16, 20, 21, 24	0	0	1
60	16, 20, 21, 23, 24, 27	0	0	1

Table 11.1 summarizes the performance through the 24-hour period of the proposed hierarchical architecture for resilient operation of distribution grid. Performance metrics include the picked up loads, VR and capacitor bank position during the one hour fault-clearing period. Referring to Table 11.1, it can be inferred that proposed approach could successfully recover all critical loads via optimal selection of loads to be picked up and minimal operation of VRs and capacitor banks.

Figure 11.5a–c show the total active and reactive power contribution of DGs as well as the total amount of load picked up for the proposed and centralized approaches during the 3600 seconds of interest. It can be seen from Figure 11.5a that if DGs follow the centralized

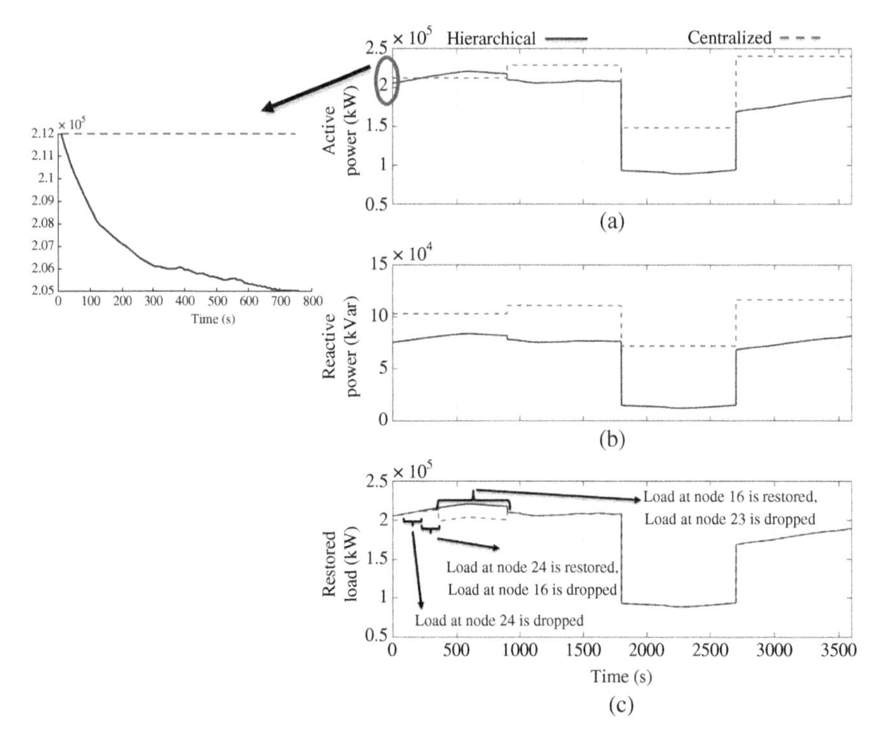

Figure 11.5 (a) Total active power contribution of DGs; (b) total reactive power contribution of DGs; (c) total amount of picked up load using the proposed and centralized approaches.

method, their total active power generation is higher than the proposed approach except for the first 900 seconds. This is mainly because the centralized method utilizes the 15-minute average data, while the proposed hierarchical method has the additional ability to reschedule DGs in layer two using finer one-second data resolution. With only centralized method, we see that the total load exceeds the total generation and results in a load drop in the time period between 223 and 900 seconds. In particular, Figure 11.5c shows that critical load at Node 24 is dropped for the time period 223–283 seconds. Then load at Node 24 is restored while load at Node 16 is dropped from 284 to 355 seconds time period. This is followed by restoration of load at Node 16 and the shedding of load at Node 23 from 356 to 900 seconds period. In contrast, with the hierarchical method, all critical loads are picked up after one second, and never dropped. Returning to the resilience definition in Definition 11.1, it follows that the resilience metric of the hierarchical method is one second, while that of the centralized method is 900 seconds. This is the main advantage of the proposed hierarchical method.

Figure 11.6a–c shows the simulations of the voltages at buses 22, 27, and 29. It can be seen that using the centralized method, voltages are higher in periods that loads have been

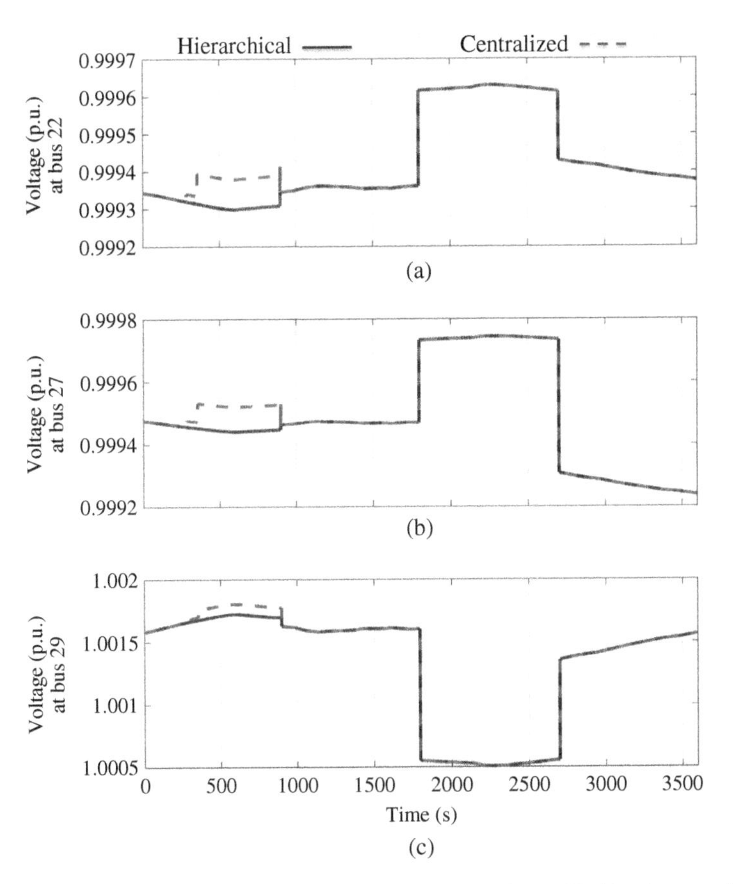

Figure 11.6 (a) voltage at Node 22; (b) voltage at Node 27; (c) voltage at Node 29 using the proposed and centralized approaches.

dropped compared to the proposed approach. This is because the proposed approach carries more loads to serve so the voltages are lower compared to the centralized approach.

11.4 Conclusions

This chapter presents a hierarchical architecture for resilient operation of distribution grids in the presence of high penetration of DGs. The architecture includes two layers where the top layer carries out central optimization and seeks to maximize critical loads to be restored via optimal scheduling of DGs. The bottom layer ensures load-generation balance using a distributed approach for active and reactive power provisioning of DGs. The top layer generates overall references for the lower layer optimized over the outage period and updated at 15-minute intervals using forecasted data regarding DGs and loads. The bottom layer provides finer correction of the active and reactive power injection every second based on the load-generation variation and a consensus-based algorithm. A suitable resilience metric is proposed, which corresponds to the time taken to accommodate all critical loads reliably. The proposed method is validated using a modified IEEE 34 bus test feeder in the presence of an outage, and is shown to increase the resilience by several orders of magnitude.

References

1 (2017). *Enhancing the Reslience of the Nation's Electricity System*. National Academies of Sciences, Engineering and Medicine, Washington, DC: The National Academies Press. https://doi.org/10.17226/24836. http:nap.edu/24836.

2 L. Mili and N. V. Center (2011), "Taxonomy of the characteristics of power system operating states," in 2nd NSF-VT Resilient and Sustainable Critical Infrastructures (RESIN) Workshop, Tucson, AZ, pp. 13–15.

3 Q. Zhu and T. Basar (2011), "Robust and resilient control design for cyberphysical systems with an application to power systems," in Proceeding of 50th IEEE Conference on Decision and Control and European Control Conference (CDC/ECC), Orlando, Florida, pp. 4066–4071.

4 Annaswamy, A.M., Malekpour, A.R., and Baros, S. (2016). Emerging research topics in control for a smart infrastructure. *Journal of Annual Reviews of Control* 42: 259–270.

5 Kitano, H. (2004). Biological robustness. *Nature Reviews* 5: 826–837.

6 Levin, S.A. and Lubchenco, J. (2008). Resilience, robustness, and marine ecosystem-based management. *BioScience* 58 (1): 27–32.

7 Ip, W.H. and Wang, D. (2011). Resilience and friability of transportation networks: evaluation, analysis and optimization. *IEEE Systems Journal* 5 (2): 189–198.

8 Tabuada, P., Caliskan, S.Y., Rungger, M., and Majumdar, R. (2014). Towards robustness for cyber-physical systems. *IEEE Transactions on Automatic Control* 59 (12): 3151–3163.

9 Rungger, M. and Tabuada, P. (2016). A notion of robustness for cyber-physical systems. *IEEE Transactions on Automatic Control* 61 (12): 2108–2123.

10 Yagan, O., Qian, D., Zhang, J., and Cochran, D. (2012). Optimal allocation of interconnecting links in cyber-physical systems: interdependence, cascading failures, and robustness. *IEEE Transactions on Parallel and Distributed Systems* 23 (9): 1708–1720.

11 Zhou, K., Doyle, J., and Glover, K. (1996). *Robust and Optimal Control.* Prentice Hall.

12 M.N. Albasrawi, N. Jarus, K.A. Joshi, S.S. Sarvestani. (2014). Analysis of reliability and resilience for smart grids. In: IEEE 38th Annual International Computers, Software and Applications Conference, Vasteras, Sweden, pp. 529–534.

13 Rieger, C.G., Moore, K.L., Baldwin, T.L. (2013) Resilient control systems: a multi-agent dynamic systems perspective. In: IEEE International Conference on Electro/Information Technology (EIT), Rapid City, SD, USA, pp 1–16.

14 Rieger, C.G., Gertman, D.I., and McQueen, M.A. (2009), "Resilient control systems: Next generation design research." In: Proceedings of IEEE 2nd Conference on Human System Interactions, Catania, Italy, pp. 632–636.

15 Holling, C.S. (1973). Resilience and stability of ecological systems. *Annual Review of Ecology and Systematics* 4: 1–23.

16 Holling, C.S. (1996). Engineering resilience versus ecological ressilience. *Engineering with Ecological Constraints* 31: 32.

17 Demirok, E., Gonzalez, P.C., Frederiksen, K.H.B. et al. (2011). Local reactive power control methods for overvoltage prevention of distributed solar inverters in low-voltage grids. *IEEE Journal of Photovoltaics* 1: 174–182.

18 Caldon, R., Coppa, M., Sgarbossa, R., and Turri, R. (2013). A simplified algorithm for OLTC control in active distribution MV networks. In: Proceedings of IEEE AEIT Annual Conference, Mondello, Palermo, Italy, pp. 1–6.

19 Malekpour, A.R., Pahwa, A., and Das, S. (2013). Inverter-based var control in low voltage distribution systems with rooftop solar PV. In: Proceedings of 2013 IEEE 45th North American Power Symposium (NAPS), Manhattan, KS, USA, September 2013, pp. 1–5.

20 Nair, N.-K.C. and Jing, L. (2013). Power quality analysis for building integrated PV and micro wind turbine in New Zealand. *Energy and Buildings* 58: 302–309.

21 Chen, C.-S., Tsai, C.-T., Lin, C.-H. et al. (2011). Loading balance of distribution feeders with loop power controllers considering photovoltaic generation. *IEEE Transactions on Power Systems* 26 (3): 1762–1768.

22 Borghetti, A. (2013). Using mixed integer programming for the volt/var optimization in distribution feeders. *Electric Power Systems Research* 98 (5): 39–50.

23 Malekpour, A.R. and Pahwa, A. (2016). A dynamic operational scheme for residential PV smart inverters. *IEEE Transactions on Smart Grid* 8: 1–10.

24 Lim, S.I., Lee, S.J., Choi, M.-S. et al. (2006). Service restoration methodology for multiple fault case in distribution systems. *IEEE Transactions on Power Systems* 21 (4): 1638–1644.

25 Nguyen, C.P. and Flueck, A.J. (2012). Agent based restoration with distributed energy storage support in smart grids. *IEEE Transactions on Smart Grid* 3 (2): 1029–1038.

26 Liu, C.C., Lee, S.J., and Venkata, S.S. (1988). An expert system operational aid for restoration and loss reduction of distribution systems. *IEEE Transactions on Power Systems* 3: 619–626.

27 Xu, Y., Liu, C.C., Schneider, K., and Ton, D. (2016). Optimal placement of remote controlled switches for distribution system restoration. *IEEE Transactions on Power Systems* 31: 1139–1150.

28 Jiang, Y., Liu, C.C., and Xu, Y. (2016). Smart distribution systems. *Energies* 9 (4): 1–20. 10.3390/en9040297 (Invited).

29 Jiang, Y., Liu, C.C., Diedesch, M., and Srivastava, A. (2016). Outage management of distribution systems incorporating information from smart meters. *IEEE Transactions on Power Systems* 31: 4144–4154.

30 Tsai, M.S., Liu, C.C., Mesa, V.N., and Hartwell, R. (1993). IOPADS (intelligent operational planning aid for distribution systems). *IEEE Transactions on Power Delivery* 8 (3): 1562–1569.

31 Pham, T.T.H., Bésanger, Y., and Hadjsaid, N. (2009). New challenges in power system restoration with large scale of dispersed generation insertion. *IEEE Transactions on Power Systems* 24 (1): 398–406.

32 Deshmukh, S., Natarajan, B., and Pahwa, A. (2012). Voltage/var control in distribution networks via reactive power injection through distributed generators. *IEEE Transactions on Smart Grid* 3 (3): 1226–1234.

33 Araujo, L.R., Penido, D.R.R., Carneiro, S., and Pereira, J.L.R. (2013). A three-phase optimal power-flow algorithm to mitigate voltage unbalance. *IEEE Transactions on Power Delivery* 28 (4): 2394–2402.

34 Bruno, S., Lamonaca, S., Rotondo, G. et al. (2011). Unbalanced three-phase optimal power flow for smart grids. *IEEE Transactions on Industrial Electronics* 58 (10): 4504–4513.

35 Zhu, Y. and Tomsovic, K. (2007). Optimal distribution power flow for systems with distributed energy resources. *International Journal of Electrical Power & Energy Systems* 29 (3): 260–267.

36 Senjyu, T., Miyazato, Y., Yona, A. et al. (2008). Optimal distribution voltage control and coordination with distributed generation. *IEEE Transactions on Power Delivery* 23 (2): 1236–1242.

37 Malekpour, A.R. and Niknam, T. (2011). A probabilistic multi-objective daily volt/var control at distribution networks including renewable energy sources. *Energy* 36 (5): 3477–3488.

38 Valverde, G. and Van Cutsem, T. (2013). Model predictive control of voltages in active distribution networks. *IEEE Transactions on Smart Grid* 4 (4): 2152–2161.

39 Paudyal, S., Canizares, C.A., and Bhattacharya, K. (2011). Optimal operation of distribution feeders in smart grids. *IEEE Transactions on Industrial Electronics* 58 (10): 4495–4503.

40 Nedic, A. and Ozdaglar, A. (2009). Distributed subgradient methods for multiagent optimization. *IEEE Transactions on Automatic Control* 54 (1): 48–61.

41 Nedic, A., Ozdaglar, A., and Parrilo, P.A. (2010). Constrained consensus and optimization in multi-agent networks. *IEEE Transactions on Automatic Control* 55 (4): 922–938.

42 Olfati-Saber, R., Fax, J.A., and Murray, R.M. (2007). Consensus and cooperation in networked multi-agent systems. *Proceedings of the IEEE* 95 (1): 215–233.

43 Olshevsky, A. and Tsitsiklis, J.N. (2009). Convergence speed in distributed consensus and averaging. *SIAM Journal on Control and Optimization* 48 (1): 33–55.

44 Baran, M.E. and Wu, F.F. (1989). Optimal sizing of capacitors placed on a radial distribution system. *IEEE Transactions on Power Delivery* 4 (1): 735–743.
45 Turitsyn, K., Sŭlc, P., Backhaus, S., and Chertkov, M. (2010) Local control of reactive power by distributed photovoltaic generators. In: Proceedings of 1st IEEE International Conference on Smart Grid Communications, Gaithersburg, MD, USA, October 2010, pp. 79–84.
46 Turitsyn, K., Sŭlc, P., Backhaus, S., and Chertkov, M. (2011). Options for control of reactive power by distributed photovoltaic generators. *Procedings of the IEEE* 99 (6): 1063–1073.
47 Robbins, B.A., Hadjicostis, C.N., and Dominguez-Garćia, A.D. (2012). A two stage distributed architecture for voltage control in power distribution systems. *IEEE Transactions on Power Systems* 28: 1470–1482.
48 Horn, R.A. and Johnson, C.R. (1987). *Matrix Analysis*. Cambridge, U.K.: Cambridge University Press.
49 Mwakabuta, N. and Sekar, A. (2007). Comparative study of the IEEE 34 node test feeder under practical simplifications. In: Proceedings of 2007 IEEE 39th North American Power Symposium (NAPS), Las Cruces, NM, USA, September 2007, pp. 484–491.
50 eGauge: Energy Metering Systems, Retrieved January 25, 2023, from: http://egauge360.egaug.es/5D813/.
51 Sengupta, M. and A. Andreas. (2010). Oahu Solar Measurement Grid (1-Year Archive): 1- Second Solar Irradiance; Oahu, Hawaii (Data), NREL Report No. DA-5500-56506, National Renewable Energy Laboratory.

Biographies

Dr. Ahmad Reza Malekpour (M'16) received the PhD degree in electrical engineering from Kansas State University, Manhattan, in 2016. He was a postdoctoral associate within Active-Adaptive Control Laboratory, Massachusetts Institute of Technology (MIT), Cambridge. Currently he is the Senior Team Lead within Grid Analytics Product team, GE Digital, San Ramon, CA. His research interests include applied optimization and machine learning in electrical power grids, energy management of microgrids, renewable and distributed energy resources (DERs) integration.
Dr. Anuradha Annaswamy received her PhD in Electrical Engineering from Yale University in 1985. She has been a member of the faculty at Yale, Boston University, and MIT where currently she is the director of the Active-Adaptive Control Laboratory and a senior research scientist in the Department of Mechanical Engineering. Her research interests pertain to adaptive control theory and applications to aerospace, automotive, and propulsion systems, cyber physical systems science, and CPS applications to Smart Grids, Smart Cities, and Smart Infrastructures.
Dr. Jalpa Shah received her PhD in Electrical Engineering from University of Minnesota, twin cities in 2011. She is currently a specialist engineer with Corporate Research and Technology, Eaton Corporation leading technology development for intelligent power grid controls. She has priorly worked with John Deere and Rockwell Automation in the domain of industrial motor control and automation and on/off road vehicle electrification. Her research interests include advanced controls for power electronics and electrical power systems.

Part III

Real-World Case Studies

12

A Resilience Framework Against Wildfire

Dimitris Trakas[1], Nikos Hatziargyriou[1], Mathaios Panteli[2], and Pierluigi Mancarella[3]

[1] *Electric Energy Systems Laboratory, School of Electrical and Computer Engineering, National Technical University of Athens, Athens, Greece*
[2] *Department of Electrical and Computer Engineering, University of Cyprus, Nicosia, Cyprus*
[3] *Department of Electrical and Electronic Engineering, The University of Melbourne, Parkville, Melbourne, Australia*

12.1 Introduction

Electrical power systems, as critical infrastructure, are key for the sustainability and growth of modern societies, since they support several critical services and infrastructures, such as transportation, communication, and health systems. Hence, given these high and complex interdependencies between these critical infrastructures, a disruption in the electricity network can have catastrophic consequences.

However, despite the efforts of keeping the power flowing and the lights on under any *credible* events, power systems (and particularly distribution networks) are occasionally exposed to *extreme weather* and *natural hazards* (e.g. wildfires, storms, and earthquakes), which as evidenced worldwide can be so intense that they can cause the collapse of power systems, leading to large and sustained power disruptions with great economic and social impacts. The threats of a power system can be broadly categorized as *credible* or "*typical*" *power system outages* and more *extreme events*, driven mainly by *natural disasters and extreme weather*, whose frequency and severity might increase as a direct impact of climate change [1]. Table 12.1 shows the distinct differences between these two categories of events [2].

Power networks have been traditionally designed and operated to be *reliable* to the more typical threats. Nevertheless, experiences around the world are now signifying the increasing importance for power networks to also achieve high levels of *resilience* to natural hazards and extreme weather [3–5], the so-called high-impact low-probability events, in order to mitigate the impacts of such events and quickly recover. Table 12.2 shows some of the key features that set the concept of reliability apart from the one of resilience [6].

Resiliency of Power Distribution Systems, First Edition.
Edited by Anurag K. Srivastava, Chen-Ching Liu, and Sayonsom Chanda.

Table 12.1 Comparison of typical power system outages and extreme events.

Typical power system outage	Extreme event
• Low impact, high probability	• High impact, low probability
• Preventive and corrective control measures portfolio in place	• No control measures are in place (typically)
• Random location and time of occurrence	• Spatiotemporal correlation between faults and event
• Supported by contingency analysis and optimization tools	• Limited mathematical tools
• Limited number (single or double) of faults due to component failures	• Multiple simultaneous faults
• Small portion of the network is damaged/collapsed	• Large portion of the network is damaged/collapsed
• Quick restoration	• More time and resources consuming/longer restoration

Table 12.2 Comparison of resilience and reliability.

Reliability	Resilience
• High-probability, low-impact	• Low-probability, high-impact
• Based on average indicators, e.g. loss of load frequency	• Based on risk profile, e.g. conditional expectation
• Shorter-term, typically static	• Longer-term, adaptive, ongoing
• Evaluates the power system states	• Evaluates the power systems states and the state transitions
• Concerned mainly with customer interruption time	• Concerned with customer interruption time and infrastructure recovery time

12.2 The Hazard of Wildfires

Wildfires is one of the major natural disasters that power networks are exposed to, particularly in countries with high temperatures. These harsh environmental conditions result in the high probability of wildfires, which makes the portions of the power systems passing through or adjacent to areas with dense vegetation, trees, and forests highly prone to faults due to wildfires every summer season. This is becoming of increasing concern as a result of longer and warmer seasons, mainly driven by climate change. In fact, climate change is believed to contribute to many catastrophic wildfires around the world, for example, United States, Alaska, and Canada [7]. The wildfires can cause significant direct (e.g. burning of distribution poles) and indirect (e.g. impact on line sag) damage to power systems, leading to long power outages. During wildfires, the uninterrupted supply of electrical energy is of major importance, since the water pump with electric motors is critical for extinguishing the fire. It is therefore highly critical to consider the likelihood and impact of wildfires when defining the design and operation of power systems in countries that are affected by wildfires, as well as the emergency procedures and practices, to minimize the effect of these catastrophic events on the resilience of these power systems.

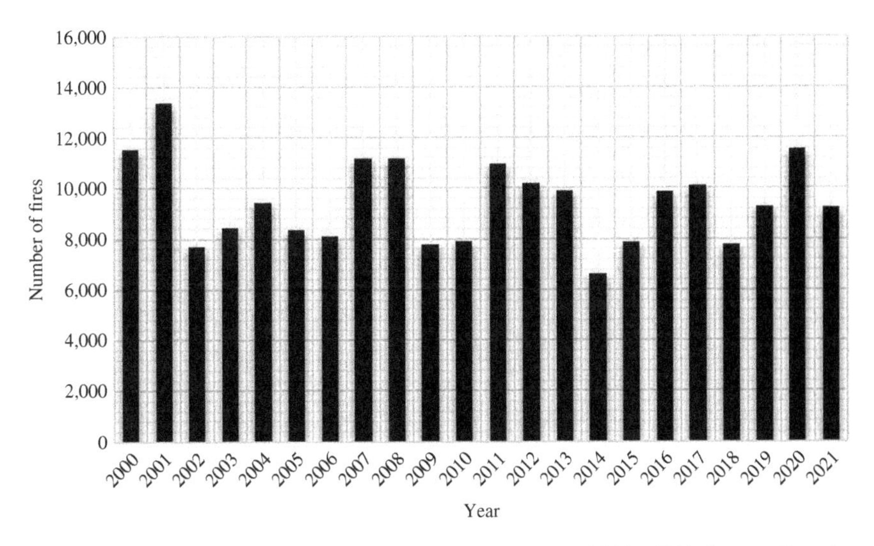

Figure 12.1 Number of forest fires in Greece for the years 2000–2021. Source: Data from [8].

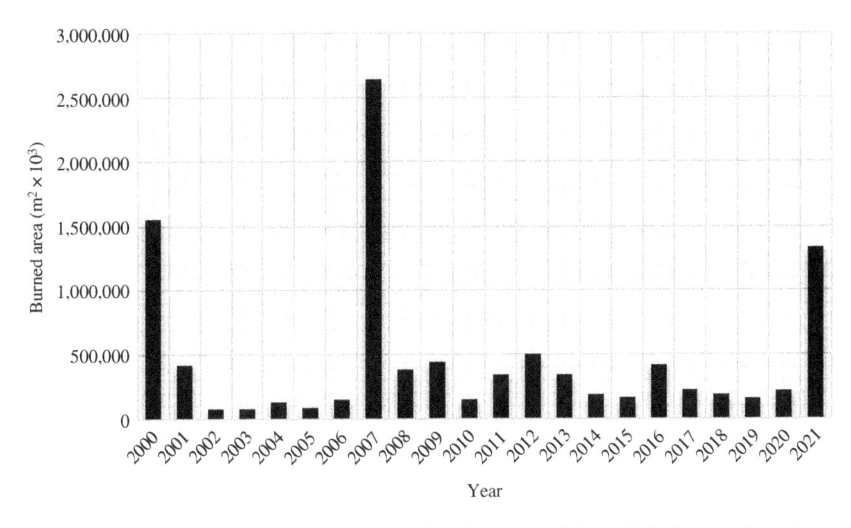

Figure 12.2 Burned forest area in Greece for the years 2000–2021. Source: Data from [8].

As an example of the severity and impact of wildfires, Figure 12.1 shows the number of fires in forests, forested areas, grassy areas, reed swamps, agricultural areas and crop residues in Greece for the years 2000–2021. It is observed that the number of fires is extremely high. The corresponding burned area is presented in Figure 12.2. For instance, in 2007, about $2,643 \times 10^6$ m^2 were burned due to wildfires. The largest burned area was located in Peloponnese. Approximately 2500 poles were burned and 400 power outages occurred over a period of one week, resulting in a serious disruption in the electricity supply of 90,000 people. 490 technicians needed to repair the damages and more than 20% of the people affected had their electricity supply restored in more than five days. Figure 12.3 shows some examples of the damages to the distribution network caused by

Figure 12.3 Damages to the distribution system in Peloponnese due to the wildfire in 2007.

the wildfire in Peloponnese in 2007. The total cost of the Peloponnese power distribution system restoration due to wildfires in 2007 amounted to €6,300,000.

A wildfire is able to cause direct physical damage to the power system components or decrease the thermal rating of the lines due to the increase in surface temperature of the conductor [9]. In this chapter, optimal distribution system operation for enhancing resilience is formulated and solved considering the varying conditions during the spread of a progressing wildfire and its impact on the system. In [10] the optimum placing of switches in certain branches of the distribution system considering the vulnerability of power system components to a wildfire threat is decided. In case of a line failure, the distribution system is sectionalized into self-sufficient microgrids to prevent a possible spread of the fault with the minimum load shedding of the islanded portion of the distribution system. In [11] a methodology to quantify the damage caused by a wildfire to the distribution system of a city is proposed. However, no measures are proposed for addressing the threat of fire. The impact of a progressing wildfire on line ratings of a transmission system is proposed in [9]. Moreover, an optimal power flow method for minimizing the generation cost is applied considering reduced line capacities due to the wildfire. In [12, 13] a method for an optimal distribution system operation against a progressing wildfire is presented and the contribution of microgrids and demand response is studied. In [9, 12, 13], the steady-state heat balance equation is used, and according to IEEE Std 738 [14], it is applied when the electrical current, conductor temperature, and weather conditions are assumed constant for all times.

12.3 Modeling and Quantifying the Resilience of Distribution Networks to Wildfires

In this section, a stochastic programming approach is presented to determine the resilient operation of a distribution system exposed to an approaching wildfire. The proposed method is applied, when the danger of a wildfire is perceived [15]. Dynamic line rating (DLR) of the overhead lines is considered in order to model the impact of the wildfire on the conductor's temperature. The non-steady-state heat balance is used to calculate conductor temperature for weather conditions that vary over the time horizon. When the wildfire is within close distance from the conductors, it is assumed that it disrupts their operation for the rest of the time horizon of the study due to the high rise of their temperature and the violation of their maximum permissible temperature. This does not necessarily mean physical damage, but it can include unacceptable line sags that lead to the outage of the line.

The proposed model assumes that distribution system operators (DSOs) have full control of distributed generators (DGs). As stated in [16, 17], system defense plans and restoration schemes are put in action during emergency situations, to avoid power interruptions and to restore electricity, when interruptions are not avoided. Close coordination between all actors playing an active role in the defense and restoration schemes is required. DSOs are responsible to supervise and control the implementation of emergency actions. Therefore, it is assumed that during the progression of the wildfire the DSO has full control of DGs and energy storage systems (ESSs). In fact, according to the German Guideline for generating plants [18], in case of potential danger to system operation security, DSOs are entitled to request a temporary limit of the DGs power or disconnect the facility. Furthermore, DSOs in Great Britain have to prepare and maintain plans for mitigating the effects of an emergency, such as a natural disaster (e.g. wildfire), that may likely disrupt the electricity supply [19]. In this case, the Distribution Operating Code sets out Contingency Planning procedures to enable coordination between all the users of the distribution system. In addition, in case of campus or military base, the assumption that the operator is the owner of system components and responsible for their operation is fully applicable [20].

Distribution systems consisting of ESSs and DGs are considered in this chapter. The production of wind turbines (WTs) and photovoltaic panels (PVs) are modeled as stochastic, based on probabilistic forecasting techniques [21], while microturbines (MTs) are modeled as controllable sources.

12.3.1 Uncertainties

12.3.1.1 Stochastic Parameters

Wind speed and direction are considered uncorrelated. Weibull and von Mises distributions are used for modeling wind speed and direction [12]. Solar radiation is modeled using beta distribution and normal distribution is used for load demand modeling [22]. The parameters of beta distribution and Weibull distribution are calculated by using the mean value and standard deviation of the predicted solar radiation and wind speed according to [23, 24], respectively. The values of the stochastic parameters between adjacent times are considered uncorrelated.

12.3.1.2 Scenario Generation and Reduction Algorithm

Monte Carlo simulation (MCS) is used to produce a large number of scenarios for the stochastic parameters based on forecasted values and the typical distribution of each parameter. The use of MCs, instead of simple mean values, is able to provide results covering a whole spectrum of probable scenario values including worst-case probable conditions, as will be shown in Section 12.4.2. The Backward Scenario Reduction algorithm is then used in order to reduce the number of scenarios in a tractable size [25]. The algorithm is applied iteratively. Based on Kantorovich distance, a scenario is removed at each iteration and its probability is assigned to the closest scenario. The algorithm is terminated when the desired number of scenarios is reached.

12.3.2 Wildfire Model

The heat from wildfire is transferred through radiation and convection. The convective transfer is neglected, since it affects the conductors' temperature, only when the fire is directly under the overhead line. It is most likely, however, that a line will already be out of order when the wildfire is within close distance [12].

For a large fire, the simplified heat flux model of [26] is used. The radiative heat flux Φ^f emitted from the fire to a conductor is computed according to:

$$\Phi^f_{ij,\omega,t} = \frac{\tau\varepsilon^f B T^{f^4}}{2}\sin\left(\delta^f_{ij,\omega,t}\right), \ \forall (i,j) \in \mathscr{L}, \omega \in \mathscr{O}, t \in \mathscr{T} \tag{12.1}$$

$$\delta^f_{ij,\omega,t} = \tan^{-1}\left(\frac{L^f \cos(\gamma^f)}{r^f_{ij,\omega,t} - (L^f \sin(\gamma^f))}\right), \ \forall (i,j) \in \mathscr{L}, \omega \in \mathscr{O}, t \in \mathscr{T} \tag{12.2}$$

where δ^f is the view angle between the flame and the object threatened by the fire for a large fire front. For a better understanding of (12.1) and (12.2), the reader is referred to the schematic geometry used in [26].

The wildfire is moving across the forested area with a specific rate of spread V^f that depends on wind speed V^w and the vegetation of the crossing area:

$$V^f_{\omega,t} = \frac{k\left(1 + V^w_{\omega,t}\right)}{\rho^b}, \ \forall \omega \in \mathscr{O}, t \in \mathscr{T} \tag{12.3}$$

where k is equal to 0.07 kg/m^3 for a wildland fire [9]. The bulk density ρ^b is equal to 40 kg/m^3 along a forest floor. The distance of the wildfire from the conductors at time t depends on wind direction φ^w and the rate of spread V^f:

$$r^f_{ij,\omega,t} = r^f_{ij,\omega,t-1} - V^f_{\omega,t}\Delta t \cos\varphi^w_{ij,\omega,t}, \ \forall (i,j) \in \mathscr{L}, \omega \in \mathscr{O}, t \in \mathscr{T} \tag{12.4}$$

12.3.3 Dynamic Model of Overhead Lines

The calculation of overhead line temperature is based on the method presented in IEEE Std 738 [14]. The heat transferred to the conductor from the wildfire is added to the sources that increase the conductor temperature. The change in conductor temperature during the time interval Δt is calculated using the non-steady-state heat balance equation:

$$(T_{ij,\omega,t+1} - T_{ij,\omega,t}) = \frac{\Delta t}{mC^p}\begin{pmatrix} q^l_{ij,\omega,t} + q^s_{ij,\omega,t} + q^f_{ij,\omega,t} \\ -q^c_{ij,\omega,t} - q^r_{ij,\omega,t} \end{pmatrix}, \ \forall (i,j) \in \mathscr{L}, \omega \in \mathscr{O}, t \in \mathscr{T} \tag{12.5}$$

Note that the three first terms, that cause the temperature rise of the conductor, are the heat gain rates due to ohmic losses q^l, solar radiation q^s and radiative heat flux q^f emitted from the fire. These three terms are given by the following equations, respectively:

$$q^l_{ij,\omega,t} = R(T_{ij,\omega,t})|I_{ij,\omega,t}|^2, \ \forall (i,j) \in \mathscr{L}, \omega \in \mathscr{O}, t \in \mathscr{T} \tag{12.6}$$

$$q^s_{ij,\omega,t} = D_{ij}K_{ij}\Phi^s_{ij,\omega,t}, \ \forall (i,j) \in \mathscr{L}, \omega \in \mathscr{O}, t \in \mathscr{T} \tag{12.7}$$

$$q^f_{ij,\omega,t} = D_{ij}\Phi^f_{ij,\omega,t}, \ \forall (i,j) \in \mathscr{L}, \omega \in \mathscr{O}, t \in \mathscr{T} \tag{12.8}$$

where $R(T_{ij})$ is a function that describes the dependency of conductor resistance form its temperature and I_{ij} is the line current.

The last two terms in (12.5) cool down the conductor. The convection q^c for non-zero wind speeds and radiative heat loss q^r rates are calculated according to:

$$q^c_{ij,\omega,t} = \max\begin{pmatrix} K^{angle}_{ij,\omega,t}\left[1.01 + 1.35\left(N^{Re}_{ij,\omega,t}\right)^{0.52}\right]\left(T_{ij,\omega,t} - T^a_{ij,t}\right) \\ K^{angle}_{ij,\omega,t}0.754\left(N^{Re}_{ij,\omega,t}\right)^{0.6}k^a\left(T_{ij,\omega,t} - T^a_{ij,t}\right)\end{pmatrix}, \quad \forall (i,j) \in \mathcal{L}, \omega \in \mathcal{O}, t \in \mathcal{T} \tag{12.9}$$

$$q^r_{ij,\omega,t} = 17.8D_{ij}\varepsilon\left[\left(\frac{T_{ij,\omega,t}}{100}\right)^4 - \left(\frac{T^a_{ij,t}}{100}\right)^4\right], \quad \forall (i,j) \in \mathcal{L}, \omega \in \mathcal{O}, t \in \mathcal{T} \tag{12.10}$$

The magnitude of the convective heat loss q^c depends on a dimensionless number N^{Re} known as Reynolds number and the wind direction factor K^{angle} that are given by:

$$N^{Re}_{ij,\omega,t} = \frac{D_{ij}\rho^a V^w_{\omega,t}}{\mu^a}, \quad \forall (i,j) \in \mathcal{L}, \omega \in \mathcal{O}, t \in \mathcal{T} \tag{12.11}$$

$$K^{angle}_{ij,\omega,t} = 1.194 - \cos\left(\varphi^w_{ij,\omega,t}\right) + 0.194\cos\left(2\varphi^w_{ij,\omega,t}\right) + 0.368\sin\left(2\varphi^w_{ij,\omega,t}\right), \quad \forall (i,j) \in \mathcal{L}, \omega \in \mathcal{O}, t \in \mathcal{T} \tag{12.12}$$

Equation (12.9) is used for non-zero winds. In the present paper, the burst and progression of a wildfire caused by high wind speeds are modeled, thus zero-wind speed is not considered.

Equations (12.1)–(12.4), (12.11), and (12.12) depend only on problem parameters and are calculated after the application of the generation and scenario reduction algorithm, in order to be used as inputs to the optimization problem. In addition, the generation of WTs and PVs that is used as input to the optimization problem is calculated according to [22].

12.3.4 Problem Formulation

12.3.4.1 Optimization Problem for Resilient Operation Against Approaching Wildfire

A stochastic programming formulation is used for enhancing distribution system resilience against an approaching wildfire. The optimization problem is expressed as:

$$\begin{aligned}\min \sum_{t=1}^{N_T}\sum_{\omega=1}^{N_\Omega}\pi_\omega\sum_{i=1}^{N_B}\left(\text{VoLL}p^{\text{shed}}_{i,\omega,t} - c^D p^D_{i,\omega,t}\right) + \sum_{t=1}^{N_T}\sum_{\omega=1}^{N_\Omega}\pi_\omega\sum_{i=1}^{N_B}\left(c^G_i p^G_{i,\omega,t}\right) \\ + \sum_{t=1}^{N_T}\sum_{\omega=1}^{N_\Omega}\pi_\omega c^{\text{UP}}_t\left(p^{\text{UP}_\text{B}}_{\omega,t} - p^{\text{UP}_\text{S}}_{\omega,t}\right) + \sum_{t=1}^{N_T}\sum_{i=1}^{N_B}\left(\text{su}^G_{i,t} + \text{sd}^G_{i,t}\right)\end{aligned} \tag{12.13}$$

$$p^G_{i,\omega,t} + p^{WT}_{i,\omega,t} + p^{PV}_{i,\omega,t} + p^{\text{ST}-}_{i,\omega,t} - p^{\text{ST}+}_{i,\omega,t} - p^D_{i,\omega,t} = \sum_{j=1}^{N_B} pf^P_{ij,\omega,t}, \quad \forall i \in \mathcal{B}, \omega \in \mathcal{O}, t \in \mathcal{T} \tag{12.14}$$

$$q^G_{i,\omega,t} + q^{\text{ST}}_{i,\omega,t} - q^D_{i,\omega,t} = \sum_{j=1}^{N_B} pf^Q_{ij,\omega,t}, \quad \forall i \in \mathcal{B}, \omega \in \mathcal{O}, t \in \mathcal{T} \tag{12.15}$$

$$\underline{p_i^G} u^G_{i,t} \le p^G_{i,\omega,t} \le \overline{p_i^G} u^G_{i,t}, \quad \forall i \in \mathcal{B}, \omega \in \mathcal{O}, t \in \mathcal{T} \tag{12.16}$$

$$\underline{q_i^G} u^G_{i,t} \le q^G_{i,\omega,t} \le \overline{q_i^G} u^G_{i,t}, \quad \forall i \in \mathcal{B}, \omega \in \mathcal{O}, t \in \mathcal{T} \tag{12.17}$$

$$su_{i,t}^{G} \geq 0, su_{i,t}^{G} \geq c_{i}^{SU}\left(u_{i,t}^{G} - u_{i,t-1}^{G}\right), \quad \forall i \in \mathscr{B}, t \in \mathscr{T} \tag{12.18}$$

$$sd_{i,t}^{G} \geq 0, sd_{i,t}^{G} \geq c_{i}^{SD}\left(u_{i,t-1}^{G} - u_{i,t}^{G}\right), \quad \forall i \in \mathscr{B}, t \in \mathscr{T} \tag{12.19}$$

$$T_{ij,\omega,t} \leq T_{ij}^{\max} + \left(1 - u_{ij,\omega,t}^{l}\right) M_1, \quad \forall (i,j) \in \mathscr{L}, \omega \in \mathscr{O}, t \in \mathscr{T} \tag{12.20}$$

$$-M_2 u_{ij,\omega,t}^{l} \leq pf_{ij,\omega,t}^{P} \leq M_2 u_{ij,\omega,t}^{l}, \quad \forall (i,j) \in \mathscr{L}, \omega \in \mathscr{O}, t \in \mathscr{T} \tag{12.21}$$

$$-M_2 u_{ij,\omega,t}^{l} \leq pf_{ij,\omega,t}^{Q} \leq M_2 u_{ij,\omega,t}^{l}, \quad \forall (i,j) \in \mathscr{L}, \omega \in \mathscr{O}, t \in \mathscr{T} \tag{12.22}$$

$$\begin{aligned} -M_3\left(1 - u_{ij,\omega,t}^{l}\right) &\leq pf_{ij,\omega,t}^{P} - (g_{ij}(v_{i,\omega,t} - v_{j,\omega,t} - \zeta_{ij,\omega,t} + 1) - b_{ij}(\vartheta_{i,\omega,t} - \vartheta_{j,\omega,t})) \\ &\leq M_3\left(1 - u_{ij,\omega,t}^{l}\right), \quad \forall (i,j) \in \mathscr{L}, \omega \in \mathscr{O}, t \in \mathscr{T} \end{aligned} \tag{12.23}$$

$$\begin{aligned} -M_3\left(1 - u_{ij,\omega,t}^{l}\right) &\leq pf_{ij,\omega,t}^{Q} - (-b_{ij}(v_{i,\omega,t} - v_{j,\omega,t} - \zeta_{ij,\omega,t} + 1) - g_{ij}(\vartheta_{i,\omega,t} - \vartheta_{j,\omega,t})) \\ &\leq M_3\left(1 - u_{ij,\omega,t}^{l}\right), \quad \forall (i,j) \in \mathscr{L}, \omega \in \mathscr{O}, t \in \mathscr{T} \end{aligned} \tag{12.24}$$

$$\zeta_{ij,\omega,t} = \sum\nolimits_{y=1}^{N_Y}\left(\kappa_{ij,\omega,t,y}\Delta\vartheta_{ij,\omega,t,y} + u_{ij,\omega,t,y}^{\zeta}\xi_{ij,\omega,t,y}\right), \quad \forall (i,j) \in \mathscr{L}, \omega \in \mathscr{O}, t \in \mathscr{T} \tag{12.25}$$

$$\sum\nolimits_{y=1}^{N_Y} u_{ij,\omega,t,y}^{\zeta} = 1, \quad \forall (i,j) \in \mathscr{L}, \omega \in \mathscr{O}, t \in \mathscr{T} \tag{12.26}$$

$$\Delta\vartheta_{\min}^{y} u_{ij,\omega,t,y}^{\zeta} \leq \Delta\vartheta_{ij,\omega,t,y} \leq u_{ij,\omega,t,y}^{\zeta}\Delta\vartheta_{\max}^{y}, \quad \forall (i,j) \in \mathscr{L}, \omega \in \mathscr{O}, t \in \mathscr{T}, \ y \in \mathscr{Y} \tag{12.27}$$

$$\sum\nolimits_{y=1}^{N_Y} \Delta\vartheta_{ij,\omega,t,y} = \vartheta_{i,\omega,t} - \vartheta_{j,\omega,t}, \quad \forall (i,j) \in \mathscr{L}, \omega \in \mathscr{O}, t \in \mathscr{T} \tag{12.28}$$

$$u_{ij,\omega,t}^{l} \leq u_{ij,\omega,t-1}^{l}, \quad \forall (i,j) \in \mathscr{L}, \omega \in \mathscr{O}, t \in \mathscr{T} \tag{12.29}$$

$$\underline{v_i} \leq v_{i,\omega,t} \leq \overline{v_i}, \quad \forall i \in \mathscr{B}, \omega \in \mathscr{O}, t \in \mathscr{T} \tag{12.30}$$

$$0 \leq p_{i,\omega,t}^{\text{shed}} \leq P_{i,\omega,t}^{D}, \quad \forall i \in \mathscr{B}, \omega \in \mathscr{O}, t \in \mathscr{T} \tag{12.31}$$

$$q_{i,\omega,t}^{\text{shed}} = p_{i,\omega,t}^{\text{shed}}\frac{Q_{i,\omega,t}^{D}}{P_{i,\omega,t}^{D}}, \quad \forall i \in \mathscr{B}, \omega \in \mathscr{O}, t \in \mathscr{T} \tag{12.32}$$

$$\text{soc}_{i,\omega,t}^{\text{ST}} = \text{soc}_{i,\omega,t-1}^{\text{ST}} + \frac{n_i^{\text{ST}} p_{i,\omega,t}^{\text{ST}+}\frac{\Delta t}{3600}}{E_i^{\text{ST}}} - \frac{p_{i,\omega,t}^{\text{ST}-}\frac{\Delta t}{3600}}{n_i^{\text{ST}}E_i^{\text{ST}}}, \quad \forall i \in \mathscr{B}, \omega \in \mathscr{O}, t \in \mathscr{T} \tag{12.33}$$

$$\underline{\text{SOC}_i^{\text{ST}}} \leq \text{soc}_{i,\omega,t}^{\text{ST}} \leq \overline{\text{SOC}_i^{\text{ST}}}, \quad \forall i \in \mathscr{B}, \omega \in \mathscr{O}, t \in \mathscr{T} \tag{12.34}$$

$$0 \leq p_{i,\omega,t}^{\text{ST}+} \leq \overline{p_i^{\text{ST}+}} u_{i,\omega,t}^{\text{ST}}, \quad \forall i \in \mathscr{B}, \omega \in \mathscr{O}, t \in \mathscr{T} \tag{12.35}$$

$$0 \leq p_{i,\omega,t}^{\text{ST}-} \leq n_i^{\text{ST}}\overline{p_i^{\text{ST}-}}\left(1 - u_{i,\omega,t}^{\text{ST}}\right), \quad \forall i \in \mathscr{B}, \omega \in \mathscr{O}, t \in \mathscr{T} \tag{12.36}$$

$$\underline{q_i^{\text{ST}}} \leq q_{i,\omega,t}^{\text{ST}} \leq \overline{q_i^{\text{ST}}}, \quad \forall i \in \mathscr{B}, \omega \in \mathscr{O}, t \in \mathscr{T} \tag{12.37}$$

$$\mathrm{soc}^{\mathrm{ST}}_{i,\omega,t_{end}} \geq \mathrm{soc}_{\mathrm{thres}}, \quad \forall i \in \mathcal{B}, \omega \in \mathcal{O}, t \in \mathcal{T} \tag{12.38}$$

$$p^{\mathrm{UP}}_{\omega,t} = p^{\mathrm{UP_B}}_{\omega,t} - p^{\mathrm{UP_S}}_{\omega,t}, \quad \forall \omega \in \mathcal{O}, t \in \mathcal{T} \tag{12.39}$$

$$0 \leq p^{\mathrm{UP_B}}_{\omega,t} \leq \overline{P^{\mathrm{UP_B}}} u^{\mathrm{UP}}_{\omega,t}, \quad \forall \omega \in \mathcal{O}, t \in \mathcal{T} \tag{12.40}$$

$$0 \leq p^{\mathrm{UP_S}}_{\omega,t} \leq \overline{P^{\mathrm{UP_S}}} \left(1 - u^{\mathrm{UP}}_{\omega,t}\right), \quad \forall \omega \in \mathcal{O}, t \in \mathcal{T} \tag{12.41}$$

$$p^{D}_{i,\omega,t} = P^{D}_{i,\omega,t} - p^{\mathrm{shed}}_{i,\omega,t}, \quad \forall i \in \mathcal{B}, \omega \in \mathcal{O}, t \in \mathcal{T} \tag{12.42}$$

$$q^{D}_{i,\omega,t} = Q^{D}_{i,\omega,t} - q^{\mathrm{shed}}_{i,\omega,t}, \quad \forall i \in \mathcal{B}, \omega \in \mathcal{O}, t \in \mathcal{T} \tag{12.43}$$

and (12.5)–(12.10).

The objective function (12.13) aims to minimize the expected social cost, expressed as a minimum load shedding during an emergency situation, such as an approaching wildfire, for the next hours, in the most efficient way and respecting the operating limits of the system. Although resilience is directly associated with minimum load shedding [5], the method also considers operating costs in order to provide the most economical solution. In (12.13), the first term represents the load shedding cost minus the retailers' revenue for selling energy to customers. The second term represents the generation cost of MTs units. The third term describes the cost of power exchange with the upstream system. The fourth term represents start-up and shut-down costs of MTs units. The MTs are controlled according to the proposed scheduling and therefore the MTs' status is the same in each scenario. Thus, the start-up and shut-down costs of MTs are not associated with the scenarios. The value of the lost load is set large enough, in order to prioritize the demand satisfaction. It is noted that the revenue from selling energy to customers and the cost of power exchange with the upstream system are attributed to retailers. The DSO does not benefit from system operation and cannot make revenues. In practice and in certain jurisdictions, the DSO may be subject to penalties for customer interruptions, customer minutes lost, etc., and this can be used as a proxy for VoLL and used in the above cost–benefit analysis approach.

Constraints (12.14) and (12.15) guarantee the active and reactive power balance at each bus, including the power exchange at the point of common coupling. Constraints (12.16)–(12.19) represent the active and reactive output limits of MTs and their start-up and shut-down costs, respectively. Constraint (12.20) sets the conductor out of order when its temperature exceeds the maximum permissible temperature $T^{\max}$ (if $u^{l} = 0$, the conductor is set out of order). In this case, it is assumed that the wildfire is very close, and therefore the conductor is outaged by the fire. Constraints (12.21)–(12.28) represent power flow equations. Constraints (12.21) and (12.22) allow the power to flow through the line only when the line is functional. The selection of M_2 should allow the maximum flow of the line. Equations (12.23) and (12.24) are converted to equalities when the line is functional and relate voltage magnitude and angle to power flow. The linearization of power flow equations proposed in [20] is used, where $\zeta_{ij,\omega,t}$ represents the piecewise linear approximation of $\cos(\vartheta_i - \vartheta_j)$. More details about the linearization of power flow equations and their accuracy can be found in [27, 28]. Constraints (12.25)–(12.28) describe the proposed piecewise linearization approximation. The number of pieces is selected as a trade-off between computation time and accuracy of the linearized power flow equations. M_1 and M_3 are activated when the lines are outaged and are suitably selected in order to allow the conductor to reach high temperature due to the fire and the calculation of voltages and angles of the buses at the ends of lines set out of order. Constraint (12.29) guarantees that if a line is out of service due to the fire, it will remain offline for the rest of the time horizon. Constraint (12.30) guarantees that bus voltages are within limits. Constraints (12.31) and (12.32) represent the load shedding bounds considering constant power factor loads. Equations (12.33)–(12.38) describe the operation and limits of ESSs. The state of charge (SOC) of ESSs is calculated by (12.33). Equation (12.34) guarantees that the SOC of ESSs is within the permissible limits. Equations (12.35) and (12.36) represent the active power limits of ESSs depending on their operation mode. Constraint (12.37) represents the reactive output limits of ESSs. Constraint (12.38) is used in order to maintain the SOC of ESSs above a prespecified threshold soc_{thres} at the end of the time horizon t_{end}. Constraints (12.39)–(12.41) model the power exchange with the upstream system. Constraint (12.39) represents the amount of exchanged power with the upstream system. Constraints (12.40) and (12.41) describe the power exchange limits depending on the mode of the distribution system (exporting or importing electricity). Constraints (12.42) and (12.43) express the actually served load, which is equal to the scenario load demand minus the load shed. The problem expressed in (12.13) is highly non-linear and non-convex due to (12.6), (12.9), and (12.10).

12.3.4.2 Convexification of Non-convex Terms

This section presents the procedure for transforming the problem into a mixed integer problem with quadratic constraints. To convexify the non-convex terms (12.6), (12.9), and (12.10), they are transformed according to [29]. The heat gain rate due to ohmic losses, given in (12.6), is equal to the product of the square of the current flow and its resistance. The resistance of the conductor is given by:

$$R_{ij,\omega,t}(T_{ij,\omega,t}) = R_{ij,ref}(1 + d_{ij}(T_{ij,\omega,t} - T_{ij,ref})), \quad \forall (i,j) \in \mathscr{L}, \omega \in \mathcal{O}, t \in \mathscr{T} \tag{12.44}$$

where $R_{ij,\mathrm{ref}}$ is the conductor resistance at the reference temperature $T_{ij,\mathrm{ref}}$ and d_{ij} is the conductor thermal resistivity coefficient. Considering that the resistance of the conductor is constant and equal to the one obtained for its maximum permissible temperature $T_{ij}^{\max}$ and the voltage is close to 1 p.u., (12.6) can be transformed to an inequality given by:

$$q^{l}_{ij,\omega,t} \geq R_{ij,\omega,t}\left(T_{ij}^{\max}\right)\left(\left|pf^{P}_{ij,\omega,t}\right|^2 + \left|pf^{Q}_{ij,\omega,t}\right|^2\right), \quad \forall (i,j) \in \mathscr{L}, \omega \in \mathcal{O}, t \in \mathscr{T} \tag{12.45}$$

The heat loss rate, given in (12.10), is linearized and is given by:

$$q^{r}_{ij,\omega,t} = \lambda\, T_{ij,\omega,t} + \beta, \quad \forall (i,j) \in \mathscr{L}, \omega \in \mathcal{O}, t \in \mathscr{T} \tag{12.46}$$

The convection heat loss rate in (12.9) can be expressed as a function of the difference between the conductor temperature and the ambient temperature, multiplied by a slope. The maximum calculated slope in (12.9) is used as input to the optimization problem. In this way, the optimization problem is transformed into a mixed integer problem with quadratic constraints and can be solved using a commercial solver.

12.3.4.3 Overview of the Proposed Method

As a first step, MCS are used to produce a large number of scenarios, that contain a time series of the stochastic parameters. The typical distributions of stochastic parameters are

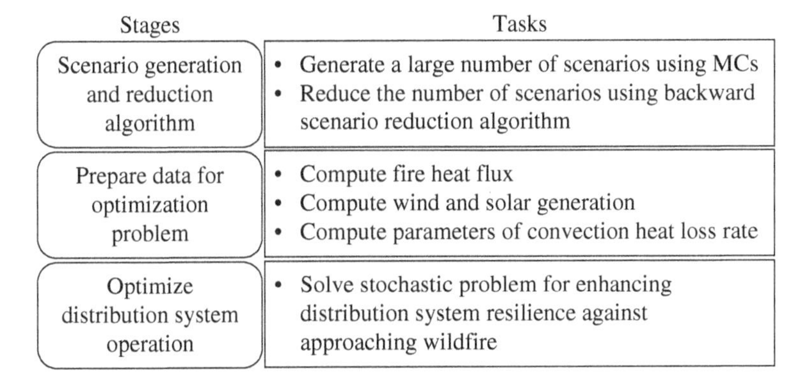

Stages
Tasks
Scenario generation and reduction algorithm
• Generate a large number of scenarios using MCs
• Reduce the number of scenarios using backward scenario reduction algorithm
Prepare data for optimization problem
• Compute fire heat flux
• Compute wind and solar generation
• Compute parameters of convection heat loss rate
Optimize distribution system operation
• Solve stochastic problem for enhancing distribution system resilience against approaching wildfire

Figure 12.4 A framework of the proposed approach.

used for each scenario. The Backward Scenario Reduction algorithm is then applied to reduce the number of scenarios and subsequently the computation efforts. In the next step, using the generated scenarios, the radiative heat flux emitted from the fire, the generation of stochastic generators, and the parameters of convection heat loss rate are computed in order to be used as inputs to the optimization problem. Finally, the optimization problem for the resilient operation against approaching wildfire is solved. The overall proposed methodology is illustrated in Figure 12.4.

12.4 Case Study Application

12.4.1 Test Network and Simulation Data

This section presents the results of the proposed method applied to a modified IEEE 33-bus distribution system [30]. The system is assumed balanced and is represented by an equivalent single-phase circuit. The single-line diagram of the modified network and the progression of the wildfire are presented in Figure 12.5. The peak active demand is equal to 15 MW, constant power factor loads are considered and the load profile for the summer season is adopted from [31]. The distribution system extends to a small geographical region, therefore its components are exposed to similar weather conditions. The location and capacity of the system components are shown in Table 12.3. The data of MTs and ESSs are shown in Tables 12.4 and 12.5, respectively. For WTs, the cut-in, cut-out, and rated wind speeds are 4, 20, and 12 m/s, respectively. The rated illumination intensity of the PVs is 1000 W/m^2. The standard deviation is considered equal to 15% of the mean value for wind speed and solar radiation and 5% of the mean value for the loads. A k-factor of 2 is assumed for the von Mises distribution. The weather parameters and wildfire data are taken from [9]. A time step of 30 minutes is selected, which according to [29] is a tolerable time step for DLR modeling. The wind and solar data are obtained from [32, 33], respectively.

A summer day with high wind speed and ambient temperature that facilitate the development and spread of a wildfire is considered. It is assumed that the fire breaks out at $t = 1$, approaches and affects only the line between buses 1 and 2 (dashed line in Figure 12.5). The proposed method can be applied at any time during the day at which the threat of the wildfire is perceived. Therefore, in light of a wildfire threat, the scheduled operation changes to enhance distribution system resilience. The initial distance from the line is $r^f_{t=1}$=1200 m.

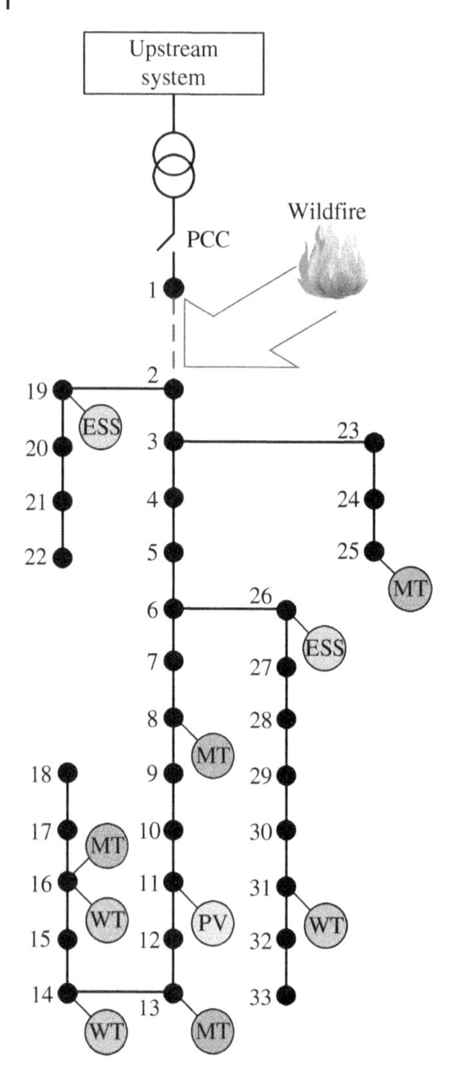

Figure 12.5 Single-line diagram of the simulated distribution system and wildfire progression, PCC = point of common coupling.

The rest of the lines are considered at a safe distance from the fire. The selected line is the one that connects the distribution system with the upstream network and therefore the power exchange between the two systems is affected by the fire. The proposed method is however general and it can be applied for the study of any line or lines approached by the fire. Assuming that the spatial data of the distribution system are known, for example, by a Geographic Information System (GIS), the distance between the fire and the conductors can be estimated, during the progression of the wildfire. Based on the estimated distance, the impact of the wildfire on the temperature of any line can be computed.

It is considered that at the end of the time horizon, the wildfire is extinguished, but the affected line remains out of service until it is repaired. Thus, the SOC of the ESSs is required to remain greater than 30% of the maximum capacity, in order to contribute to demand satisfaction for the next hours after the study. An aluminum conductor steel-reinforced

Table 12.3 Location and capacity of distribution system components.

Type	Bus	Capacity (MW)
WTs	14/16/31	0.8/0.8/0.8
PV	11	0.5
MTs	8/13/16/25	3/2/2/3
ESSs	19/26	0.5/0.5

Table 12.4 Parameters of microturbines.

Microturbine	$\overline{p^G}$ (MW)	$\underline{p^G}$ (MW)	$\overline{q^G}$ (MVAr)	$\underline{q^G}$ (MVAr)	c^G (\$/MWh)
MT1 (bus 8)	3	0.21	2.1	−2.1	90
MT2 (bus 13)	2	0.19	1.9	−1.9	90
MT3 (bus 16)	2	0.19	1.9	−1.9	90
MT4 (bus 25)	3	0.22	2.2	−2.2	90

Table 12.5 Parameters of energy storage systems.

Energy storage system	E^{ST} (MWh)	$\overline{p^{ST+/-}}$ (MW)	$\overline{q^{ST}}$ (MVAr)	$\underline{q^{ST}}$ (MVAr)
ESS (bus 19)	1.5	0.5	0.3	−0.3
ESS (bus 26)	1.5	0.5	0.3	−0.3

(ACSR)-type conductor is considered. The diameter and the maximum permissible temperature of the line are assumed equal to 21 mm and 353 K, respectively.

Twenty pieces are used for the piecewise linear approximation of $\cos(\vartheta_i - \vartheta_j)$ in the linearized power flow equations. One thousand scenarios have been generated using MCS. The appropriate number of representative scenarios has been selected taking into account the range of optimization results (their impact on the objective function) and the required computation time. In this application, 10 representative scenarios are the best choice, since a higher number has an insignificant effect on the results and the computation time becomes unacceptably high for the online application of the method.

The mean value of active demand and solar radiation are presented in Figure 12.6. Figure 12.7 presents the mean values of wind speed and wind direction. The angle is measured with respect to the axis defined by the fire and the conductor. Wind speed $V^{w,\mathrm{WT}}$ in Figure 12.7 refers to the height L^{WT} of WTs. The wind speed at L^l is given by [34]:

$$V^{w,l} = V^{w,\mathrm{WT}} \left(\frac{L^l}{L^{\mathrm{WT}}} \right)^{0.143} \tag{12.47}$$

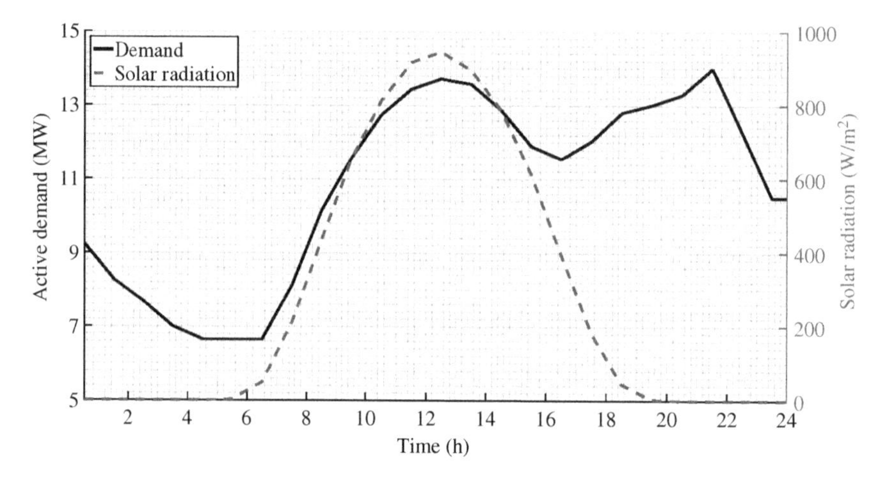

Figure 12.6 Mean value of active demand and solar radiation.

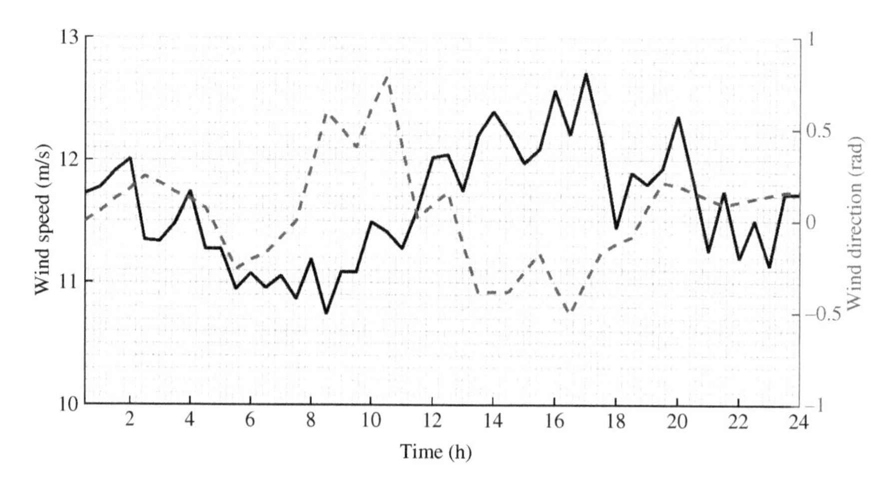

Figure 12.7 Mean value of wind speed at L^{WT} and wind direction.

The ambient temperature and the price of energy exchange with the upstream system are presented in Figure 12.8. The price of buying and selling energy from and to the upstream system is considered equal and is adopted from [35].

The proposed model was solved using the IBM CPLEX solver. A PC with Intel Core i7 CPU @3.40 GHz and 4 GB RAM was used. The computation time was 973.074 seconds.

12.4.2 Operation Against Approaching Wildfires

The proposed method is applied to optimize the distribution system operation for enhancing its resilience against the approaching wildfire for the next 24 hours. Figure 12.9 shows the distance between the fire and lines 1–2. It is observed that the fire approaches the line during the day reaching zero distance in the last hours. Figure 12.10 presents the conductor

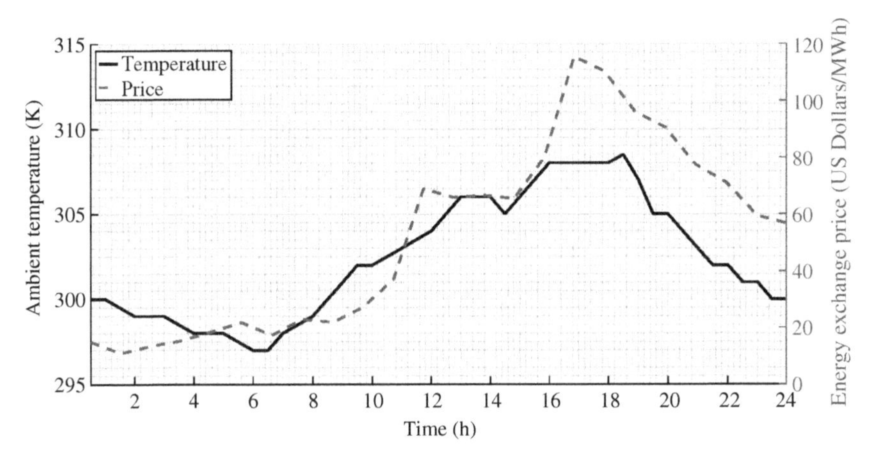

Figure 12.8 Ambient temperature and power exchange price.

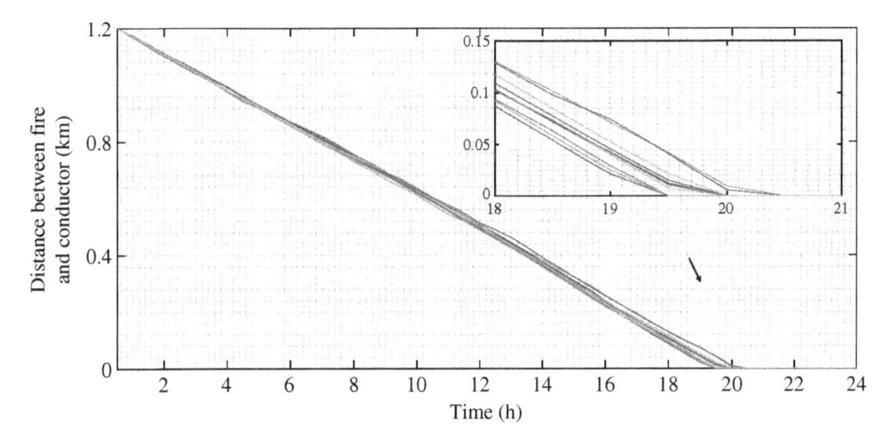

Figure 12.9 Distance between fire and lines 1–2 in the reduced number of scenarios and magnified sub-figure.

heat gain rate from the fire. By comparing Figures 12.9 and 12.10, it is observed that the conductor heat gain rate from the fire gets values greater than zero when the distance of the fire from the line is less than 300 m and increases exponentially, as the fire approaches the line. Figures 12.9 and 12.10 also show the added value of the stochastic analysis. In Figure 12.9, for example, it is observed that the wildfire reaches a close distance from the line (less than 15 m), in different time steps for each examined scenario. As a result, the application of a deterministic optimization based on mean forecasted values, would not consider the line outage at an earlier, yet probable time step.

Two cases are considered in order to evaluate the effectiveness of the proposed method. In Case I, the distribution system operation is optimized to minimize social costs without considering the wildfire progression. In Case II, the wildfire progression is considered and the resilient operation of the distribution system is optimized.

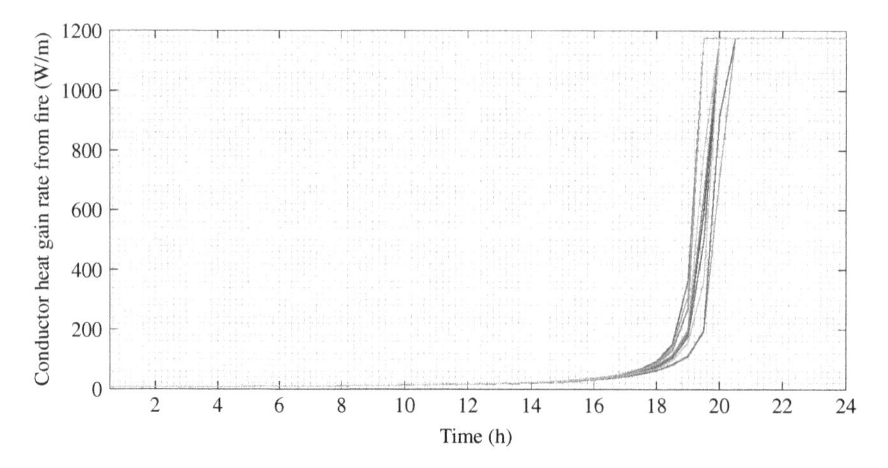

Figure 12.10 Conductor 1–2 heat gain rate from the fire in the reduced number of scenarios.

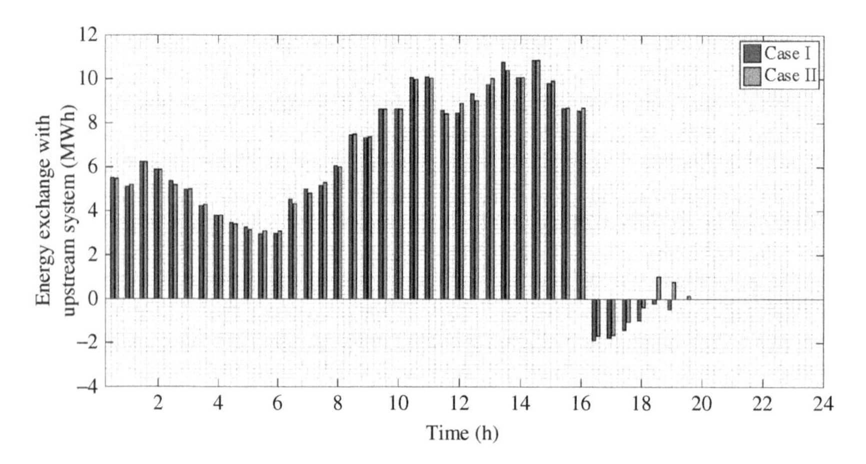

Figure 12.11 Expected energy exchange with the upstream system for Cases I and II.

The expected energy exchange with the upstream system for both cases is presented in Figure 12.11. It is observed that until 16:00 hour, the energy exchange is almost identical for both cases and energy is bought from the upstream system since the price of power bought is lower than the cost of the local MTs and the power from renewable generators does not cover the demand. After 16:00 hour and until 18:00, the power exchange price has high values and therefore energy is sold to the upstream system. For Case I, the exchanged energy is higher because as shown in Figure 12.12 the ESS injects active power equal to its higher capacity in order to maximize the revenues from selling energy to the upstream system. For Case II, as shown in Figure 12.13, the discharging power of ESSs is lower since it is desired to maintain the SOC at the necessary level in order for the discharging power of ESSs to be used for minimizing load shedding when the connection with the upstream system will be disrupted due to the wildfire. After 18:00 hour and until 19:30, the power exchange price is lower but still higher than the cost of the local MTs. For Case I, the operator continues to

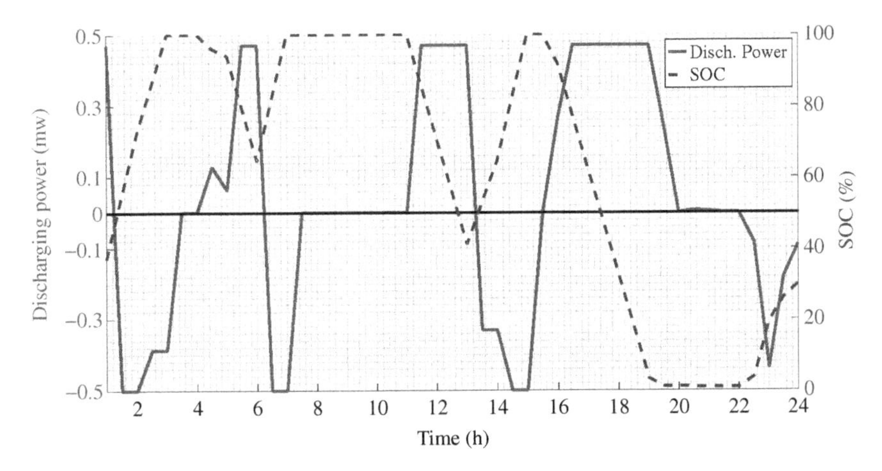

Figure 12.12 Expected discharging power and SOC of ESS at bus 19 for Case I.

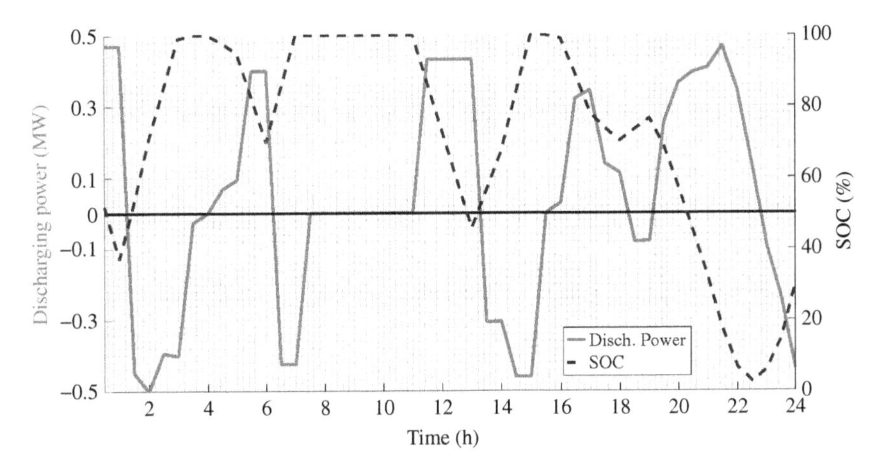

Figure 12.13 Expected discharging power and SOC of ESS at bus 19 for Case II.

sell energy to the upstream, while for Case II energy is bought for charging the ESSs before lines 1–2 is outaged due to the wildfire.

After this time, the power exchange with the upstream system drops to zero, since line 1–2 is set out of order due to the violation of its maximum temperature. As a result of the strategy that was followed in Case I, the SOC of the ESS at bus 19 is at its lowest level. Meanwhile, for Case II the ESSs are charged to the desired levels in order to serve the demand during the time the distribution system is disconnected from the upstream system and minimize the load shedding.

Figure 12.14 presents the expected amount of load that is shed to keep the balance between generation and demand, when line 1–2 is on outage. Until 19:00 hour, the load is not shed in both cases. Between 19:30 and 22:30, an amount of load is shed due to the disconnection of line 1–2 and the inability of the DGs and ESSs to meet the demand. For Case II, the amount of load shedding is much lower due to the proper operation of the ESSs before the disconnection of line 1–2. After 23:00 hour, the demand is met by the DGs,

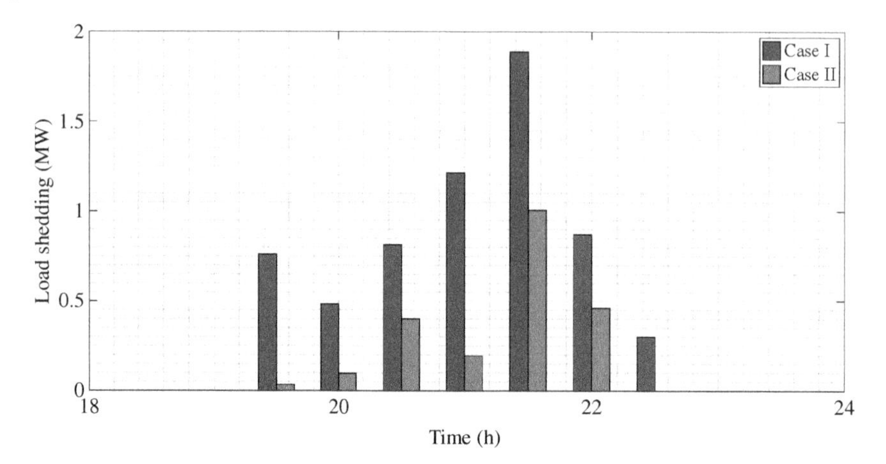

Figure 12.14 Expected load shedding for cases I and II.

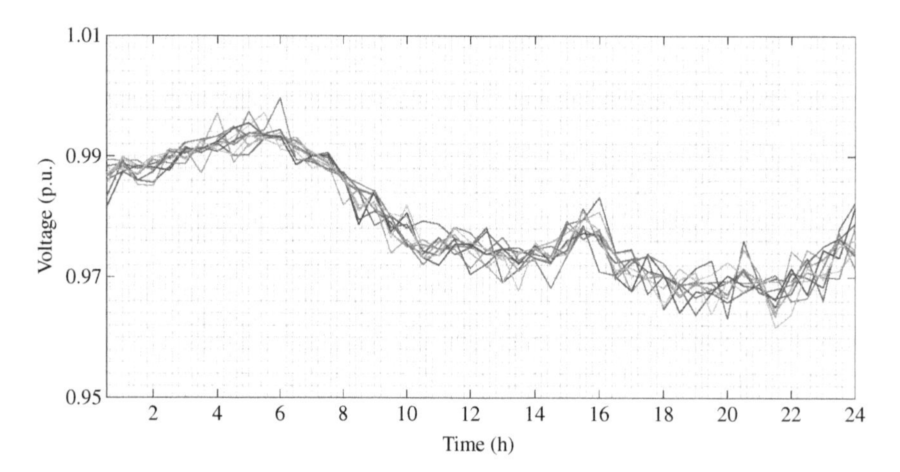

Figure 12.15 Bus 33 voltage in the reduced number of scenarios for case II.

and load shedding is not recorded for both cases. Moreover, the ESSs are charged to reach the required levels.

Figure 12.15 shows the voltage at bus 33 for Case II (minimum recorded voltage). The minimum voltage is recorded during the disconnection of line 1–2. It is observed that the reactive power injected by the MTs and ESSs is able to keep the voltage within permissible limits ($\pm5\%$) during the isolation of the distribution system. After the 22:00 hour, the voltage increases, as the load decreases (see Figure 12.6).

Table 12.6 presents the load shedding cost, the revenue from selling energy to customers, the cost of power exchange with the upstream system, and the generation cost for both cases. Furthermore, Case III is considered, where the wildfire progression is taken into account, but only load shedding is minimized without consideration of other costs (Eq. (12.13)). The value of lost load VoLL is set at 1000 USD/MWh [12, 36], in order to prioritize demand satisfaction. Load shedding cost for Cases II and III is identical since for

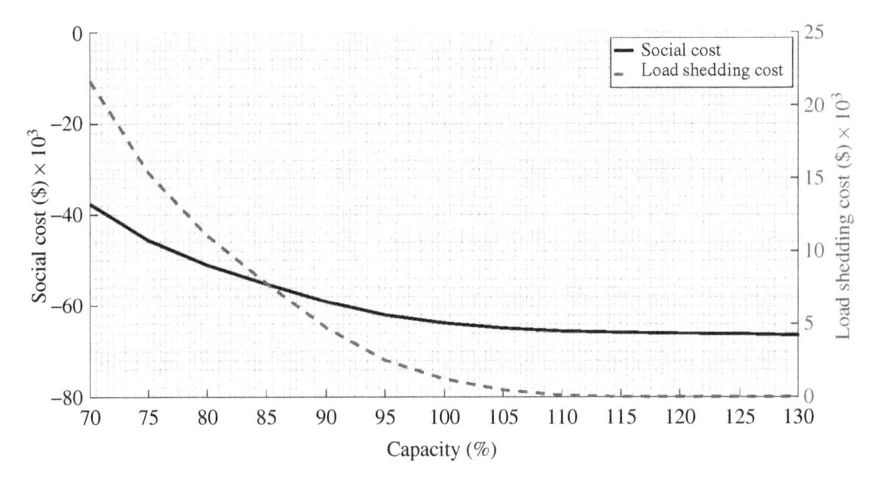

Figure 12.16 Objective function value (social cost) and load shedding cost considering the variation of installed power ratings.

Table 12.6 Revenues and costs for cases I, II, and III ($\$ \times 10^3$).

Cases	Load shedding cost	Generation cost	Power exchange cost	Revenue from selling energy to customers
I	3.16	8.24	4.29	77.16
II	1.23	8.29	4.43	77.75
III	1.23	17.82	0.71	77.75

both cases the priority is to minimize load shedding, while for Case I load shedding cost is higher, as expected. The higher load shedding in Case I results in lower revenue from selling energy to customers. For Case II, the power exchange cost and generation cost are slightly higher than Case I due to the energy used to charge the ESSs before disconnection of line 1–2. Generation cost for Case III is much higher due to the formulation of the objective function.

12.4.3 Effects of Capacity and Location of DGs and ESSs

The rated powers of the system components in Table 12.3 are varied from 70% to 130% in steps of 10% compared to Case II. The value of the objective function (social cost) and the load shedding cost are presented in Figure 12.16. As expected, the value of the objective function and the load shedding cost decrease as the installed capacities increase. When the overall capacity exceeds 110%, load shedding cost reaches zero and the value of the objective function decreases at a much lower rate. Further decrease in the social cost is mainly due to the increase of the energy provided to the upstream system before the outage of line 1–2.

The location of the distribution system components, in combination with the affected lines by the wildfire, has a high influence on load shedding. In particular, the presence of DGs and ESSs in an isolated part of the distribution system after the wildfire is critical for maintaining demand supply and minimizing load shedding. For example, in case the wildfire affects the line between busses 19 and 20, the demand for busses 20, 21, and 22 will not be satisfied following the outage of line 19–20, since there are no DGs and ESSs in the isolated network.

In order to show the dependence of spatial load shedding on the affected lines by the wildfire and the location of DGs and ESSs, an additional case is considered. In Case IV, line 6–26 (and not line 1–2) is considered as the affected line by the fire. The same progression of the wildfire, as in Case II, is considered. Figure 12.9 corresponds to the distance between the fire and line 6–26. In this case, the isolated part of the distribution system comprises an ESS on bus 26 and a WT on bus 31.

The expected spatial load shedding for Cases II and IV are presented in Figures 12.17 and 12.18, respectively. The only difference between the two cases is the affected line. Given that, in Case II, the affected line 1–2 isolates the distribution system from the upstream system, the location of DGs and ESSs will not have a critical influence on spatial load shedding, unless network lines are congested. In case IV, load shedding is recorded at the isolated buses. At the rest of the system that is connected with the upstream system, the load demand is met by the DGs, the ESS at bus 19, and the power imported from the upstream system. At the isolated part, the ESS at bus 26 and the WT at bus 31 are not able to satisfy the total demand and therefore load has to be shed. If an MT, similar to the one connected at bus 8, was connected at bus 28, no shedding would be recorded in the isolation part. Therefore, it

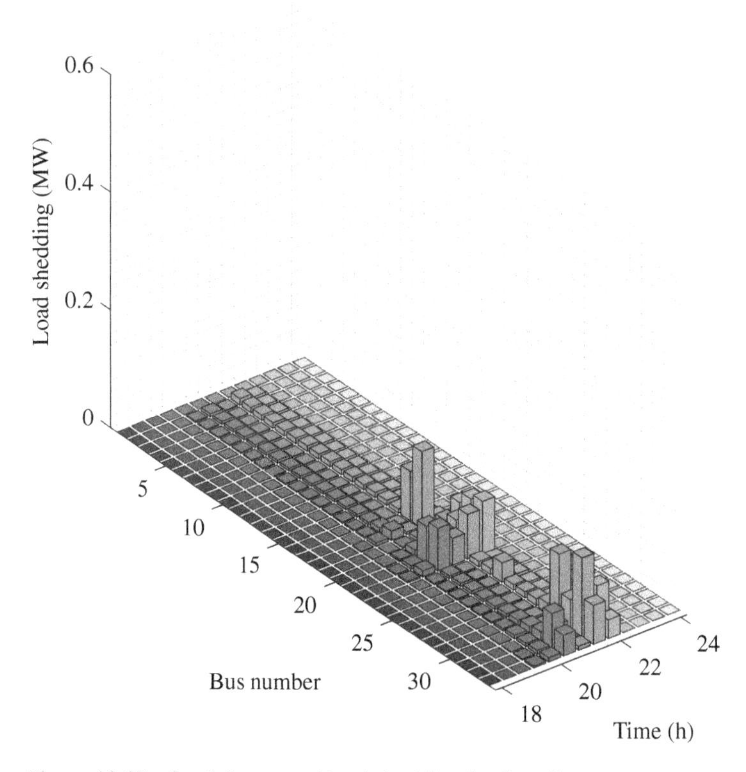

Figure 12.17 Spatial expected load shedding for Case II.

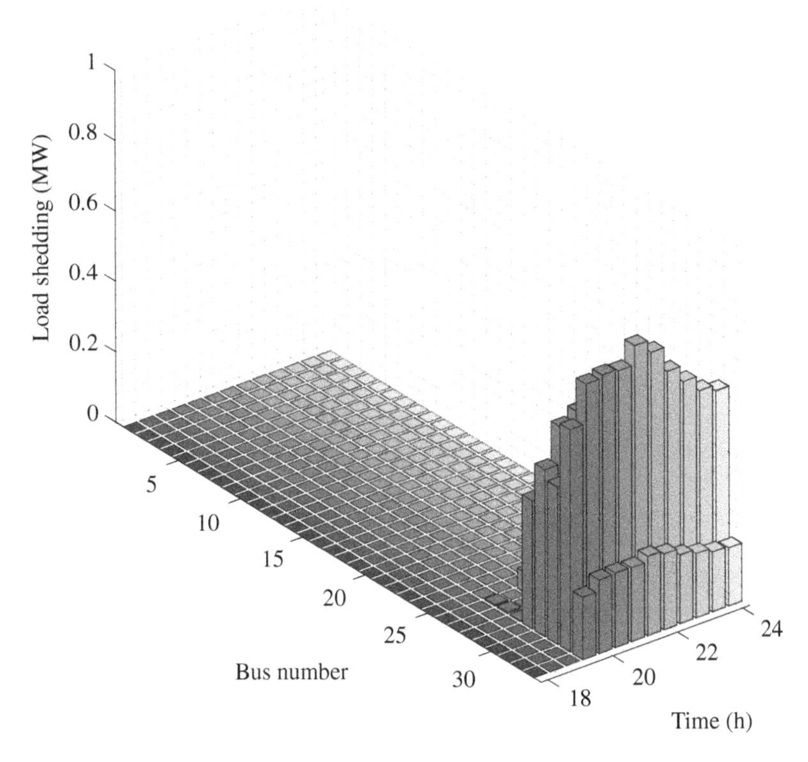

Figure 12.18 Spatial expected load shedding for Case IV.

is recognized that the spatial load shedding depends critically on which lines are affected by the wildfire.

12.5 Summary

This chapter has presented a framework for assessing the resilience of a distribution system exposed to emergency situations, where the exposure of a distribution system to a wildfire is considered as an illustrative case study application. A stochastic mixed integer programming with quadratic constraints is presented to enhance the resilience of the distribution system by minimizing load shedding. The progress of the wildfire based on wind speed and direction is modeled and a DLR model of the overhead lines is integrated into the framework in order to take into account the effect of the wildfire on the distribution system operation. The radiative heat flux emitted from the wildfire affects the lines in close proximity by reducing their capacity or by disrupting their operation due to the high rise of their temperature. The approach has been illustrated using a modified IEEE 33-bus distribution system. The results show that the resilience of a distribution system against an approaching wildfire can be enhanced by considering the progression of the wildfire during the operation of a distribution system. This is essential due to the dependence of the spatial load shedding on the combination of the affected lines by the wildfire and the location of distributed system components, mainly the DGs and ESSs. Although the operating cost is higher when considering the resilience operation of the system, the potential load shedding cost decreases significantly. This is achieved mainly due to the proper operation of the ESSs, which are charged to the desired levels in order to serve the demand during the time the distribution system is disconnected from the upstream system. These results also highlight that there may be a need to reconsider current operation and planning approaches to also incorporate resilience in systems analysis.

Nomenclature

Indices and Sets

i, j	indices for buses (1 to N_{B}).
(i,j)	distribution line between buses i and j.
t	index for time periods (1 to N_{T}).
ω	index for scenarios (1 to N_{Ω}).
y	index of linearization model of $\cos(\vartheta_i - \vartheta_j)$ (1 to N_{Y}).
$\overline{}, \underline{}$	symbols for upper and lower bounds.
$\mathscr{L}$	Set of distribution system lines.
$\mathscr{B}$	Set of buses.
$\mathscr{O}$	Set of examined scenarios.
$\mathscr{T}$	Set of time intervals for the examined horizon.
$\mathscr{Y}$	Set of pieces used for linearization model of $\cos(\vartheta_i - \vartheta_j)$.

Parameters

Flame

L^f	flame length (m).
γ^f	flame tilt angle (rad).
T^f	flame zone temperature (K).
ε^f	flame zone emissivity.
ρ^{b}	fuel bulk density (kg/m^3).

Weather and Atmospheric Conditions

V^{w}	wind speed (m/s).
φ^{w}	angle between the wind direction and the conductor axis (rad).
Φ^{s}	solar radiation (W/m^2).
T^{a}	ambient temperature (K).
τ	dimensionless atmospheric transmissivity.
k^{a}	thermal conductivity of air (W/(m $\cdot$ K)).

ρ^{a} density of air (kg/m^3).
μ^{a} absolute (dynamic) viscosity of air ($kg/(m \cdot s)$).

Conductor-Line

mC^{p} total heat capacity of conductor ($J/(m \cdot K)$).
K solar absorptivity.
D conductor diameter (m).
ε conductor emissivity.
L^{l} vertical position of the line (m).
g, b conductance and suspectance of line (p.u.).
T^{max} conductor maximum permissible temperature (K).

Prices, Costs, and Values

c^{UP} price for buying/selling energy from/to the upstream system ($/MWh).
c^{D} price for selling energy to consumers ($/MWh).
c^{G} generation cost of controllable generators ($/MW).
c^{SU}/c^{SD} start-up/shut-down cost of controllable generators ($).
VoLL value of lost load ($/MWh).

Energy Storage System

n^{ST} conversion efficiency of energy storage systems.
E^{ST} energy capacity of energy storage systems (MWh).

Coefficients, Constants, and Bounds

λ, β coefficients used in the linearization of radiated heat loss rate.
κ, ξ coefficients used in the piecewise linearization of $\cos(\vartheta_i - \vartheta_j)$.
B Stefan–Boltzmann constant ($W/(m^2 \cdot K^4)$).
$\Delta\vartheta^{y}_{min/max}$ lower/upper bounds of segments for piecewise linearization of $\cos(\vartheta_i - \vartheta_j)$.

Others

π probability of scenario.
Δt duration of time intervals (s).
$M_{1,2,3}$ sufficiently big positive numbers.

Variables

Flame

V^{f} fire rate of spread (m/s).
r^{f} distance of fire from the line (m).

Φ^f	radiative heat flux emitted from fire (W/m^2).
δ^f	view angle between the flame and the conductor (rad).

Conductor-Line

T	conductor temperature (K).
$pf^{P/Q}$	line active/reactive power flow (p.u.).

Heat Gain and Loss Rates

q^l	resistive heat gain rate (W/m).
q^s	heat gain rate from sun (W/m).
q^f	heat gain rate from fire (W/m).
q^c	convection heat loss rate (W/m).
q^r	radiated heat loss rate (W/m).

Upstream System, Distributed Generation, Energy Storage Systems, and Loads

p^{UP}	active power to exchange with the upstream system (p.u.).
$p^{UP_{B/S}}$	active power to buy/sell from/to the upstream system (p.u.).
$p^{WT/PV/G}$	active power generation of wind/solar/controllable generators (p.u.).
$p^{ST+/-}$	charging/discharging active power for energy storage systems (p.u.).
P^D, Q^D	active and reactive power for demand (p.u.).
p^D/q^D	served active/reactive power (p.u.).
p^{shed}/q^{shed}	active/reactive power shedding (p.u.).
q^G	reactive output of controllable generators (p.u.).
q^{ST}	reactive output of energy storage systems (p.u.).
su^G, sd^G	controllable generator start-up and shut-down cost.
soc^{ST}	state of charge of energy storage systems.

Binary Variables

u^l	binary variable for determining the status of line.
u_{ij}^{ζ}	binary variable for linear approximation of $\cos(\vartheta_i - \vartheta_j)$.
u^{UP}	binary variable for determining the buying or selling energy from/to the upstream system.
u^{ST}	binary variable for determining the charging or discharging status of energy storage systems.
u^G	binary variable for determining the status of controllable units.

Bus

v	bus voltage magnitude (p.u.).
ϑ	bus voltage angle (rad).

Others

K^{angle} wind direction factor.
N^{Re} dimensionless Reynolds number.
ζ_{ij} piecewise linear approximation of $\cos(\vartheta_i - \vartheta_j)$.

For the calculations, the base values of power and voltage are used.

References

1 Panteli, M. and Mancarella, P. (2015). Influence of extreme weather and climate change on the resilience of power systems: impacts and possible mitigation strategies. *Electric Power Systems Research* 127: 259–270.

2 Wang, Y., Chen, C., Wang, J., and Baldick, R. (2016). Research on resilience of power systems under natural disasters – a review. *IEEE Transactions on Power Systems* 31 (2): 1604–1613.

3 Panteli, M., Pickering, C., Wilkinson, S. et al. (2016). Power system resilience to extreme weather: fragility modelling, probabilistic impact assessment, and adaptation measures. *IEEE Transactions on Power Systems*, Early Access 32 (5): 3747–3757.

4 Panteli, M., Mancarella, P., Trakas, D. et al. (2017). Metrics and quantification of operational and infrastructure resilience in power systems. *IEEE Transactions on Power Systems*, Early Access 32 (6): 4732–4742.

5 Panteli, M., Trakas, D.N., Mancarella, P., and Hatziargyriou, N.D. (2017). Power systems resilience assessment: hardening and smart operational enhancement strategies. *Proceedings of the IEEE* 105 (7): 1202–1213.

6 Panteli, M. and Mancarella, P. (2015). The grid: stronger, bigger, smarter?: presenting a conceptual framework of power system resilience. *IEEE Power and Energy Magazine* 13 (3): 58–66.

7 Kezunovic, M., Dobson, I., and Dong, Y. (2008). Impact of extreme weather on power system blackouts and forced outages: new challenges. In: *Balkan Power Conference*, Šibenik Croatia (September 2008).

8 (2017). Fire Department of Greece. https://www.fireservice.gr/el_GR/synola-dedomenon.

9 Choobineh, M., Ansari, B., and Mohagheghi, S. (2015). Vulnerability assessment of the power grid against progressing wildfires. *Fire Safety Journal* 73: 20–28.

10 Venizelos, E.C., Trakas, D.N., and Hatziargyriou, N.D. (2017). Distribution system resilience enhancement under disastrous conditions by splitting into self-sufficient microgrids. In: *2017 IEEE Manchester PowerTech*, Manchester, pp. 1–6.

11 Bagchi, A., Sprintson, A., and Singh, C. (2009). Modeling the impact of fire spread on the electrical distribution network of a virtual city. In: *41st North American Power Symposium*, Starkville, MS, USA, pp. 1–6.

12 Mohagheghi, S. and Rebennack, S. (2015). Optimal resilient power grid operation during the course of a progressing wildfire. *International Journal of Electrical Power & Energy Systems* 73: 843–852.

13 Ansari, B. and Mohagheghi, S. (2015). Optimal energy dispatch of the power distribution network during the course of a progressing wildfire. *International Transactions on Electrical Energy Systems* 25: 3422–3438.

14 IEEE Standard for Calculating the Current-Temperature Relationship of Bare Overhead Conductors. (2013). In: *IEEE Std 738-2012 (Revision of IEEE Std 738-2006 - Incorporates* IEEE *Std 738-2012 Cor 1-2013)*, IEEE, New York, USA, pp. 1–72 (23 December 2013).

15 Trakas, D.N. and Hatziargyriou, N.D. (2018). Optimal distribution system operation for enhancing resilience against wildfires. *IEEE Transactions on Power Systems* 33 (2): 2260–2271.

16 European Distribution System Operators for Smart Grids. (2022). Coordination of transmission and distribution system operators: a key step for the Energy Union. https://www.edsoforsmartgrids.eu/wp-content/uploads/public/Coordination-of-transmission-and-distribution-system-operators-May-2015.pdf.

17 Nikos Hatziargyriou, Thierry Van Cutsem, Jovica Milanović, Pouyan Pourbeik, Costas Vournas et al. (2017). Contribution to bulk system control and stability by distributed energy resources connected at distribution network. *PES Tech. Rep. TR-22.*

18 Bartels, W., Ehlers, F., and Heidenreich, K. (2022). Generating plants connected to the medium-voltage network. Technical Guideline of BDEW. https://erranet.org/download/generating-plants-connected-to-medium-voltage-network/?wpdmdl=33264&refresh=636e6330e17d21668178736.

19 (2017). The distribution code of licensed distribution network operators of Great Britain. http://www.dcode.org.uk/.

20 Gholami, A., Shekari, T., Aminifar, F., and Shahidehpour, M. (2016). Microgrid scheduling with uncertainty: the quest for resilience. *IEEE Transactions on Smart Grid* 7 (6): 2849–2858.

21 Sideratos, G. and Hatziargyriou, N.D. (2012). Probabilistic wind power forecasting using radial basis function neural networks. *IEEE Transactions on Power Systems* 27 (4): 1788–1796.

22 Liu, Z., Wen, F., and Ledwich, G. (2011). Optimal siting and sizing of distributed generators in distribution systems considering uncertainties. *IEEE Transactions on Power Delivery* 26 (4): 2541–2551.

23 Arefifar, S.A., Mohamed, Y.A.R.I., and El-Fouly, T.H.M. (2012). Supply-adequacy-based optimal construction of microgrids in smart distribution systems. *IEEE Transactions on Smart Grid* 3 (3): 1491–1502.

24 Li, W. (2005). *Risk Assessment of Power Systems: Models, Methods, and Applications.* New York: Wiley.

25 Heitsch, H. and Römisch, W. (2003). Scenario reduction algorithms in stochastic programming. *Computational Optimization and Applications* 24 (2, 3): 187–206.

26 Rossi, J.L., Simeoni, A., Moretti, B., and Leroy-Cancellieri, V. (2011). An analytical model based on radiative heating for the determination of safety distances for wildland fires. *Fire Safety Journal* 46: 520–527.

27 Trodden, P.A., Bukhsh, W.A., Grothey, A., and McKinnon, K.I.M. (2014). Optimization-based islanding of power networks using piecewise linear AC power flow. *IEEE Transactions on Power Systems* 29 (3): 1212–1220.

28 Coffrin, C. and Hentenryck, P.V. (2014). A linear-programming approximation of AC power flows. *INFORMS Journal on Computing* 26 (4): 718–734.

29 Nick, M., Alizadeh-Mousavi, O., Cherkaoui, R., and Paolone, M. (2016). Security constrained unit commitment with dynamic thermal line rating. *IEEE Transactions on Power Systems* 31 (3): 2014–2025.

30 Wang, C. and Cheng, H.Z. (2008). Optimization of network configuration in large distribution systems using plant growth simulation algorithm. *IEEE Transactions on Power Systems* 23 (1): 119–126.

31 Sfikas, E.E., Katsigiannis, Y.A., and Georgilakis, P.S. (2015). Simultaneous capacity optimization of distributed generation and storage in medium voltage microgrids. *International Journal of Electrical Power & Energy Systems* 67: 101–113.

32 Papaioannou, D.I., Papadimitriou, C.N., and Dimeas, A.L. (2014). Optimization & sensitivity analysis of microgrids using HOMER software – a case study. In: *MedPower 2014*, Athens, pp. 1–7.

33 (2022)NREL, PVWatts Calculator. http://pvwatts.nrel.gov.

34 Greenwood, D.M., Ingram, G.L., and Taylor, P.C. (2017). Applying wind simulations for planning and operation of real-time thermal ratings. *IEEE Transactions on Smart Grid* 8 (2): 537–547.

35 Khodaei, A. (2014). Resiliency-oriented microgrid optimal scheduling. *IEEE Transactions on Smart Grid* 5 (4): 1584–1591.

36 Farzin, H., Fotuhi-Firuzabad, M., and Moeini-Aghtaie, M. (2017). Stochastic energy management of microgrids during unscheduled islanding period. *IEEE Transactions on Industrial Informatics* 13 (3): 1079–1087.

13

Super Microgrid in Inner Mongolia

Jian Xu, Siyang Liao, and Yuanzhang Sun

School of Electrical Engineering and Automation, Wuhan University, Wuhan, China

13.1 Definition and Significance of the Super Microgrid

Renewable energy, which is represented by wind and solar power, has been developing rapidly because it is clean and environmentally friendly. However, due to the stochastic fluctuation of the output of renewable energy, the development of renewable energy is facing many challenges [1]. In China, for example, renewable energy resources such as wind energy and solar energy are mainly distributed in "three northern areas" (North China, Northeast China, and Northwest China) which are far from load centers. There are insufficient local loads and power supply to quickly respond to the power fluctuation of renewable energy in these areas so the capacity to consume large-scale renewable energy is not enough. Meanwhile, the power grid framework in these "three northern areas" is weak, and these areas have a weak ability in trans-provincial or interregional power transmission. It is difficult to meet the power transmission requirements after the large-scale integration of renewable energy. In 2016, China's average utilization time of wind power equipments was only 1724 hours per year, representing an increase of 14 hours year-on-year [2]. In 2016, wind curtailment was even more serious in these "three northern areas" which are rich in wind resources. The average utilization time of wind power equipments in Jilin, Gansu, and Xinjiang provinces was only 1333, 1080, and 1290 hours, respectively. It indicates that there is a great amount of waste in clean renewable energy. At the same time, being affected by electricity price, environmental, and other factors, the high energy-consuming industrial capacity shows an obvious trend of moving westward. There is an example related to high energy-consuming loads. In 2012, the electrolytic aluminum output of two central provinces Henan and Shandong accounted for 18.5% and 10% of the total national output, respectively, and the output of the northwest province Xinjiang only accounted for 4.6%. In 2015, the electrolytic aluminum output of Xinjiang rose rapidly, accounting for 18.6% of the total national output of electrolytic aluminum, ranking second in China. For electrolytic aluminum, it is concentrated in areas where renewable energy resources are rich, such as Xinjiang, Gansu, and Inner Mongolia. This provides a good condition for high energy-consuming loads to locally consume renewable energy. In order to realize efficient utilization of large-scale renewable energy, a new

Resiliency of Power Distribution Systems, First Edition.
Edited by Anurag K. Srivastava, Chen-Ching Liu, and Sayonsom Chanda.

concept "Super Microgrid" has been created [3, 4]. This chapter describes the composition of the super microgrid and its role in consuming renewable energy.

13.1.1 Composition of the Super Microgrid

The super microgrid is a power system with a megawatt-level installed capacity and consists of thermal power units, renewable energy, and high energy-consuming loads. High energy-consuming loads refer to industrial loads consuming a lot of power, including electrolytic loads (such as electrolytic aluminum, electrolytic copper, and electrolytic zinc), electric arc loads (such as submerged arc furnaces and electric arc furnaces), and industrial rotary loads (such as steel rolling mills). Based on the large installed capacity of each industrial load, efficient use of renewable energy can be realized if the industrial load can locally consume renewable energy such as wind power and photovoltaic power. Due to the stochastic fluctuation of the output of renewable energy, the super microgrid usually contains a certain capacity of thermal power units. In general, the voltage level of the transmission line of the super microgrid is 110 and 220 kV. The length of the transmission line can be up to 40 km or above. Therefore, the super microgrid can be called an industrial power system integrated with wind power/photovoltaic power, thermal power units, and loads.

13.1.2 Typical Architecture of the Super Microgrid

Tongliao, Inner Mongolia, China has rich wind energy and coal resources. The local government has used them to develop the electrolytic aluminum industry and has established a super microgrid consisting of thermal power units, wind farms, and electrolytic aluminum loads. The structure of this super microgrid is shown in Figure 13.1 [3]. The thermal power units and wind farms constitute the power sources, and the electrical loads are relatively simple, mainly including electrolytic aluminum loads, heating loads, and a dynamic load. The electrolytic aluminum loads account for more than 90%. The composition of the power sources and loads is shown in Table 13.1:

13.1.3 Analysis on Operation and Control of the Super Microgrid

The lack of power support from the large power grid results in small inertia of the super microgrid system. Thus, the power fluctuation of renewable energy and system failure can have a significant impact on the operation and control of the super microgrid, threatening the safe and stable operation of the system.

The primary frequency regulation of thermal power units is a conventional means to stabilize system power disturbance. However, a super microgrid has a limited number of thermal power units, and the primary frequency regulation capacity of thermal power units is limited by factors such as boiler pressure. Therefore, the primary frequency regulation capacity of the super microgrid is limited. Once power disturbance goes beyond this regulation capacity, other control measures are in critical need.

A security and stability control system is the key to ensuring the safe operation of the power system. When the system fails or the frequency exceeds a certain limit, the security and stability control system protect system stability by means of generator-shedding and

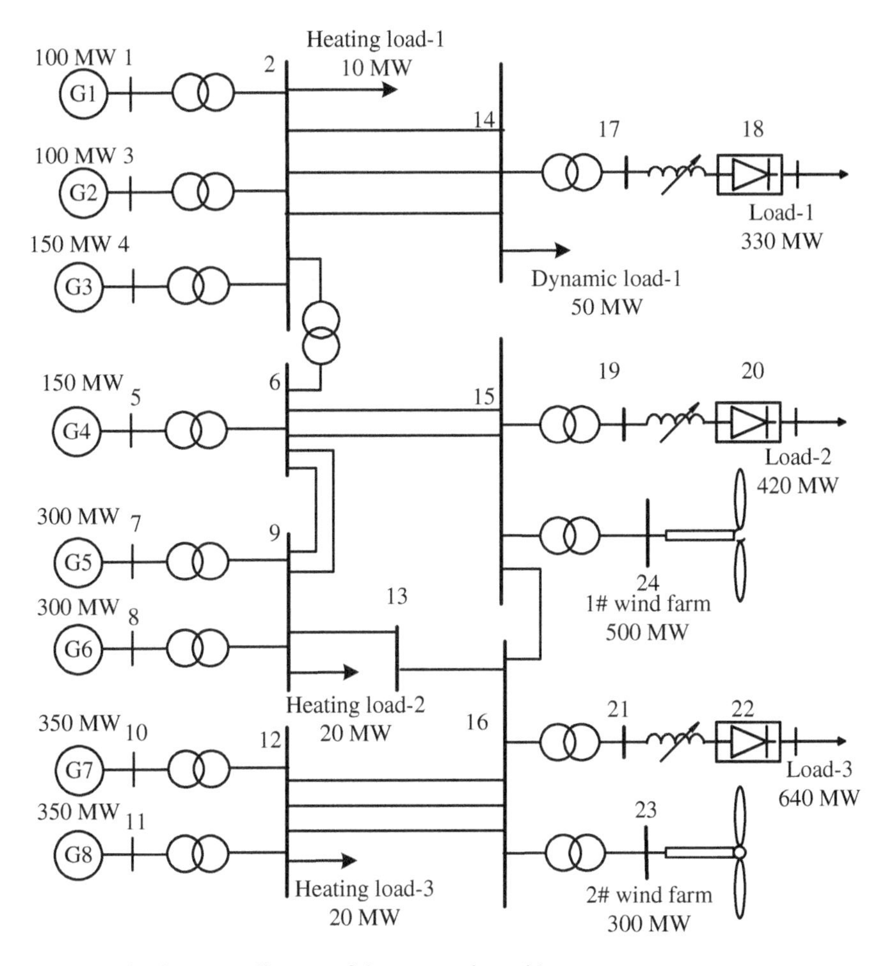

Figure 13.1 Structure diagram of the super microgrid system.

Table 13.1 Power sources and loads of the super microgrid.

Power sources	Loads
Thermal power unit 100 MW ×2	Aluminum plant 1# 330 MW
Thermal power unit 150 MW ×2	Aluminum plant 2# 420 MW
Thermal power unit 300 MW ×2	Aluminum plant 3# 640 MW
Thermal power unit 350 MW ×2	Dynamic load 50 MW
Wind farms 300 MW (already built) +500 MW (to be built)	Heating loads 50 MW

load-shedding. However, for a super microgrid composed of high energy-consuming loads, due to the simple types and large individual capacity of loads, the load curtailment amount and the disturbance amount cannot be matched properly. In addition, the wind power is fluctuating continuously. Even if the fluctuation of wind power exceeds the threshold of the security and stability control system, it cannot be recognized by this system smoothly.

Once the wind power fluctuation goes beyond the primary frequency regulation capacity of the thermal power unit, the system frequency drops rapidly, triggering the under-frequency load shedding action. In the production of electrolytic aluminum loads, all electrolytic tanks are connected in series for electrolysis, and there is no load differential between electrolytic aluminum loads. If under-frequency load shedding acts, the whole series of electrolytic aluminum loads could be cut. As a result, there is a huge difference between the shed load power and wind power change, indicating serious over-shedding. It can eventually lead to system collapse.

Therefore, there are many difficulties in frequency control of the super microgrid. The solution is discussed in the next section.

13.2 Applying Load Control Technology to the Super Microgrid

Since traditional frequency regulation methods are difficult to maintain a safe and stable operation of the super microgrid, a method for controlling the loads in the super microgrid to participate in frequency control is introduced in this section. According to this method, power disturbance in the super microgrid can be stabilized with the coordinated control of load power control and the primary frequency regulation of thermal power units. The electrolytic aluminum load as a typical high energy-consuming load is studied in this section.

13.2.1 Modeling of Electrolytic Aluminum Load Characteristics and Active Power–Voltage External Characteristics

In terms of the production process, electrolytic aluminum loads depend on a large current of several hundred kilo-amperes to electrically smelt alumina to produce aluminum. The cryolite-alumina molten salt electrolysis method is used for production. Molten cryolite serves as a solvent; alumina serves as solute; carbon serves as anode; and molten aluminum serves as a cathode. After a powerful direct current goes through and at the temperature of 950 °C–970 °C, electrochemical reaction takes place on two electrodes in the electrolysis tank [5]. The schematic diagram of the production process of electrolytic aluminum is shown in Figure 13.2.

The electrolytic aluminum load is a typical small-voltage high-current DC load. In the production process of electrolytic aluminum, all the electrolytic tanks are connected in series for electrolysis. In the rectification part, all the rectifier units are connected in parallel for rectification. The electrolytic tanks can be equivalent to a series resistance R and a counter electromotive force E. The equivalent circuit model of electrolytic aluminum is shown in Figure 13.3, where V_{AH} is the HV-side voltage of the load bus; V_{AL} is the LV-side voltage of the load bus; k is the transformation ratio of the step-down transformer of the aluminum plant; and L_{SR} is the inductance value of the saturation reactor.

The equivalent resistance R and the counter electromotive force E are vital to the active power control of the electrolytic aluminum load. In order to obtain these parameters, the electrolytic aluminum load is field tested. Formula (13.1) characterizes the linear

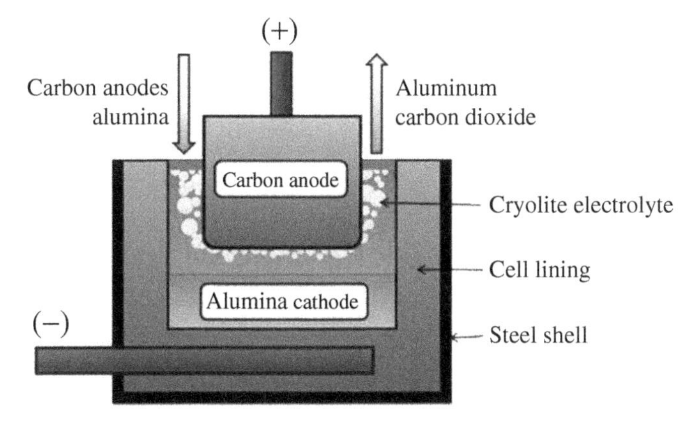

Figure 13.2 Schematic diagram of the production process of electrolytic aluminum.

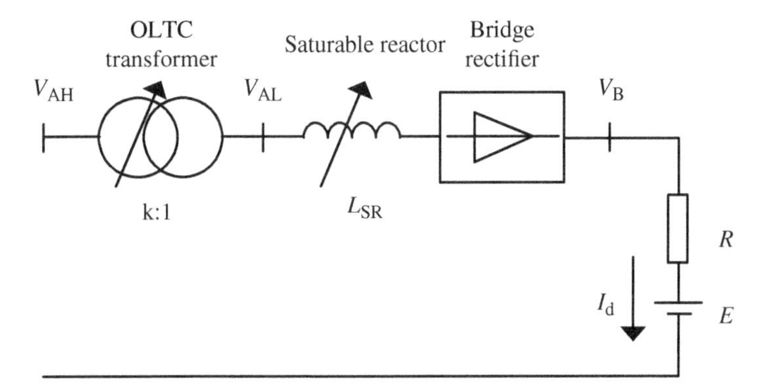

Figure 13.3 Equivalent circuit of the electrolytic aluminum load.

relationship between the DC voltage V_B and the DC current I_d of electrolytic tanks. The deformation formula can be obtained:

$$E = V_B - I_d R \tag{13.1}$$

In formula (13.1), E and R are unknown parameters; V_B and I_d are measurable data, which can be read directly from the main monitoring station of the aluminum plant. By adjusting the on-load tap changer (OLTC) tap to change the transformer ratio k, the DC-side bus voltage V_B can be changed, and the corresponding DC current I_d can be monitored. In order to avoid the influence of the saturation reactor, the current stabilization control of the saturation reactor is withdrawn from the tests. The inductance of the saturation reactor inductance is a constant $L_{SR} = 3.0\,\mu H$ [6]. With this method, many sets of voltage and current values of the DC-side bus can be gotten. The test results are shown in Table 13.2.

Based on the least square method adopted in formula (13.1), the equivalent resistance R and the counter electromotive force E of this electrolytic aluminum load can be identified as: $R = 2.016\,m\Omega$, and $E = 354.6\,V$. According to the equivalent circuit in Figure 13.3, the

Table 13.2 Field test results of DC-side voltage and current.

No.	V_B(V)	I_d(kA)	No.	V_B(V)	I_d(kA)
1	997	310.4	8	976	306.6
2	993	309.5	9	964	298.1
3	990	311.1	10	959	300.4
4	989	311.1	11	955	300.6
5	985	303.7	12	953	299.9
6	983	307.2	13	951	298.6
7	978	304.9	14	948	297.4

active power of the electrolytic aluminum load can be expressed by formula (13.2):

$$P_{\mathrm{Load}} = V_{\mathrm{B}} I_{\mathrm{d}} = \frac{V_{\mathrm{B}}(V_{\mathrm{B}} - E)}{R} = \frac{V_{\mathrm{B}}(V_{\mathrm{B}} - 354.6)}{2.016} \times 10^{-3}(\mathrm{MW}) \quad (13.2)$$

where P_{Load} is the active power of the electrolytic aluminum load. R and E are related to the electrolyte composition, electrolytic tank temperature, and the electrode polar distance, and they can be considered as constants for any given electrolytic tank. Therefore, the active power of the electrolytic aluminum load P_{Load} has a strong coupling relationship with its DC voltage V_{B}.

13.2.2 Active Power Control Method for the Electrolytic Aluminum Load

It can be seen from formulas (1.2–1.4) that the active power of the electrolytic aluminum P_{Load} has a strong coupling relationship with its DC voltage V_{B}. How to adjust V_{B} in real-time in order to control the electrolytic aluminum load to participate in the frequency control of the super microgrid is the research focus of this section. The DC-side voltage of the electrolytic aluminum load is related to the voltage drop of the saturation reactor and the HV-side bus voltage. By controlling the voltage drop of the saturation reactor and the HV-side bus voltage, the active power of the electrolytic aluminum load can be controlled.

13.2.2.1 Electrolytic Aluminum Load Active Power Control Method Based on the Saturation Reactor

The structure of the saturation reactor in the electrolytic aluminum load is shown in Figure 13.4, where CW is the control winding and PW is the power winding. When D1 is turned on, since a large current goes through the power winding, the iron core IC of the saturation reactor is in the saturation zone. When D1 is turned off, the current in the power winding is 0, and the iron core IC of the saturation reactor is in the unsaturated zone.

While the iron core is located at point N and point I, the corresponding magnetic flux of the iron core is different. Thus, as the diode on the branch, where the saturation reactor is located, is turned from off to on, the saturation reactor consumes some voltage that affects the DC output voltage of the rectifying circuit. By adjusting the current in the power winding, the position of point I can be adjusted, so that the difference in magnetic flux at point N and point I can be adjusted, and further, the voltage drop of the saturation reactor can be adjusted.

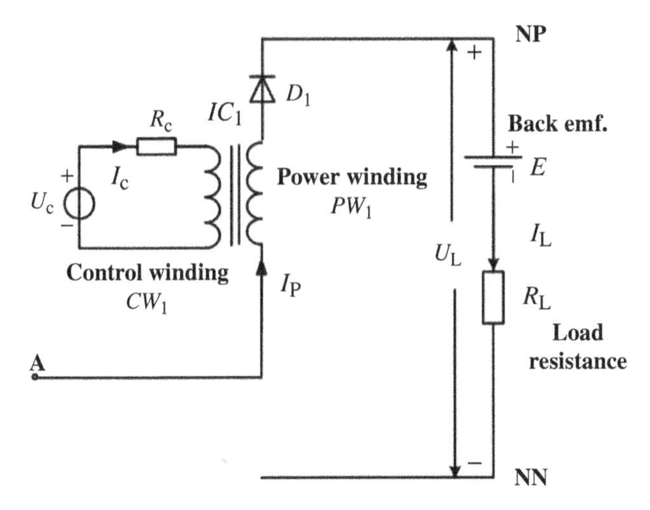

Figure 13.4 Structure of the saturation reactor.

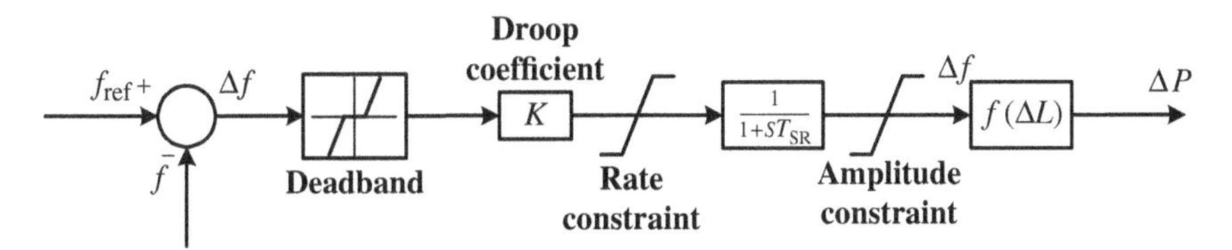

Figure 13.5 Structure of the controller for the saturation reactor to participate in system frequency control. Source: Adapted from Jiang et al. [6].

The saturation reactor can adjust the DC-side output voltage of the rectifying circuit. Figure 13.5 shows the structure of the controller for the saturation reactor to participate in system frequency control. Its basic principle is similar to that of the generator governor: if the deviation between the system frequency f and the reference frequency f_{ref} exceeds a certain deadband, the deviation value could be amplified and then the control current of the control winding can be changed, changing the power of the electrolytic aluminum load, and enabling it to participate in system frequency control.

13.2.2.2 Electrolytic Aluminum Load Active Power Control Method Based on Voltage Control of the Excitation System

The quantitative relationship between the DC voltage V_B and the HV-side bus voltage V_{AH} is given by formula (13.3) [7–9]

$$V_B = \left(\frac{1.35V_{AH}}{k} + \frac{3\omega}{2\pi} \frac{L_{SR}}{R} \cdot E \right) \bigg/ \left(1 + \frac{3\omega}{2\pi} \frac{L_{SR}}{R} \right) \tag{13.3}$$

where ω is the angular frequency of the system. Since the frequency change is small during the frequency process, such change is ignored in formula (13.3). The saturation reactor is considered to have withdrawn from current stabilization control, and the saturation reactor inductance is considered as a constant. Therefore, there is only one variable V_{AH} in the right side of formula (13.3), and V_B and V_{AH} have a linear relationship.

Table 13.3 Active power–voltage regulation capacity of the electrolytic aluminum load.

V_B (p.u.)	0.95	0.90
ΔP_{Load} (p.u.)	12.8%	24.7%

Substitute formula (13.3) into formula (13.2), and the quantitative relationship between P_{Load} and V_{AH} can be obtained. Table 13.3 lists the active power–voltage regulation capacity of the electrolytic aluminum load when the HV-side bus voltage is adjusted to 0.05 p.u. and 0.1 p.u. The reference value of the DC-side bus voltage is 1 kV.

As shown in Table 13.3, when the DC-side bus voltage of three electrolytic aluminum loads drops to 0.95 p.u. and 0.90 p.u., the total regulation capacity of the electrolytic aluminum loads can be 0.128 p.u. and 0.247 p.u., respectively. As a thermal storage load, the electrolytic aluminum load can operate in a heat preservation way for four hours with a 25% reduction in active power [10, 11]. Based on this, the upper limit of the load voltage offset ΔV_{AH} is set to be 0.1 p.u. in this section, so that the requirements for operating in a heat preservation way can be met during the regulation process of the electrolytic aluminum load. In fact, during actual operation, when unit maintenance or under-load operation is necessary, it is not uncommon to adjust the transformer tap to reduce the DC voltage to 0.90 p.u. or even lower to adjust the active power of the load, so that the operation requirements in the field can be met. However, the tap adjustment action takes too long time, and it does not work for the frequency control discussed in this section.

The relationship between P_{Load} (the active power of the electrolytic aluminum load) and V_{AH} is shown by the curve in Figure 13.6.

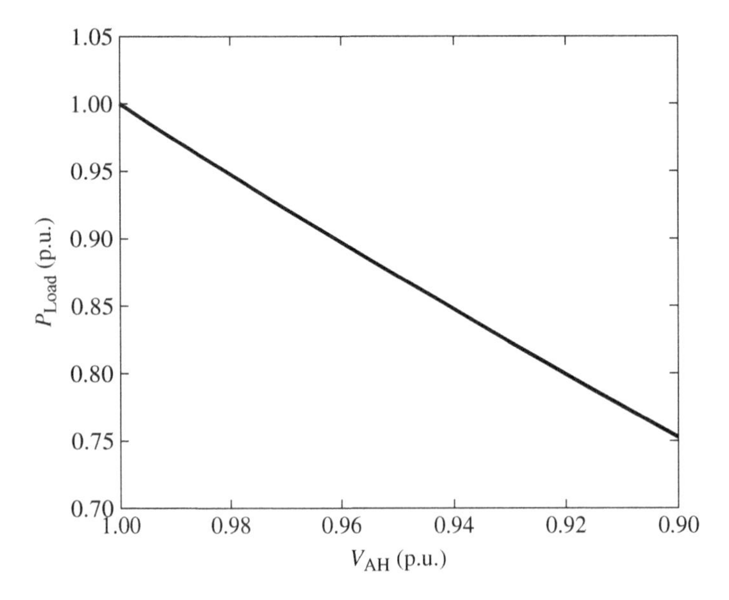

Figure 13.6 Active power–voltage relationship of the electrolytic aluminum load. Source: Adapted from Xu et al. [7].

13.3 Research on Load–Frequency Control Methods for the Super Microgrid

13.3.1 Design of the WAMS-Based Load-Frequency Open-Loop Control System

According to the characteristics of the electrolytic aluminum load, the active power of the electrolytic aluminum load can be significantly adjusted by adjusting the bus voltage of the electrolytic aluminum load. This section presents a frequency and voltage coordinated control method with the excitation control method, which is utilizing PMU to acquire system frequency precisely in real-time. It takes the frequency deviation Δf as the feedback information of the reference voltage V_{ref} of the excitation regulator in order to control the electrolytic aluminum load to participate in frequency control when system frequency control means are limited, and eliminate the power imbalance existing in the system.

13.3.1.1 Power Disturbance Monitoring and Power Imbalance Calculation

For the external power disturbance P_{step}, based on the SFR model, the frequency response of the system should be able to be calculated from formula (13.4) [12]:

$$\Delta f(t) = \frac{RP_{\text{Step}}}{2\pi(DR + K_{\text{m}})}\left[1 + \alpha e^{-\zeta\omega_n t}\sin(\omega_r t + \varphi)\right] \tag{13.4}$$

For this super microgrid, when P_{step} is set to be a different value, the corresponding frequency response can be obtained by simulation. The equivalent parameters of the super microgrid can be identified by a series of simulation data, and they can be applied to the online identification of system power disturbance.

At time $t = t_0$, a disturbance occurs in the system, and the initial change rate of frequency is as shown in formula (13.5), where H is the system inertia constant:

$$\left.\frac{\mathrm{d}f}{\mathrm{d}t}\right|_{t=t_0} = \frac{\alpha R f_n P_{\text{step}}}{DR + K_m}\sin\varphi_1 = \frac{P_{\text{step}}}{4\pi H} \tag{13.5}$$

The initial change rate of frequency $\mathrm{d}f/\mathrm{d}t|_{t=t0}$ can be monitored by means of a wide-area measuring device, so that the system power imbalance P_{step} can be identified online according to formula (13.5).

When power disturbance takes place in the system, the thermal power unit can adjust its output level with its primary frequency regulation capacity to stabilize the power disturbance. However, due to the limited frequency regulation capacity of the thermal power unit, the system power imbalance ΔP may still occur when the system is subjected to large power disturbance. ΔP can be calculated by formula (13.6):

$$\Delta P = P_{\text{step}} - P_{\text{res}} \tag{13.6}$$

$$\begin{cases} P_{\text{res}} = \begin{cases} P_{\text{res-up}}, & \text{if } P_{\text{step}} > 0 \\ P_{\text{res-down}}, & \text{if } P_{\text{step}} < 0 \end{cases} \\ P_{\text{res-up}} = \sum_{j=1}^{N}(P_{\text{G}j\max} - P_{\text{G}j}) \\ P_{\text{res-down}} = \sum_{j=1}^{N}(P_{\text{G}j\min} - P_{\text{G}j}) \end{cases} \tag{13.7}$$

P_{res} is the reserve thermal capacity available for all thermal power units; $P_{res\text{-}up}$ and $P_{res\text{-}down}$ are the maximum up and down-regulation capacity of all thermal power units; P_{Gj}, $P_{Gj\,max}$, and P_{Gjmin} are the current output power, maximum output power, and minimum output power of the jth generator; and N is the number of thermal power units in this super microgrid.

13.3.1.2 Power Imbalance Elimination Method Based on Voltage Adjustment

V_{AH} can be adjusted by changing the reference voltage of the generator bus. The reference voltage of the generator bus can be calculated with the voltage sensitivity method proposed in Reference [12].

When the system is in normal operation, the variation between different bus voltages is always proportional [13], that is:

$$\Delta V_i = a_{ij}\Delta V_j \tag{13.8}$$

$$a_{ij} = \frac{\partial V_i}{\partial Q_j} \Big/ \frac{\partial V_j}{\partial Q_j} \tag{13.9}$$

α_{ij} is the voltage sensitivity of bus i to bus j.

Nevertheless, the voltage sensitivity calculated by formulas (13.8) and (13.9) only applies to the case where the electrical distance between bus i and bus j is small. To avoid errors caused by electrical distance, formula (13.8) can be written into formula (13.10):

$$\Delta V_i = \frac{\sum(a_{ij}\Delta V_j/d_{ij})}{\sum 1/\mathrm{d}_{ij}} \tag{13.10}$$

$$d_{ij} = -\ln(\alpha_{ij}\alpha_{ji}) \tag{13.11}$$

d_{ij} is the electrical distance between bus i and bus j.

Using the above method, the voltage sensitivity of the generator bus to the load-side bus can be obtained, so that the reference voltage of the generator bus can be obtained. The frequency control based on voltage adjustment is realized in the following steps:

Step 1: Obtaining the parameters of each component of the super microgrid system, such as generators, transformers, and the transmission line;
Step 2: Identifying the equivalent parameters of the SFR model;
Step 3: Monitoring the initial change rate of frequency of the system, and use the equivalent parameters obtained in Step 2 to calculate the system power disturbance value;
Step 4: Monitoring the generator's active power output and calculate the available reserve thermal reserve capacity, P_{res}, of the super microgrid;
Step 5: Calculating the power imbalance of the system through Step 3;
Step 6: If P_{step} is greater than zero and ΔP is less than zero, or P_{step} is less than zero and ΔP is greater than zero, Step 3 is performed. In other cases, Step 7 is performed;
Step 7: Calculating the regulation capacity of each aluminum load, namely:

$$\Delta P_{Load\text{-}i} = \Delta P \times P_{Load\text{-}i} \Big/ \sum_{i=1}^{3} P_{Load\text{-}i} \tag{13.12}$$

$P_{Load\text{-}i}$ is the current active power consumption of the ith aluminum load;

Step 8: Calculating the target value of the bus voltage of each aluminum load according to formulas (13.6), (13.7), and (13.12);

Step 9: Calculating the bus voltage reference of each generator, V'_{Grefj}, through formulas (13.7)–(13.11);

Finally, by adjusting the bus voltage of each thermal power unit, the system power shortage can be compensated and the system frequency can be controlled.

13.3.2 Design of the WAMS-Based Load-Frequency Closed-Loop Control System

This section presents a load damping control method based on frequency feedback control, which can further solve the problem of power imbalance elimination and frequency stabilization when the system is subjected to continuous wind power fluctuation. In addition, with the time-varying damping control strategy, the frequency offset of the power grid during the transient process can also be significantly reduced.

13.3.2.1 Electrolytic Aluminum Load Damping Control Method

In this section, the electrolytic aluminum load uses the static load model based on reference [14], and the voltage dependence of its active power demand can be expressed by formula (13.13):

$$\Delta P_{\mathrm{Load}} = f(\Delta V_{\mathrm{Load}}) = 2.4\ \Delta V_{\mathrm{Load}} \tag{13.13}$$

The load bus voltage can be adjusted by changing the generator terminal voltage. The relationship between the load bus voltage and the generator terminal voltage can be expressed as:

$$\boldsymbol{\Delta V}_{\mathrm{L}} = \boldsymbol{K}_{\mathrm{sens}} \cdot \boldsymbol{\Delta V}_{\mathrm{G}} \tag{13.14}$$

where $\boldsymbol{\Delta V}_{\mathrm{L}}$ and $\boldsymbol{\Delta V}_{\mathrm{G}}$ are the load bus voltage vector and the generator terminal voltage vector, respectively, and $\boldsymbol{K}_{\mathrm{sens}}$ is the voltage sensitivity matrix.

The voltage sensitivity $K_{Li\text{-}Gj}$ between bus i and bus j can be calculated by formula (13.15) [13]:

$$K_{Li-Gj} = a_{ij} = \frac{\partial V_{\mathrm{Load}\text{-}i}}{\partial Q_j} \Big/ \frac{\partial V_{Gj}}{\partial Q_j} \tag{13.15}$$

where α_{ij} is the voltage sensitivity coefficient of bus i and bus j, and α_{ij} can be calculated from the elements in the Jacobian matrix. Based on the voltage dependence of load active power, a closed-loop control method is proposed to introduce frequency deviation feedback, so that the power of the electrolytic aluminum load can respond to the frequency change of the system, as shown in Figure 13.7:

13.3.2.2 Equivalent Load Damping Characteristics Control Method

The SFR model of the super microgrid can be equivalent to the structure shown in Figure 13.8:

The equivalent load damping coefficient D_{K} in Figure 13.8 can be obtained from the following formula:

$$D_{\mathrm{K}} = \boldsymbol{f}(\boldsymbol{\Delta V}_{\mathrm{L}})^{\mathrm{T}} \cdot \boldsymbol{K}_{\mathbf{sens}} \cdot \boldsymbol{K}. \tag{13.16}$$

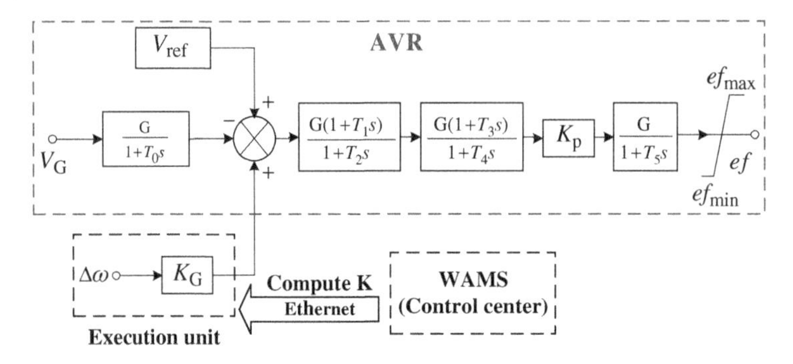

Figure 13.7 Block diagram of the closed-loop control excitation system to introduce frequency deviation feedback.

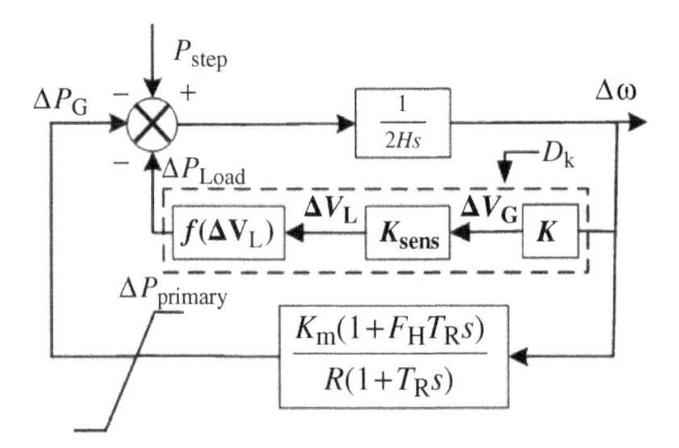

Figure 13.8 Block diagram of the SFR equivalent model considering closed-loop feedback control of the excitation system.

where

$$\boldsymbol{K} = [K_{G1}\ K_{G2}\ \cdots\ K_{G8}]^{\mathrm{T}} \tag{13.17}$$

$$\mathbf{f}(\boldsymbol{\Delta V}_{\mathrm{L}}) = \left[f(\Delta V_{\text{Load-1}})\ f(\Delta V_{\text{Load-2}})\ f(\Delta V_{\text{Load-3}})\right]^{\mathrm{T}} \tag{13.18}$$

13.3.2.3 Calculation of the Load Damping Coefficient

In the SFR equivalent model shown in Figure 13.8, the steady-state frequency offset $\Delta\omega_s$ can be calculated according to formula (13.19) [12]

$$\Delta\omega_s = \frac{RP_{\text{step}}}{D_{\mathrm{K}}R + K_{\mathrm{m}}} \tag{13.19}$$

The steady-state frequency offset determines the increment of the generator output. In order to make full use of the primary frequency regulation capacity of the generator set, it can be assumed here that the increment of the power of the generator has reached the upper limit $\Delta P_{\text{primary}}$, and combining with the primary frequency regulation formula of the generator, the control target of the steady-state frequency offset $\Delta\omega s$ can be obtained as follows:

$$\Delta\omega_s = \frac{R}{K_m}\Delta P_{\text{primary}} \tag{13.20}$$

Substitute formula (13.19) into formula (13.20), the following formula can be gotten:

$$D_{\mathrm{K}} = \frac{(P_{\mathrm{step}} - \Delta P_{\mathrm{primary}})K_m}{R\Delta P_{\mathrm{primary}}} \tag{13.21}$$

The target steady-state frequency offset $\Delta\omega$ and the load damping coefficient D_{K} can be calculated, respectively, by formulas (13.20) and (13.21), and the proportional feedback coefficient $K_{\mathrm{G}j}$ of each generator in the super microgrid can be obtained by the following steps:

For a given frequency deviation $\Delta\omega$, the terminal voltage offset $\Delta V_{\mathrm{G}j}$ of the generator is:

$$\Delta V_{Gj} = K_{Gj}\Delta\omega \tag{13.22}$$

The corresponding electrolytic aluminum load bus voltage change $\Delta V_{\mathrm{Load}-i}$ is:

$$\Delta V_{\mathrm{Load}-i} = \sum_{j=1}^{8} \Delta V_{Gj}\alpha_{ij} = \left(\sum_{j=1}^{8} K_{Gj}\alpha_{ij}\right)\Delta\omega \tag{13.23}$$

Therefore, the total adjustment ΔP_{Load} of the load active power is:

$$\Delta P_{\mathrm{load}} = \sum_{i=1}^{3} \Delta P_{\mathrm{load}\text{-}i} = 2.4\Delta\omega \sum_{i=1}^{3}\sum_{j=1}^{8} K_{Gj}\alpha_{ij} \tag{13.24}$$

Substituting $D_K = \Delta P_{\mathrm{Load}}/\Delta\omega$ into formula (13.24), it can be obtained as follows:

$$\sum_{i=1}^{3}\sum_{j=1}^{8} K_{Gj}\alpha_{ij} = \frac{D_k}{2.4} \tag{13.25}$$

Since the parameters of generators $G1$ and $G2$, generators $G3$ and $G4$, generators $G5$ and $G6$, and generators $G7$ and G8 are, respectively, the same, in order to simplify the calculation, there are:

$$\begin{cases} K_{G1} = K_{G2} \\ K_{G3} = K_{G4} \\ K_{G5} = K_{G6} \\ K_{G7} = K_{G8} \end{cases} \tag{13.26}$$

The power regulation amount of the loads in the aluminum plant is allocated according to their rated power, that is:

$$\begin{cases} \sum_{j=1}^{8} K_{Gj}\alpha_{1j} \Big/ \sum_{j=1}^{8} K_{Gj}\alpha_{2j} = P_{\mathrm{Load\text{-}1}}/P_{\mathrm{Load\text{-}2}} \\ \sum_{j=1}^{8} K_{Gj}\alpha_{1j} \Big/ \sum_{j=1}^{8} K_{Gj}\alpha_{3j} = P_{\mathrm{Load\text{-}1}}/P_{\mathrm{Load\text{-}3}} \end{cases} \tag{13.27}$$

For the control target $\Delta\omega_{\mathrm{s}}$, when the system frequency reaches a steady state, the total load adjustment amount should be able to exactly compensate for the power imbalance in the system, that is:

$$2.4\Delta\omega_{\mathrm{s}} \sum_{i=1}^{3}\sum_{j=1}^{8} K_{Gj}\alpha_{ij} = P_{\mathrm{step}} - \Delta P_{\mathrm{primary}} \tag{13.28}$$

According to formulas (13.27)–(13.28), the proportional feedback coefficient K_{Gj} of each generator can be obtained.

13.3.2.4 Time-Varying Load Damping Control Method

In order to reduce the maximum frequency offset during the transient process, a two-stage time-varying load damping control scheme that can be applied online is proposed, as shown in Figure 13.9. The specific implementation steps are as follows:

The First Stage of Control In order to reduce the frequency offset during the transient process, the first-stage load damping coefficient $D_{\text{K-transient}}$ should be so set that the maximum transient offset of its corresponding frequency response $\Delta\omega_{\text{max-tranient}}$ is equal to the steady-state offset of the target frequency $\Delta\omega_s$, as shown by the green line in Figure 13.9, namely:

$$\Delta\omega_{\text{max-transient}} = \frac{R}{K_m}\Delta P_{\text{primary}} \tag{13.29}$$

Meanwhile, $\Delta\omega_{\text{max-tranient}}$ can also be expressed as a function of $D_{K\text{-transient}}$. The relationship is as follows:

$$\Delta\omega_{\text{max-transient}} = \frac{RP_{\text{Step}}}{D_{K-\text{transient}}R + K_m}\left[1 + \alpha e^{-\varsigma\omega_n t_z}\sqrt{1-\varsigma^2}\right] \tag{13.30}$$

where $\alpha,\zeta,\omega_n,t_z$ are all functions of $D_{K\text{-transient}}$. The details can be seen in Reference [12].

Substituting formula (1.29) into formula (13.30), after deformation, $D_{K\text{-transient}}$ can be obtained by formula (13.31).

$$D_{K\text{-transient}} = \frac{K_m P_{\text{Step}}\left[1 + \alpha e^{-\varsigma\omega_n t_z}\sqrt{1-\varsigma^2}\right] - K_m \Delta P_{\text{primary}}}{R\Delta P_{\text{primary}}} \tag{13.31}$$

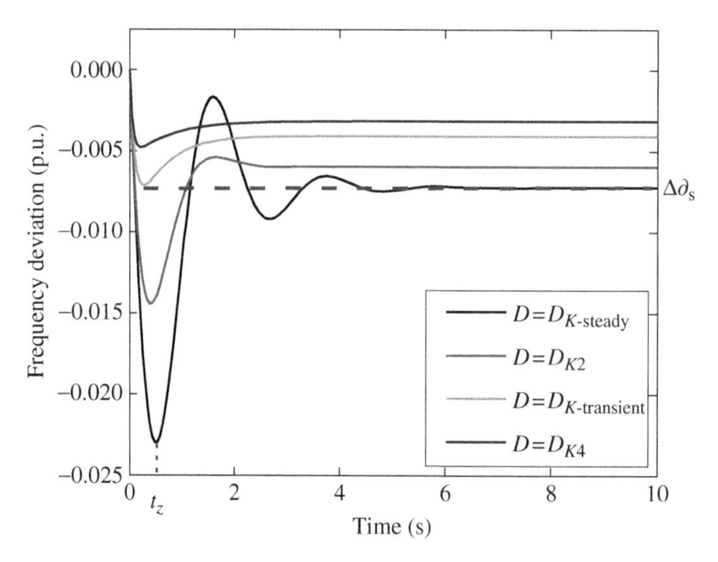

Figure 13.9 Curves of corresponding frequency response of different load damping coefficients ($D_{K\text{-steady}} < D_{K2} < D_{K\text{-transient}} < D_{K4}$).

The Second Stage of Control Suppose that at $t = t_Z$, the theoretical frequency offset of the system reaches the maximum value. At this time, the value of the load damping coefficient D_K is stepped from $D_{K\text{-transient}}$ to $D_{K\text{-steady}}$ to achieve the second stage of control. t_Z can be calculated by formula (1.32) [12].

$$t_z = \frac{n\pi - \phi_1}{\omega_r} = \frac{1}{\omega_r}\tan^{-1}\left(\frac{\omega_r T_R}{\varsigma\omega_n T_R - 1}\right) \tag{13.32}$$

Therefore, the time-varying damping coefficient control method can be expressed in the form of the following function:

$$D_K = \begin{cases} \dfrac{K_m P_{\text{Step}}\left[1 + \alpha e^{-\varsigma\omega_n t_z}\sqrt{1-\varsigma^2}\right] - K_m \Delta P_{\text{primary}}}{R\Delta P_{\text{primary}}}, (t \le t_z) \\ \dfrac{(P_{\text{step}} - \Delta P_{\text{primary}})K_m}{R\Delta P_{\text{primary}}}, (t > t_z) \end{cases} \tag{13.33}$$

The above two-stage time-varying load damping control scheme can greatly reduce the maximum transient frequency offset of the system, and it can maintain the steady-state value of the frequency if necessary.

13.4 Implementation of the Load–Frequency Control Method for the Super Microgrid

13.4.1 Architecture Design of the Wide-Area Information-Based Control System

This section mainly introduces the control flow and system architecture of the wide-area information-based super microgrid real-time control system. The conventional wide area measurement system (WAMS) has an uplink channel for synchronous data and the master station can parse synchronous data, but it does not have the calculation function for control instructions and the downlink channel for control instructions. In order to realize the WAMS control function proposed in this section and control the electrolytic aluminum load to participate in frequency control of the super microgrid, WAMS control functions are extended based the conventional WAMS tasks in this section, so that the control system has a downlink channel for control instructions, and the control instructions of the WAMS master station can be issued through the downlink channel to the controlled devices of the power system, thus realizing WAMS real-time control function.

The overall architecture of the designed WAMS control system is shown in Figure 13.10. The WAMS master control station is independent of the conventional WAMS master station. The data acquired by PMU is packaged with the TCP protocol to be sent to the conventional WAMS master station for analysis and application. Meanwhile, it is also packaged with the user datagram protocol (UDP) to be sent to the WAMS master control station dedicated for real-time control of advanced applications, serving as the input data of the WAMS control.

The master control station equipped with advanced WAMS control applications calculates the control instructions for the analyzed data according to the proposed control method. It provides a downlink channel for issuing WAMS control instructions. The master station sends digital signals to the network control unit (NCU) via Ethernet. Then

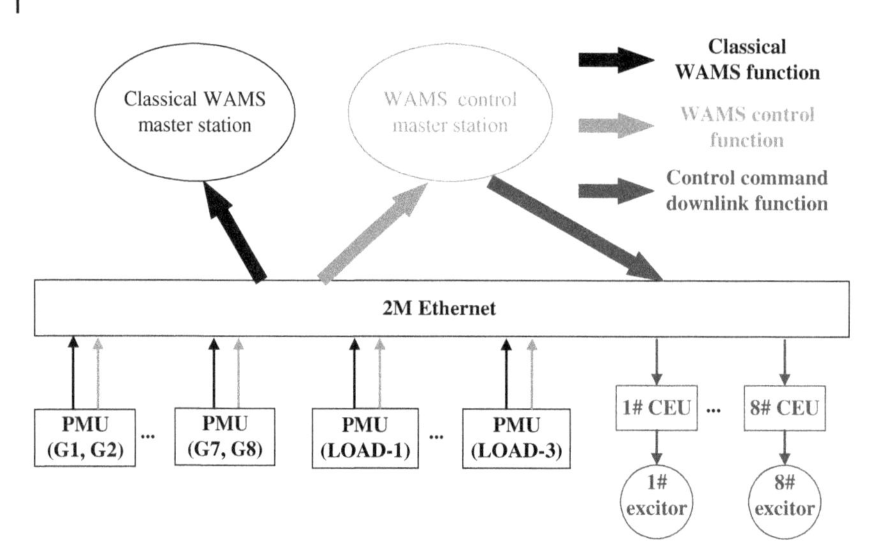

Figure 13.10 The overall architecture of the wide-area information-based control system. Source: Adapted from Sun et al. [15].

NCU, via digital-to-analog conversion, sends control instructions to the analog input channel of the excitation regulator, which is the controlled object in the power system, so that they can act on the controlled object to realize the WAMS control function.

13.4.2 Controlled Device Interface Definition Method

NCU is used to control the generator excitation system. According to the proposed control strategy, NCU has the functions of frequency signal acquisition and real-time communication. It can convert digital quantity into physical quantity output and has a logic calculation function. The NCU output port is connected to the reference voltage superimposition point of the generator excitation system, and it takes acquired frequency signals as feedback signals to input. When power disturbance occurs in the system, the frequency variation is output by the control section of NCU to be superimposed to the reference voltage superimposition point of the excitation system of the thermal power unit, so that the excitation voltage is changed to change the AC-side voltage of the electrolytic aluminum load. Thus, the active power of the electrolytic aluminum load is adjusted to stabilize the power imbalance of the system. The WAMS main control station contains advanced applications to monitor the operating status of the system and determine the fault conditions. Then it calculates corresponding control parameters of NCU based on system faults and issues them to NCU. In order to shorten the electric wire distance of the analog transmission channel so as to avoid control error caused by attenuation, NCU is installed in the cabinet of the excitation regulator. The physical installation of NCU is shown in Figure 13.11, where the excitation regulator in the figure uses the ABB UNITROL-5000 model. The control constructions of the WAMS master control station are sent to NCU via the network receiving port shown in Figure 13.11, and the corresponding

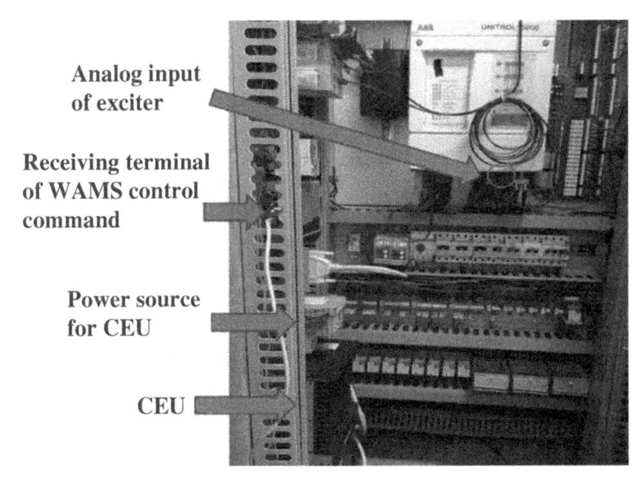

Figure 13.11 Physical installation of the NCU unit in the excitation regulator. Source: Sun et al. [15]/John Wiley & Sons, Inc.

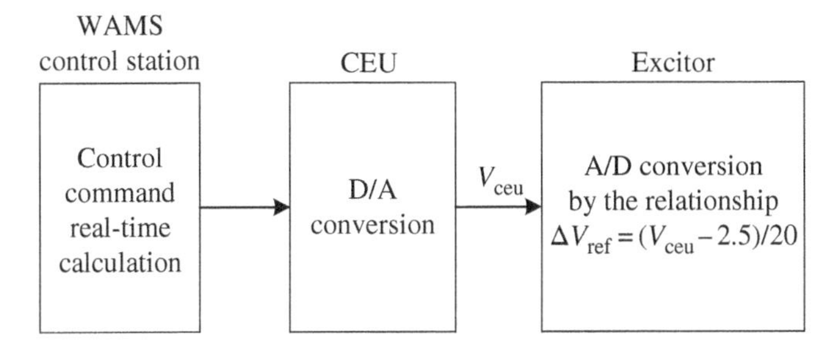

Figure 13.12 Data flow of control instructions in the downlink channel of the WAMS control system.

control instructions are sent to the analog input channel of the excitation regulator to complete WAMS control through digital–analog conversion.

In summary, the data flow of control instructions in the downlink channel of the WAMS control system can be represented by Figure 13.12. In the process, NCU is used for digital/analog conversion. Its input signal is the network digital signal of the WAMS main station and its output is the 0–5 V DC-voltage analog signal, respectively. The analog input channel of the excitation regulator is used for analog/digital conversion, namely, receiving the analog instructions of NCU, converting them to digital signals, and superimposing them onto the superposition point of the excitation transfer function block diagram.

13.5 Operation of the Super Microgrid

According to the smooth operation of the wide-area information-based super microgrid real-time control system, the local consumption of large-scale renewable energy can be achieved with the regulation of high energy-consuming loads. The actual operation

Figure 13.13 The operation interface of the wide-area information-based super microgrid real-time control system.

interface of the system is shown in Figure 13.13. As the project is completed and put into production, the electricity cost is greatly saved compared with purchasing electricity from the power grid. In 2014, the thermal power generation was 2 billion kilowatt hours, saving an electricity purchase cost of 240 million CNY. In 2015 and the next few years, the thermal power generation will reach 2 billion kilowatt hours, representing an annual saving of 600 million CNY, and comprehensively utilizing more than 3 million tons of low-quality coal. Meantime, this project organically combines power generation and power consumption, realizes local transformation, and saves 150 million CNY of wind power tariff subsidy for China each year. If all the China's electrolytic aluminum production capacity is replaced with the "aluminum production by wind power" industrial chain, China could save an annual wind power tariff subsidy of 20 billion CNY and 40 million tons of standard coal, and reduce the emission of 100 million tons of carbon dioxide and the discharge of more than 10 million tons of pollutants.

13.6 Summary

Using the super microgrid consisting of renewable energy and high energy-consuming loads to consume renewable energy is a highly effective way to solve the difficult problem of consuming renewable energy. However, the safe and stable operation of such a microgrid is facing great challenges. In this chapter, for the difficulties in the operation and control of a super microgrid, the key technology of consuming large-scale wind power by high energy-consuming industrial loads in the super microgrid is studied. Based on the production process characteristics and active power–voltage regulation characteristics of loads in the super microgrid, the excitation regulator of the frequency deviation control generator can be used. Then based on feedback control, the active power of the electrolytic aluminum load can automatically respond to the frequency deviation of the system so as to continuously regulate the active power of the electrolytic aluminum load, quickly respond to the fluctuation of wind power, quickly eliminate the power imbalance existing in the system, and achieve the coordination between voltage control and frequency control. A real-time super microgrid control system based on wide-area information is also developed. When this system is put into use, each device works normally and the operation is stable and reliable. The system can correctly implement the frequency-voltage coordinated control strategy and control the electrolytic aluminum load to participate in frequency control of the super microgrid based on the saturation reactor to control loads, and use the load-side response capability of high-energy-consuming loads to stabilize wind power fluctuation. Its regulation capacity provides an effective way to eliminate power disturbance and solve the problem of rapid ramping scenarios caused by extreme wind power fluctuations in the super microgrid. Additionally, it provides technical and equipment support for the new model which is high-energy-consuming loads to locally consume large-scale wind power.

References

1 Hart, E.K., Stoutenburg, E.D., and Jacobson, M.Z. (2012). The potential of intermittent renewables to meet electric power demand: current methods and emerging analytical techniques. *Proceedings of the IEEE* 100: 322–334.

2 National Energy Administration. Wind power utilization hours of provinces in China (in Chinese) (2016).

3 Sun, Y.Z., Lin, J., Song, Y.H. et al. (2012). An industrial system powered by wind and coal for aluminum production: a case study of technical demonstration and economic feasibility. *Energies* 5 (11): 4844–4869.

4 Sun Y Z, Bao Y, Xu J, et al. Discussion on strategy of stabilizing high proportion of renewable energy power fluctuation (in Chinese). *Chinese Science Bulletin*, 2016, doi:https://doi.org/10.1360/N972016-00484.

5 Principles of the Hall-Héroult process [EB/OL]. (2015-11-05). http://www.peter-entner.com/e/theory/PrincHH/PrincHH.aspx.

6 Jiang, H., Lin, J., Song, Y. et al. (2014). Demand side frequency control scheme in an isolated wind power system for industrial aluminum smelting production. *Power Systems, IEEE Transactions on* 29 (2): 844–853.

7 Xu, J., Liao, S.Y., Sun, Y.Z. et al. (2015). An isolated industry system driven by wind-coal power for aluminum productions: a case study of frequency control using voltage adjusting. *IEEE Transactions on Power Apparatus and Systems* 30 (1): 471–483.

8 S. Y. Liao, J. Xu, Y. Z. Sun, W. Gao, X. Y. Ma, M. Zhou, Y. Qu, X. Li, J. Gu, J. Dong, Load-damping characteristic control method in an isolated power system with industrial voltage-sensitive load, *IEEE Transactions on Power Apparatus and Systems*, vol. 31, no. 2, pp. 1118-1128, 2015, 2016.

9 Cui, T., Lin, W., Sun, Y. et al. (2015). Excitation voltage control for emergency frequency regulation of island power system with voltage-dependent loads. *IEEE Transactions on Power Apparatus and Systems* 31 (2): 1204–1217.

10 Paulus, M. and Borggrefe, F. (2011). The potential of demand-side management in energy-intensive industries for electricity markets in Germany. *Applied Energy* 88 (2): 432–441.

11 Babu, C.A. and Ashok, S. (2008). Peak load management in electrolytic process industries. *Power Systems, IEEE Transactions on.* 23 (2): 399–405.

12 Anderson, P.M. and Mirheydar, M. (1990). A low-order system frequency response model. *IEEE Transactions on Power Apparatus and Systems* 5 (3): 720–729.

13 Lagonotte, P., Sabonnadiere, J.C., Leost, J.Y. et al. (1989). Structural analysis of the electrical system: appli-cation to secondary voltage control in France. *Power Systems, IEEE Transactions on* 4 (2): 479–486.

14 Concordia, C. and Ihara, S. (1982). Load representation in power system stability studies. *IEEE Transactions on Power Apparatus and Systems* PAS-101 (4): 969–977.

15 Sun, Y., Liao, S., Xu, J. et al. (2016). Industrial implementation of a wide area measurement system based control scheme in an isolated power system driven by wind–coal power for aluminium productions. *IET Generation, Transmission and Distribution* 10 (8): 1877–1882.

14

Technology and Policy Requirements to Deliver Resiliency to Power System Networks

Mani Vadari[1], Gerald Stokes[2], and John (JD) Hammerly[3]

[1] *Modern Grid Solutions, Redmond, WA, USA*
[2] *Stony Brook University, Long Island, NY, USA*
[3] *The Glarus Group, Spokane, WA, USA*

14.1 Introduction

While informed people can argue about the impact of climate change and its relationship to storms and other severe weather-related incidents, a recent analysis from The Weather Channel has identified the 2017 Atlantic hurricane season as among the top 10 most active in recorded history [1]. Another interesting fact is that these storms are becoming more regular and increasing in intensity year after year. The remnants of the disasters caused by Harvey and Irma can still be seen in places on both the Atlantic and Gulf coasts as well as in Puerto Rico.

Let's look at the damage caused by some of the storms in recent history, with a focus on the damage to the electrical grid and the impact on grid resiliency (Figure 14.1).

14.1.1 Superstorm Sandy

Superstorm Sandy was one of the most significant storms in the history of the US power sector. But, it was not a worst-case scenario. It was not the most expensive. And it was not the deadliest [2]. When the storm finally dissipated over the Great Lakes, it left 8.5 million people without power in 21 states – the highest outage total for any US extreme weather event in history. Sandy did not cause the same level of displacement and economic destruction as Hurricane Katrina had, seven years earlier, but it was the second-costliest hurricane ever to hit America. It caused 65 billion dollars in damage and took the lives of 181 people in the US Sandy highlighted some glaring vulnerabilities in the way utilities manage the electrical system and communicate with customers. But it also helped build momentum for another round of an increase in spending on the part of utilities and governmental agencies on innovative technologies.

Although the wind-induced damage was widespread and substantial, saltwater flooding caused the most significant restoration delay after Sandy made landfall. Facilities below grade flooded and water had to be removed to allow crews access before the restoration could begin. After the water was pumped out and mud and debris removed, the disassembly,

Resiliency of Power Distribution Systems, First Edition.
Edited by Anurag K. Srivastava, Chen-Ching Liu, and Sayonsom Chanda.

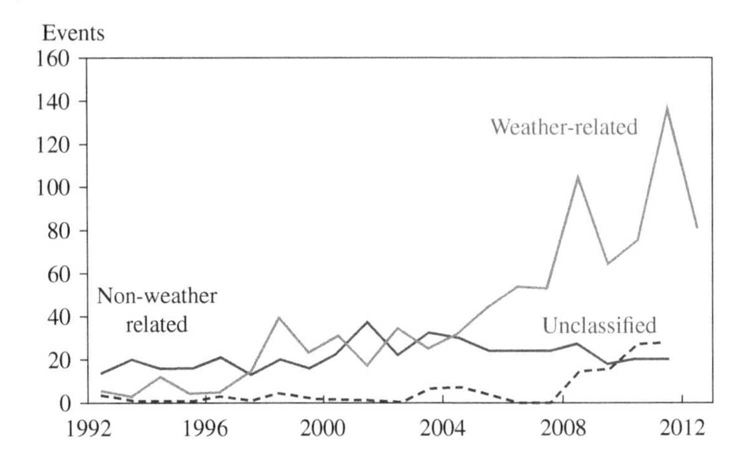

Figure 14.1 Observed outages to the bulk electric system, 1992–2012. Source: EIA/Public Domain. Documenting the surge in weather-related "grid disturbances."

cleaning, drying, and reassembly of the substation equipment could begin. Only then could it be tested, repaired, and put back into service. Although underground distribution lines are more resilient than overhead lines, many underground cables failed after an extended period of saltwater submersion.

Many substations had 50+-year-old equipment damaged, and replacement parts in depots were unavailable. This required cannibalization of the needed parts from other locations or companies. In many cases, replacement parts were simply no longer available. Full replacement often required circuit redesign "on the fly." In total, Consolidated Edison spent nearly 600 million dollars, PSE&G in Newark spent 350 million dollars, and estimates for all impacted utilities were greater than 15 billion dollars. A portion of this cost-focused on a redesign to increase the hardening of the electric infrastructure when the next "Sandy" comes ashore.

14.1.2 Hurricane Irma

When Hurricane Irma side-swiped South Florida, almost every home fell into darkness and surrendered to the tropical heat. Tens of thousands of customers suffered a week or more without power. Overall, nearly 4.5 million of Florida Power & Light's 4.9 million customers had their power fail. The widespread outages happened despite FPL spending nearly three billion dollars over the past decade to "harden" its electrical grid against the next monster storm.

The investor-owned utility – which by law makes a guaranteed profit for shareholders somewhere between 9.6% and 11.6% – stated it responded quickly to restore outages and that its storm-hardening efforts were working [3]. Puerto Rico fared even worse with full power not expected to come back on for several months.

At the peak of Irma's damage, 13 million customers in South Florida and the Keys were without electricity. Most of the electric infrastructure in the Florida Keys was blown over and required complete rebuilding. Most of the damage to substations in the Florida Keys resulted from airborne debris and saltwater flooding from the storm surge. Florida Power and Light and Duke Energy combined had 30,000 staff and workers from other utilities outside of Florida working during the restoration. The estimated cost to restore the electric system from Irma in Florida is four billion dollars.

14.1.3 Hurricane Katrina

Hurricane Katrina was one of the five worst storms in US history and the worst hurricane of the 2005 season. It was also the country's most expensive natural disaster. The hurricane formed over the Bahamas and gained strength in the Gulf of Mexico. It affected the states along the Gulf, from Florida to Texas, but most notably Louisiana, before continuing up the east coast. Hurricane Katrina made landfall three times, in Florida and Louisiana between August 23 and August 31, when it finally dissipated. In total, it affected an area of roughly 90,000 mi^2. At the time, Hurricane Katrina was the strongest hurricane recorded in the Gulf region. At one point, the wind speed approached 175 mi/h.

Katrina destroyed most of the electric system in multiple locations within 40 mi of the coast across a wide swath of its landfall. The wind damage was extensive and widespread, bringing down tens of thousands of poles and a surprising number of steel towers. Although the storm surge reached more than 20 ft in some areas, many utilities, due to frequent hurricanes, had built elevated substations. Some are as much as 3 m above grade. Most of the elevated substation equipment survived but could only be manually operated because telecommunications failed. Like many utilities, several of those impacted by Katrina relied on third-party carriers for landline connections to remotely monitor and operate field equipment. Many of the third-party-supplied landlines in urban and suburban areas ran underground, where severe flooding occurred and water remained for weeks.

Although damage from Katrina's wind and surge was massive, utilities were impacted by more than solely those forces. Entergy, based in New Orleans, was hardest hit. All of the windows in their corporate headquarters were blown out, and Entergy could not reoccupy the building for almost six months. Katrina also impacted Entergy's control centers their major distribution control center serving New Orleans survived and stayed in operation but could only be supported with food, fuel, and relief staff by boat for nine weeks.

In all, damage from Katrina required 6500 crews from outside the region to be deployed and cost nearly six billion dollars to restore the electric system. And, Entergy paid out three billion dollars on their own.

14.1.4 Summarizing the impact of Storms on the Electric Grid

With most storms, the transmission system stays intact. Much of the impact of storms and the need for resiliency is in the distribution systems. There is an extreme option that can be taken to improve resiliency – put the entire distribution system underground and network them so that, like the transmission system, every location would have multiple paths of power delivery. This option would be prohibitively expensive, however, so, what are other options?

A new buzzword has emerged as a framework for the next round of investment: "resiliency." The National Association of Regulatory Utility Commissioners defines resilience as "robustness and recovery characteristics of utility infrastructure and operations, which avoid or minimize interruptions of service during an extraordinary and hazardous event."

The term is not new for the power sector. But it is taken on a more urgent meaning within utility boardrooms, regulatory proceedings, government agencies, and technology companies. It also partially drives the business plans of East Coast utilities – consequently spreading throughout the United States to influence the broader power sector. Although resiliency is only one lens through which power companies view the world, it has arguably become central for many of them [4].

Resiliency is not about preventing damage when a storm such as Superstorm Sandy hits the area. It is about minimizing the impact of the damage to the electric grid so that facilities can continue operating or are designed to enable a rapid return to normal operations after damage and outages occur.

Making the power distribution system more resilient starts with three major aspects: bringing new technologies into the grid, changing the design of the grid, or a combination of the two. More automation, which includes intelligent switches supported by enhanced communication connectivity, can detect a combination of short circuits and disconnected feeders. This, in turn, blocks power flows to the affected area, communicates with other nearby switches, and reroutes power to keep as many customers energized as possible. Automatically executing these actions or executing through centralized control of a distribution system operator reduces restoration time and restores the power to customers much faster than current capabilities. It may also be advantageous to split the macrogrid into smaller circuits and reexamine the circuit capabilities to repair feeders quickly [5].

In this chapter, the authors focus on the use of existing, new, and still undeveloped technologies, laying the groundwork for research and including assets such as drones to make the networks more resilient and, if damaged, how to restore them as fast as possible.

14.2 A Broad Perspective on the Need to Apply Technology

Since the 1970s, while much of the developed world's transmission grid was built, transmission systems enjoyed substantial investment in sensors, communications, and analytic solutions. These transmission grid assets have aged, with many now at or past their expected lifespans. Utilities must decide which will require replacement, refurbishment, and/or continued maintenance. The cost of refurbishing or replacement has become a constraint. Addressing this dilemma required the utility industry to sharpen its focus on asset management beyond traditional financial tools [6, 7].

14.2.1 Asset Health

Today's asset management ambition is to maintain or improve reliability while replacing or refurbishing those assets likely to fail within a year or two. Simultaneously, the rest must be maintained as before. Further, today's asset management solutions must prioritize those assets requiring maintenance while delaying maintenance for those that can be delayed. Such decisions require asset health information derived from condition monitoring, asset class history, and predictive analysis. Advances in asset management sensors and analytics will allow utilities to maintain and improve reliability while managing their cost.

The next phase of transmission asset management will be a holistic optimization that considers the asset health, cost of replacement, the impact of the loss of each specific asset in the grid, time and skill required to do the replacement, and available resources. Over time, these analytic tools will be applied to distribution asset populations as well.

14.2.2 Operational Systems and Distribution Automation

Historically, the grid has been viewed as requiring active engagement and "hands-on" monitoring and control. Relays were initially deployed to protect grid assets, primarily transformers. Reply deployment paved the way for more automated grid reconfiguration. The penetration of microprocessor-based relays, although still protecting grid assets, also provides significant insight into the health of the grid through measurement data and provides key information such as fault location. As relays advanced, related technology led to the deployment of distributed control systems, smart switches, and reclosers, enabling a more sophisticated and complex automated grid reconfiguration in the form of distribution automation, DA.

In parallel with these technological impacts came deployment of modern monitoring, control, and analytic systems. For the transmission system, SCADA and Energy Management Systems have provided control and monitoring for decades. These systems provided data to further analyze the grid's status through state estimation, powerflow, optimal power flow, contingency analysis, and short circuit analysis. In addition, these systems supported a suite of generation applications that dispatched generators based both on demand and economics constrained only by the individual generator's capabilities.

Today, driven by data availability, new systems have appeared to improve reliability, restoration times, and resiliency. Advanced Distribution Management Systems, ADMS, which include formerly stand-alone Outage Management Systems' capabilities, are being deployed to provide monitoring and control from the transmission grid to the end-point electricity consumer. The ADMS also provide automation of key processes required to enable maintenance and grid expansion. These ADMS integrate not only traditional telemetry data but also assimilate critical data from the grid's edge, such as Meter Data Management System, MDMS. For the first time, there is visibility of what customers are without power based on empirical data from Smart Meters. These Smart Meters also provide critical insight into the quality of power each customer "sees" through instantaneous and continuous voltage monitoring. This visibility not only provided immediate awareness of the facts behind customer power quality complaints, but also enabled new applications to analyze, manage, and optimize voltage across the grid.

In the future, driven by the historically unprecedented availability of low-cost sensors and automated applications able to consume the incredible volumes of data, a self-healing grid will become the norm. By automatically converting data to information, information to action, "hands-on" operational engagement will not be required. In the future, operational management of the grid-of-the-future will mirror how communication networks are managed today. These grids will be automatically reconfigured as necessary to minimize impacts of failures and constantly optimize utilization and performance.

14.2.3 Phasor Measurement Units (PMU)

In the engineering and operations space, phaser measurement units, PMUs, have increased the quality of data and greatly expanded data quantity. PMUs provide detailed grid data stamped with high-resolution GPS time. For that reason, time skew, which historically impacted the analytic results quality, is a thing of the past. Today, PMUs can provide a more accurate representation of the state of the grid and improved insight into the order of events that impact the grid, but the capability to do predictive analysis based on PMU data has lagged.

14.2.4 Smart Meters

While transmission grids enjoy the evolution of the available technology, distribution grids, long lacking in sophisticated solutions to improve their design, operations, and maintenance, are experiencing a revolution. With the volume of assets in the distribution grid, no widespread technology solution deployment happens quickly, but because of that volume, device costs can also drop rapidly. The first example of this has made smart meters the launch point for the distribution grids.

The primary focus of smart meters and the justification to deploy them was enabling complex rate structures such as Time of Use and Critical Peak Pricing. Once deployed, however, smart meters quickly brought additional value because they provide multiple measurement channels in a single device, which directly enables distributed energy resource deployment. Further, smart meters reduce cost through remote connect and disconnect in "blue sky" operations by reducing truck rolls. Smart meters also have improved reliability and restoration times by providing the precise scope of outages and enabling sophisticated outage management analytics and the ability to identify the most likely outage source, often before customers report an outage.

14.2.5 Data Analytics

In practice, such applications of smart meter data are still in the early stages. As this data becomes more available to the operational and engineering users, the new value will be extracted. The volume of smart meter interval data offers the prospect of long-time historical analysis. This historical perspective can provide more precise load forecasting, enabling better resource management, and reducing the need and cost of extensive asset upgrades. More recently, this long-term historical data shows promise for enabling predictive equipment failure analysis of underground cables and real-time fault analysis allowing identification of fault causes and required response.

Without question, new analytics that consume, analyze, and predict events based on smart meter data will change how utility design, operate, and maintain the distribution grid. With this technology, the distribution grids can be designed to meet a realistic "need" rather than to the worst case, operating closer to capacity, maintaining what will fail, all the while increasing reliability and managing cost.

14.2.6 Grid Edge

The second wave of the distribution technology revolution is at the grid edge. This technology spans a wide variety of value propositions, from automated restoration via distribution automation to voltage optimization focused on reduction in energy and cost. The grid edge technologies are much less uniform and utility adoption is driven by specific and sometimes unique requirements, resulting in both successes and disappointments. Today's most promising and widespread deployments have focused on voltage reduction and power quality. Economics can demonstrate energy savings of a few percent without impacting customers and more widespread adoption is underway at utilities. Again, voltage optimization is in its infancy, and continued advances will bring greater savings as technological improvements change the utilities' ability to optimize voltage in real-time and with greater granularity.

14.2.7 Ubiquitous Communications

Technology deployment and utilization are data-driven, data's conversion to information requires fusion with other data, and information's conversion to action requires communications. Reliable collection of data, integration of diverse data, consumption by analytics, and long-term retention all require communications, which becomes a critical technology enabler. The electrical industry is about to experience a period of accelerating technological change. Although this change will impact every aspect of the utility, from the nature of the utilities' value proposition to how customers interact with their utility, arguably the distribution system will experience the most significant impacts.

In an always-connected world, the availability of low-cost, reliable communication bandwidth will enable an explosion of new sensor deployment, increased real-time grid visibility, the emergence of new analytic tools, and improved decision-making. Today's communication approach remains classically "hub and spoke," with analytics residing in central locations – the hub – with data being collected across the grid – the spokes. Fueled by the availability of new, low-cost sensor technology, ever-improving processing capability, miniaturization, and expanded analytics enabled by communications, more decisions will occur without human interaction, closer to the edge of the grid.

Analytics operating in substations and the grid edge, which optimize local assets within the guidelines and under the direction of system-wide considerations constraints, are central to the nature of a transformed utility. Peer-to-peer communication becomes a requirement for increased closed-loop control and autonomous grid operations. Communications technology enables these changes in both wired and wireless, but the more momentous change will be in wireless. In the next five years, the move from 4G LTE to 5G offers three orders of magnitude capacity increase, support for 100 billion connections, speeds to 10 GB, and less than a millisecond latency. In the past, wireless technology focused on increasing throughput with 5G, although increasing throughput substantially focused on ubiquitous connectivity and the ability to deliver the needed throughput to machine-type communications. Although some may argue this is just "more of the same," factually, the impacts of 5G availability to utilities are far-reaching. 5G enables secure peer-to-peer capabilities for sensors, a sensor to analytics, and autonomous decision-making with predictable latency. 5G's availability will change not just how quickly they operate but how solutions are implemented.

For the utility industry to transform itself while embracing distributed generation, deployed storage, electric vehicles, a connected customer, and a society requiring an ever-more reliable and efficient electricity supply, they must be able to adapt their practices and adopt technology quickly, deploy wisely and efficiently, and with secure communications.

14.2.8 Summary

For the utility industry to transform itself while embracing distributed generation, deployed storage, electric vehicles, a connected customer, and a society requiring an ever-more reliable and efficient electricity supply, they must be able to adapt their practices and adopt technology quickly, deploy wisely and efficiently, and with secure communications.

14.3 Use of Microgrids to Improve Resiliency Response

During SuperStorm Sandy, several locations in the affected areas still had electricity, because they had both on-site power generation and the means to operate independently of the grid. The College of New Jersey, New York University, Princeton University, and Stony Brook University each used a combined heat and power (CHP) plant on campus to keep the lights and heat on during and after Sandy. New York University's case was particularly visible because the area immediately surrounding it in downtown Manhattan was completely blacked out. Thanks to NYU's 13.4-MW CHP plant and its self-sufficient microgrid system, which distributes electricity independently of the main utility grid, 26 of its buildings still had electricity [5].

After a major storm, the typical utility approach is top-down. This approach implies restoration begins by ensuring the viability of the transmission system, followed by securing and validating the viability of the sub-transmission network. Crews go about restoring higher voltage lines and associated equipment to ensure that a centralized generation system can be brought to bear and that the supply side of the energy value chain is ready to deliver.

Mostly in parallel, distribution crews also move into action restoring the distribution system to ensure that the demand side of the energy value chain is ready to receive the energy from the transmission system. Restoration follows pre-determined utility priorities of first restoring critical loads, then focusing on the backbone distribution system until finally everyone else is restored. There are always pockets of customers in remote feeders that take much longer.

This approach has worked well in the past, but the advent of distributed energy resources (DERs) and storage created new opportunities to change all of that.

Reform the energy vision (REV) initiative in New York defines DER as a set of technologies that include photovoltaic (PV) cells, battery storage, fuel cell, wind, thermal, hydro, biogas, cogeneration, compressed air, flywheel, combustion generators, demand response (DR), and energy efficiency [8–10].

DER-based technologies are changing how electricity is generated, transmitted, and consumed. Instead of depending on large, remotely located generators to produce and transmit power, energy can now be generated, at or close to delivery voltages. This process reduces dependence on large centralized and remote generators and long transmission lines to bring power from generating centers to the load centers.

DERs are typically smaller-scale power generation technologies located close to the load being served. Locating them close to the load reduces delivery losses and additional investment in the distribution system. While the renewable options are intermittent, pairing these resources with storage helps improve their reliability profile while simultaneously reducing emissions. DERs also tend to be small and modular compared to central power plants. Due to their distributed and modular nature, as compared to large power plants they are flexible, quieter, emit less, and often require no water. Not all DERs are based on renewables, though most are. Several are also based on natural gas and other non-renewable sources.

14.3.1 Enter the Microgrid

The Galvin Electricity Initiative has defined microgrids as modern, small-scale versions of the centralized electricity system. They achieve specific local goals established by the community being served, such as reliability, carbon emission reduction, diversification of energy sources, and cost reduction. Like the bulk power grid, smart microgrids generate, distribute, and regulate the flow of electricity to consumers, but do so locally [11].

14.3.2 Why Use Microgrids to Improve Grid Resiliency?

Microgrids, which include local sources of generation accompanied by energy storage, allow operators (centralized or local) to quickly reconfigure parts of the system when portions of the grid fail. Implicit to microgrid deployment plans, is the need to ensure uninterrupted power to critical sites such as oil and gas refineries, water-treatment plants, and telecommunication networks, as well as gasoline stations, hospitals, and pharmacies. These critical or highly available parts of the infrastructure are necessary to get cities and localities up and running in the aftermath of a storm.

How the system can be restored when microgrids are present?

- In a large-scale storm, the breakdown of the system typically happens along the path of the storm, a path that is different for every storm.
- Microgrids, where they exist, will maintain their supply–demand balance for their region while disconnected from the grid. Supply–demand balancing was already evident in the example provided, in which several, university campus-based microgrids were able to continue operating independently of the distribution grid during the aftermath of Superstorm Sandy.
- Each stable microgrid will have independent sources of supply and will not be dependent on centralized sources of supply. When this happens, the number of people who will lose power for extended periods of time will be diminished.
- As the distribution level backbone of the utility is restored, the individually operating microgrids can be reconnected back to the grid. Fixing the system will be done in parallel on a microgrid by microgrid basis (instead of today's customer by customer or feeder by feeder basis), and hence, can be done much faster. Also, given the microgrid operator will have local sources of generation, it is anticipated that restoration should be faster and some customers/microgrids may never lose power at all.

Now, imagine a situation in which the distribution grid itself is redesigned with the microgrid at its core instead of today's bulk-power macrogrid. The implication is that in steady-state conditions, the system would still be functioning as today, but with additional access to cheaper and centrally sourced large generating units. It would also contain the following sets of IT and OT technologies strategically placed around the network, which can deliver support for short or extended periods as business needs dictate.

- Microgrid controllers are distributed around the grid at strategic locations that can perform localized control, which could be either centrally managed or in a fully distributed approach.
- A mechanism that has been identified as "I See You" can determine local parts of the network that the microgrid controller(s) can get electrically connected through the control of one or more switches. These will also represent local loads that demand response is a process by which electrical providers, distributors, transmitters, and customers –residential, commercial, and industrial – manage their electrical needs, particularly at times of peak usage or in response to market costs, thereby limiting, growing, or eliminating the demand for short or extended periods of time: the controller can pick up if the conditions of adequate generation/load balance can be maintained;
- Localized demand response capabilities would allow the microgrid controllers to drop non-critical load within its boundaries while simultaneously picking up a new load, thereby growing the reach and boundary of the microgrid.

Demand Response is a process by which electrical providers, distributors, transmitters, and customers – residential, commercial, and industrial – manage their electrical needs, particularly at times of peak usage or in response to market costs, thereby limiting, growing, or eliminating the demand for short or extended periods of time.

Once each microgrid has reached the fullest extent of its capabilities and operates stably, the distribution system operator will slowly get the microgrids connected to each other, thereby focusing on bringing the entire system up and running as a single, integrated macrogrid.

For this type of approach to work, a new type of architecture will be required along with some new technologies that are not yet available. Figure 14.2 identifies some key technology components necessary to build this architecture.

14.3.3 Research Ideas on the Microgrid of the Future

As identified before, some of the technologies are still in research mode and need additional work. Some of those are identified here. New and innovative control schemes – to disconnect and reconnect these microgrids in as close to real-time as necessary

- Battery-backed "I See You" mechanism that can be installed across the various parts of the grid and continue working through major storms to become reconnection enablers in the immediate aftermath.
- Designing a dynamic control mechanism, which can either function in a centralized or decentralized manner driving key areas of focus as:
  - real-time supply-demand balance,
  - synchronizing grids when interconnectivity returns,

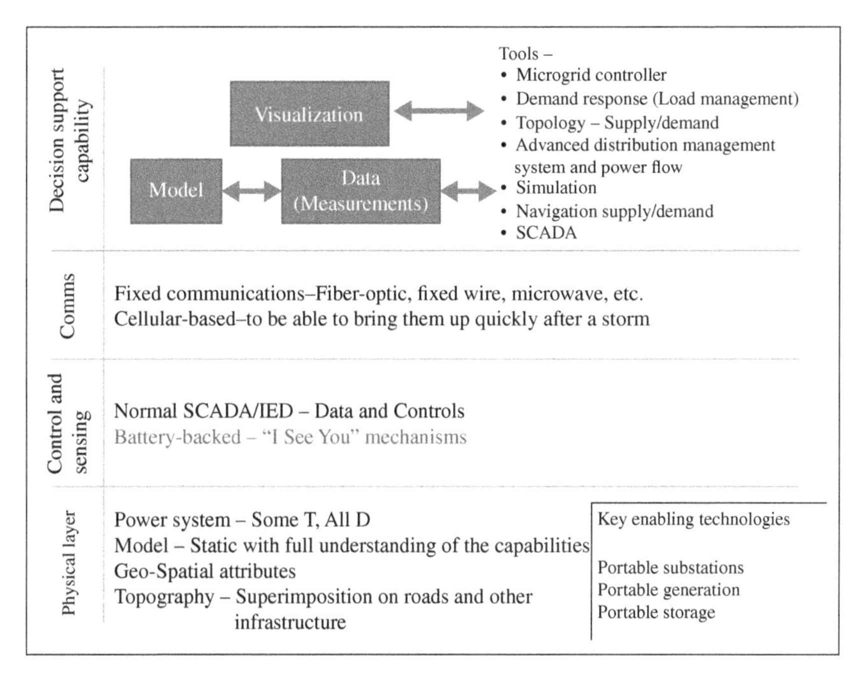

Figure 14.2 Architecture of the next-generation microgrid. Source: Copyright Modern Grid Solutions, used with permission.

- separate the system into microgrids either in a static (pre-determined) or dynamically (real-time) manner,
- drop demand (load shedding/demand response) as needed (or as available) based on local or global requirements,
- rotate available supply through periodic switching as needed to ensure microgrids continue to operate if fuel or generation sources are intermittent, and
- raise generation (regular or renewable-based) as needed (or as available) based on local or global requirements.

- Converting localized supply-demand balancing schemes to link up to a global level when connected together.
- Better modeling mechanisms and capabilities – both at the planning level and at the operational level.

While this list is not exhaustive, it does cover some of the larger areas of concern. Within these categories, several new and undiscovered specific problems are likely to arise and require additional research.

14.4 Use of Drones to Perform Advanced Damage Assessment

Utilities have developed clear processes to reduce the impact of widespread outages. The initial phase of these processes involves the deployment of staff to specific locations, having replacement equipment in geographically diverse depots, food, shelter, communications, and necessary materials to support that staff in place before a likely event. Contingency planning is critical to prepare for events of all types; however, as soon as an event begins, utilities commence their process of impact assessment. Although telemetry, smart meter status, and modern outage management analytics provide utilities valuable insight into the scope and scale of outages, until knowledgeable staff conduct "on the ground" surveys, restoration cannot take place. All outages are not equal. A narrow outage from a tornado requires less effort than a widespread outage from a hurricane or earthquake. The time, effort, and equipment required to restore a damaged substation transformer are significantly different from that required to replace a feeder span just outside the substation fence, although the number of customers impacted could be the same.

The survey provides the restoration planning process with damage severity, the estimated level of effort, equipment/spare parts required, and any logistical challenges the crews are likely to face. Historically, the survey process began as soon as it was safe for utility personnel in vehicles, primarily on the ground, to patrol the utility field assets. With access to the first survey data, utilities establish restoration priorities, and the initial restoration work begins. As more survey data become available, priorities are revised, and the order of restoration activities changes. As a result of the survey data, logistics efforts deploy replacement assets such as poles, conductors, and transformers to the areas that require rebuilding.

In the process, the initial damage on a feeder is found, repaired, and the power restored. Many times, even after the initial damage on a feeder is repaired and the feeder breaker closed, the breaker trips immediately because additional damage to the feeder exists. This common scenario slows the restoration process by requiring crews to find the next cause, repair the damage, and attempt re-energization only to discover further damage exists. Not only does this process slow restoration but also repeatedly closing in on faults can cause additional damage to substation equipment or, at a minimum, require more frequent substation maintenance. Some situations require cutting the conductor and supplying from another source if the damage is extensive or the customer's electrical supply is critical. Many of these changes require future work to return them to their original configuration once the bulk of the restoration is complete.

More than two years ago, Amazon's CEO announced a secret R&D project: "Octocopter" drones that would fly packages directly to your doorstep in 30 minutes [12]. Using these concepts, utilities could also assess the feasibility to deliver spare parts and other equipment directly to an affected area, thereby reducing the time to fix a problem.

During the restoration process, access to the utility's assets becomes a significant challenge. Roads closed by downed trees, utility poles, debris, flooding, or earthquake damage limit or delay surveying, often requiring rerouting of survey vehicles many miles. Utility vehicles require fuel; as a result of widespread outages, restoration slows if no service stations have power, requiring utilities to bring the fuel to the crews.

In the future, utilities will accomplish their initial surveys more quickly with the use of drones, either remotely piloted or autonomous. Video of the damage will be telemetered directly to those establishing the restoration priorities by drones capable of flying over impassable roads. Further, drones can loiter over damaged assets as necessary, providing high-resolution information on the extent of specific damage and identifying logistical considerations for restoration.

Using the utilities Geographic Information System data for their initial routing, these drones will follow the distribution lines providing a complete picture of the extent of damage from the substation to the last customer on the feeder. This video data can be processed to identify any employee or public safety issues quickly. It can also assess the required level of effort, specialized equipment needed, anticipated replacement equipment, and challenges for accessing the damaged locations. The completeness of the video data allows the creation of a more optimum plan to restore each specific substation, feeder, and neighborhood.

Deployment of drones providing live-streamed video allows significantly more rapid damage assessment and prioritization. For utilities, safety is paramount, both for staff and the public. After an event, drones can be deployed as soon as they can fly without placing staff at risk. Further, drones can be deployed over areas that present risk to staff, such as during flooding. Many times, crews spend significant time clearing roads just to access the additional downstream damage. If the downstream damage is known, crews can sectionalize the line and restore upstream customers before clearing roads to access the site of the damage.

The detail provided in the video from a drone allows differentiation between a conductor that may be shorted to the ground from one damaged, but which may not cause additional breaker trips if reenergized. In the latter case, an energized, damaged conductor represents a major safety concern. With this information in hand, crews can again sectionalize downstream of their repairs, restore additional customers, and continue their rebuilding without potentially closing the breaker only to discover additional problems.

The value of having a more detailed understanding of the scope and extent of the damage is that it allows a decision on how to restore more customers or more critical customers faster. The crews and resources deployed can be better matched to the extent of the damage and the size of the restoration effort. As a result, restoration becomes a more orderly process. Decisions can be better planned, and the logistics better organized, allowing crew support for meals and housing, and crews relieved as appropriate during the restoration work.

The drone video also has added value to first responders. Providing information on the condition of roads and access to locations for emergency response. While the restoration is underway, a drone could provide live video of progress and risks, and guide circumstances requiring expertise from elsewhere. Further, the video information will improve the quality and accuracy of restoration times for customers and the larger community. Once the restoration is complete, the video provides several additional benefits. First, it will improve the quality and planning of future restoration activities. Next, the video will impact distribution design decisions regarding improved resiliency. It can also focus on vegetation management activities to avoid future outages. Lastly, the video can be used to return the grid to "normal" by removing the temporary cuts, jumpers, and configuration changes.

In the future, the diversity of drone applications will no doubt expand rapidly. With miniaturization and reduced cost, crews will carry smaller, one-person deployable, drones to identify specific situations that may exist farther down the line or on long secondary service drops without having to patrol each one in a vehicle. Larger drones, capable of carrying transformers, poles, and the bundles of the conductor will bring needed assets to the damaged location to support crews. As their capability increases, even specialized equipment can be deployed precisely where and when it is needed, and easily moved to another location with fewer delays.

As drone sensors expand beyond current abilities, to include infrared and other capabilities, drones will provide precise identification about the last energized point on distribution lines before the crew is deployed. The availability of this information will likely alter the restoration tactics including crews dispatched, where work will begin, and related logistics such as feeding and housing crews. Further, even though drones will illuminate the damage, enabling video streaming sensors beyond the visible such as in darkness becomes less of a limiting factor. As drones become more widely deployed and utilized in the utility industry, restoration processes will improve, restoration time will be reduced, and the restoration process made safer. As drones become more autonomous, planned and emergency surveys will automatically occur. Swarms will sweep over the damaged lines and transmit video directly to the storm centers. Once the video is analyzed, smart restoration plans can be developed and implemented. Drones will continue to provide updates on the restoration, delivery equipment, and supplies required by the field crews.

14.5 Case Study: Lessons Learned and Forgotten. The North American Hurricane Experience

In the aftermath of Hurricane Sandy, the lessons learned focused very much on increasing the resiliency of the electric grid [5]. The usual set of hardening activities, such as burying power lines and replacing utility poles with stronger poles, were proposed. Similarly, there was an industry-wide recognition for reliable outage information as described earlier. These studies also marked a change in strategy for increasing resiliency, however. There was a distinct recognition that moving forward, traditional hardening was necessary, but not sufficient.

At the time of Sandy, the number of customers without power became a key measure of storm severity. Similarly, the restoration of service has been a measure of recovery. The litany of customers without power is very familiar – Hurricane Katrina more than two million over six states; Superstorm Sandy more than 7.5 million over the nine most affected states. In the wake of Hurricane Maria, all of Puerto Rico was without power and of the 3.5 million customers, 500,000 were still without power almost three months after the event. Storms are not the only cause of these outages. For example, Puerto Rico lost all power in 2016, because of a system problem and the island of Zanzibar in Tanzania lost all power for a month following the rupture of an undersea cable.

Outage count only tells part of the story. Increasingly, there is a focus on the power system as support for the critical infrastructure of communities. It is quite striking how much of "everyday life" depends on the electric infrastructure and how much of it, in retrospect, is viewed as critical. A 2012 report from the National Research Council, "Terrorism and the Electric Power Delivery System" delineates key elements of this critical infrastructure including [13]:

- emergency services (fire and police; 911 call centers),
- medical services (hospitals and emergency care facilities),
- communication and cyber services (cell phones, radio and TV stations, computer services, and both conventional and wireless telephones),
- water and sanitation (pumps),
- food (refrigeration),
- financial (general banking and ATMs),
- fuel (delivery and gas stations/pumps),
- non-emergency government services (schools, prisons, and security systems),
- transportation (traffic lights, air traffic control, and rail transportation),
- lighting (building emergency lighting and street lighting), and
- buildings (elevators and air conditioning).

These elements of "critical infrastructure" emphasized backup power options for individual facilities, noting that fuel for such systems becomes an issue the longer the outage period (Table 8.1 in [13]). We should note that the above list contains obvious elements such as medical facilities where respirators and other life support equipment depend on electric power. Many of the items on the list, such as gas station pumps, were identified in the wake of other storms. Yet, they were still generally without backup power in 2012. Power to non-emergency government facilities is important, as shown in the aftermath of Maria when 12,100 federal prisoners were moved from Puerto Rico to Mississippi [14]. Even street lighting and traffic signals can become a matter of life and death [15].

An earlier report commissioned by the state of Pennsylvania provided an extensive review of such systems It came out with a variety of recommendations, including "pass laws or change regulations to facilitate the construction, interconnection, and operation of distributed generation systems, and the operation of competitive microgrid systems." The recognition that the use of distributed energy and microgrids would require regulatory action was important. There is little evidence that before Superstorm Sandy much work was done on policy reviews of the kind recommended [16].

As noted in the introduction, Sandy provided a real impetus to change the microgrid approach. Specifically, in the wake of Sandy reports of islands of power from the Food and Drug Administration's White Oak research facility in Maryland, Princeton University in New Jersey, parts of the NYU campus in Manhattan, and the Stony Brook University campus on Long Island brought focus to the benefit of microgrids.

In a Forbes article written by Peter Kelly-Detwiler as Sandy approached, he described four conclusions from a microgrid meeting held in Hartford [17]:

- The technology around microgrids continues to develop, and real-world examples are already being deployed.
- Synergistic thinking is crucial to improving the economics of microgrids.
- The value of the microgrid is very much in the eye of the beholder, but there are significant benefits. Some of these are hard to value.
- Very little progress will be made in this space without the collaboration and cooperation of the utility.

In the post-Sandy era, these four points became clearer. First, and very critical for the future, the technological advancement driven by "smart grid" thinking and tools made managing a microgrid increasingly possible. In the post-Katrina era, and as manifested in the NRC report, the concept was largely one of providing backup power. Backup power was an emergency-planning tool. Even before Sandy, the Princeton facility was used to manage electricity costs for the campus when power prices were high – a real-world example that drives the second conclusion. Microgrids have a variety of possible benefits beyond backup power.

Second, the economics of microgrids are important as their role evolves beyond backup power. In general, backup power is an insurance policy and is very often mandated by regulation. For example, large hospitals and critical care facilities are required to have backup power. For facilities that run 24/7 operations, this kind of power may be used regularly to protect against short-term outages. As the desire for protection becomes more focused on longer outages, less critical facilities come into play, such as cell phone towers and gas stations, and the cost of insurance rises relative to the benefit to the owner. By finding other uses for the microgrid's power that have value to the utility, the cost of the microgrid can be reduced.

Third, these other uses may be hard to value, like voltage support or power factor management, but others, like demand response, are already valued in some markets.

Finally, the State of New York recognizing this value attempted to jump-start the adoption of microgrids for critical infrastructure through what is called the New York Prize competition https://www.nyserda.ny.gov/All-Programs/Programs/NY-Prize. New York Prize is a "first-in-the-nation competition to help communities create microgrids – standalone energy systems that can operate independently in the event of a power outage." This program funded 83 feasibility studies and 11 comprehensive engineering, financial, and commercial assessments. As became evident, the confluence of new technologies and the lessons learned in the wake of Sandy drive the consideration and adoption of microgrids in New York and other parts of the Northeastern United States. It remains to be seen as to whether the same focus can be brought to Puerto Rico and the nation-at-large in the wake of hurricanes Irma, Maria, and others in the future. A recently released plan suggests that with leadership from places like New York, Puerto Rico may well be on the path to this happening [18].

Drs. Vadari and Stokes raised a final question in the wake of Sandy. Can we move beyond the microgrids as isolated pockets of resilience to create dynamic microgrids that support resilience and facilitate recovery and restoration, maximizing the rate of recovery for non-critical customers? [19].

14.6 Bringing it All Together – Policy and Practice

The transmission and distribution of electricity is a highly regulated public good. As such, any discussion of new ways to provide electricity and electrical services must inevitably involve regulation and policy. It goes beyond the conclusion made by Kelly-Detwiler that "very little progress will be made in this space without the collaboration and cooperation of the utility."

Regulations and policies have been playing a role in electric utility for a long time. Public Utilities Regulatory Policies Act (PURPA) is one of the first major changes to hit the modern electric utility in the United States after the Public Utilities Holding Companies Act (PUHCA). The latter established the modern US investor-owned utility, and the former was implemented to promote energy conservation (reduce demand) and promote greater use of domestic energy and renewable energy (increase supply). The next change came in 1996 with the advent of FERC orders 888/889 which played a major role in opening the US electricity markets to competition [20–22]. These orders allowed in existing power pools, such as Pennsylvania/Jersey/Maryland (PJM), New York Power Pool (NYPP), and New England Power Pool (NEPool), forming Independent System Operators. Independent System Operators became Regional Transmission Operators (RTOs) with the release of FERC Order 2000, which administered the transmission grid on a regional basis across the United States and Canada.

As RTOs were settling into wholesale markets, another change appeared. The "Smart Grid" started bringing in new technologies, both OT and IT, distribution automation, microgrids, storage, DERs, and demand response.

Before the round of deregulation gripped the electric utility industry, the relationship between utilities and their customers was driven philosophically by the "regulatory compact," which he saw as "an expression of the nature and intent of the relationship between the regulated utilities and their investors, the utilities' customers, and the general public." This compact provides:

- Utilities give up certain rights for the benefit of a monopoly territory granted by the government.
- The customers of the utilities give up the right to choose the supplier of the utility commodity within that territory, for the assurance of government regulation of the price the utility may charge for that commodity.

Also, through all this change, the distribution side of the utility business and the final regulatory compact with the end-user customer has stayed the same with some changes in some states of the United States.

Today, the nature of the compact is changing – and this time not driven by the regulator or the utility. It is being driven by the customer. The following drivers are causing the change:

- **Customers are generating their own power and creating a change in electric energy consumption**. The advent of DERs allowed customers to install technologies such as solar PV on their roofs and generate their own power. This changes their consumption profile concerning the connection to the grid. In fact, in many situations, the customer is also delivering power to the grid, in effect changing their relationship with the utility.
- **Modification of the boundary between the utility and the customer**. Traditionally, the boundary between the customer and the utility has been at the meter. With a microgrid, the customer can put energy into the system and conduct activities that might affect utility operations even more than in the past. Historically, the utility has managed these requirements. A microgrid can deliver a clear value to the grid. How the customer realizes that value is important to the economics of microgrids. Early attempts at this

for distributed energy systems have had limited success, and some would argue a lack of fairness. Perhaps the most important experiment in recognizing the value in the retail market is taking place in the State of New York where their REV program is looking to build retail markets where the value can be fairly priced and traded.

- **Customers grouping together (or brought together by a commercial entity).** This concept, also known as the aggregator, is now allowing entities to bring multiple to band together – disconnected from each other and not even contiguous with respect to the grid – customers and use their new-found ability to generate power and interact at the wholesale level.

The desire to create competition in the marketplace, the basis of current deregulation efforts, changes this relationship. Beyond that, customers generating their own power reflects a fundamental shift in the relationship as well. Generalizing the regulatory compact a little, we can see that it is intended to provide the utility with a fair return with manageable risk and the customers with a fair price for the services they obtain.

In the United States, there are over 3200 utilities, and each state, the District of Columbia, and territory is an independent regulatory authority over this relationship. The issues associated with resiliency vary dramatically, however, based on where the utility's jurisdiction lies, concerning exposure to extreme weather systems. Regardless, there are some policy and regulatory issues that need to be seriously considered ensuring that the next-generation electric grid functions much better than the last one.

Invest in technology: Utilities are already investing in technologies, such as FLISR, which allow very rapid location and isolation of grid faults. As this chapter has identified, however, they also need to invest in technologies such as drones, which could quickly bring the grid back to normalcy after a major storm.

Invest in business and technical constructs such as microgrids: Microgrids should not be construed purely from a competitive viewpoint, as the potential for losing customers from their rate base. The next generation of the microgrid could be a utility construct necessary not just for resiliency, but also to provide the utility with greater flexibility in operations and better customer service. Technologies such as "I See You" also create an opportunity to have better visibility into the status and connectivity of the grid.

Invest in grid hardening: Hardening the grid is a necessity regardless of investments in other directions. Options such as moving the more vulnerable parts of the grid underground and placing key infrastructure components of the grid, such as transformers on elevated concrete pads, are still necessary.

Finally, as the industry moves toward dynamic microgrids, a whole new management scheme will likely be required. It is not clear that a recovery microgrid would necessarily follow utility line boundaries or even state boundaries. While there are provisions for the suspension of some rules during a time of emergency, setting policies and procedures for the use of this approach will require state and federal support.

14.7 Conclusions

Power system resiliency and the need to handle intense storms have received a lot of attention from utilities, technology solution providers, regulators, and policymakers. Today, there is general agreement that one of the most advanced countries in the world cannot have outages that last months, weeks, or even days, given this, funds are made available to pilot technology along with a renewed emphasis on system hardening.

Utilities have achieved tremendous improvements with the implementation of technology to combat storms. Utilities have at their disposal, systems such as smart meters, fault locators, and advanced distribution management systems, including Outage Management Systems. For example, within a week after Hurricane Irma struck in 2017, 95% of the people and businesses in Florida had their power back on, which was four times faster than after Hurricane Wilma in 2005 [23].

Much more needs to be done, however. The intensity of the electric grid and the need for it to stay on continue to increase. For example:

- The importance of the extensive use of Smartphones and their apps in daily life has increased exponentially. Mobile phones require infrastructure components such as cell towers to stay on and require direct access to electricity to charge.
  The advent of electric cars requires faster and more accessible charging stations. Otherwise, people with electric cars are stuck without personal transportation.
- The advent of increased penetration of DERs has led to new opportunities to respond to large-scale disturbances. The availability of new mechanisms, such as microgrids and nanogrids, to bring parts of the grid up and running ahead of the macrogrid.

All of this is piled on top of existing issues with loss of power, such as the need for access to light, heat, and other situational problems.

This chapter highlights the importance of technologies and policies to combat issues associated with power system resiliency. The chapter provided several examples of technologies that are already either in use, being piloted, or are on the research horizon. This is a very important area for society at large and necessary to support the resiliency of the electric grid – identified as the greatest engineering achievement of the twentieth century.

References

1 Erdman, J. (2017). 2017 Atlantic Hurricane Season Among Top 10 Most Active in History. weather.com, https://weather.com/storms/hurricane/news/2017-atlantic-hurricane-season-one-of-busiest-september (October 02, 2017).

2 Lacey, S. (2014). *Resiliency: How Superstorm Sandy Changed America's Grid*. GreenTech-Media https://www.greentechmedia.com/articles/featured/resiliency-how-superstorm-sandy-changed-americas-grid#gs.N=3Ntgo.

3 Nehamas, N. and Dahlberg, N. (2017). FPL spent $3 billion preparing for a storm. So why did Irma knock out the lights?, Miami Herald. http://www.miamiherald.com/news/weather/hurricane/article174521756.html.

4 Swartwout, R. Current Regulatory Practice from a Historical Perspective. http://lawschool.unm.edu/nrj/volumes/32/2/04_swartwout_current.pdf.

5 Abi-Samra, N.C. (2013). One Year Later: Superstorm Sandy Underscores Need for a Resilient Grid. IEEE Spectrum. https://spectrum.ieee.org/energy/the-smarter-grid/one-year-later-superstorm-sandy-underscores-need-for-a-resilient-grid.

6 Vadari, M. (2018). *Smart Grid Redefined – Transformation of the Electric Utility*. Publisher Artech House.

7 Vadari, M. (2020). *Electric System Operations – Evolution to the Modern Grid*, 2nd Edition. Publisher Artech House.
8 Members of Working Subgroup – Mani Vadari, D.L. Dr: 'Reforming the energy vision (REV) working group II: platform technology.' Department of Public Service, New York. https://www3.dps.ny.gov/W/PSCWeb.nsf/96f0fec0b45a3c6485257688006a701a/853a068321b1d9cb85257d100067b939/%26dollar%3BFILE/WG%202_Platform%20Technology_Final%20Report%20%26%20Appendices.pdf.
9 Department of Public Service, S. o.: Staff proposal distributed system implementation plan guidance. Available at http ://www3.dps.ny.gov/W/PSCWeb.nsf/All/C12C0A18F55877 E785257E6F005D533E?OpenDocument, Retrieved from Department of Public Service, State of New York, 15 October 2015
10 Interconnection of Distributed Generation in New York State: 'A utility readiness assessment'. Available at http://www3.dps.ny.gov/W/PSCWeb.nsf/96f0fec0b45a3c6485257688006a701a/dcf68efca391ad6085257687006f396 b/$FILE/83930296.pdf/EPRI%20Rpt%20-%20Interconnec tion%20of%20DG%20in%20NY%20State-complete%20-%20Sept%202015.pdf, Retrieved from New York State Department of Public Service, September 2015.
11 Initiative, G.E. (2017). What are Smart Microgrids? http://www.galvinpower.org/microgrids.
12 Bezos, J. (2013). Amazon unveils futuristic plan: Delivery by drone - Amazon's secret R&D project aimed at delivering packages to your doorstep by "octocopter" mini-drones with a mere 30-minute delivery time. CBSNews.
13 National Research Council (2012). Terrorism and the Electric Power Delivery System.
14 Gates, J.E. (2017). Mississippi to take in 1,200 prisoners from hurricane-scarred Puerto Rico. https://www.usatoday.com/story/news/nation-now/2017/10/18/mississippi-take-1-200-prisoners-hurricane-scarred-puerto-rico/777820001/.
15 College (2002). Vishwaja Muppa, Stony Brook Student, Killed In Sandy-Related Car Crash, Huffington Post. https://www.huffingtonpost.com/2012/11/02/vishwaja-muppa-stony-broo_n_2065897.html.
16 PA DEP (Pennsylvania Department of Environmental Protection) (2005). Critical Electric Power Issues in Pennsylvania: Transmission, Distributed Generation, and Continuing Services when the Grid Fails. Report prepared for the PA DEP by the Carnegie Mellon Electricity Industry Center, Carnegie Mellon University, Pittsburgh, PA.
17 Kelly-Detwiler, P. (2012) Peter Kelly-Detwiler is a Contributor to Forbes. https://www.forbes.com/sites/peterdetwiler/2012/10/26/with-all-eyes-on-hurricane-sandy-a-good-time-to-evaluate-microgrids/#2645df6576c0.
18 Wagman, D. (2017). $17 Billion Modernization Plan for Puerto Rico's Grid Is Released. IEEE Spectrum. https://spectrum.ieee.org/energywise/energy/the-smarter-grid/17-billion-modernization-plan-for-puerto-rico-is-released.
19 Vadari, M. and Stokes, G. (2013). *Utility 2.0 and the Dynamic Microgrid*. Public Utility Fortnightly.
20 United States of America 76 ferc 61,009, Federal Energy Regulatory Commission, Order Clarifying Order Nos. 888 and 889 Compliance Matters, https://www.ferc.gov/legal/maj-ord-reg/land-docs/rm95-8-0aj.txt.

21 Max Luke Why we need bigger electricity markets. World Economic Forum, Research Associate, Massachusetts Institute of Technology, https://www.weforum.org/agenda/2015/02/why-we-need-bigger-electricity-markets/.

22 Eric, P. (1997). *The Electric Utility Restructuring Debate - A Primer, Fifth Annual North American Waste To Energy Conference, Panel on Electric Utility Restructuring*. North Columbia: Research Triangle Park http://www.seas.columbia.edu/earth/wtert/sofos/nawtec/nawtec05/nawtec05-05.pdf.

23 Ebi, K. (2017) Cities that partner with utilities have a powerful advantage, from Smart Cities Council Global. http://na.smartcitiescouncil.com/article/cities-partner-utilities-have-powerful-advantage.

Index

Please note that page references to Figures will be followed by the letter '*f*'; to Tables by the letter '*t*'.

Resiliency of Power Distribution Systems, First Edition.
Edited by Anurag K. Srivastava, Chen-Ching Liu, and Sayonsom Chanda.

b

c

d

n

y